LOCAL INDUCTION

W0193014

SYNTHESE LIBRARY

MONOGRAPHS ON EPISTEMOLOGY,

LOGIC, METHODOLOGY, PHILOSOPHY OF SCIENCE,

SOCIOLOGY OF SCIENCE AND OF KNOWLEDGE,

AND ON THE MATHEMATICAL METHODS OF

SOCIAL AND BEHAVIORAL SCIENCES

Managing Editor:

JAAKKO HINTIKKA, *Academy of Finland and Stanford University*

Editors:

ROBERT S. COHEN, *Boston University*

DONALD DAVIDSON, *The Rockefeller University and Princeton University*

GABRIËL NUCHELMANS, *University of Leyden*

WESLEY C. SALMON, *University of Arizona*

VOLUME 93

LOCAL INDUCTION

Edited by

RADU J. BOGDAN

Stanford University

D. REIDEL PUBLISHING COMPANY

DORDRECHT-HOLLAND / BOSTON-U.S.A.

Library of Congress Cataloging in Publication Data

Main entry under title:

Local induction.

    (Synthese library ; 93)
    "A selected bibliography of local induction, by
Radu J. Bogdan" : p.
    Includes indexes.
    CONTENTS: Levi, I. Acceptance revisited. —
Rosenkrantz, R. Cognitive decision theory. – Goosens, W. K.
A critique of epistemic utilities. [etc.]
    1. Induction (Logic) — Addresses, essays, lectures. I.
Bogdan, Radu J.
BC91.L62   161   75-34922

ISBN 978-94-011-9801-1   ISBN 978-94-011-9799-1 (eBook)
DOI 10.1007/978-94-011-9799-1

Published by D. Reidel Publishing Company,
P.O. Box 17, Dordrecht, Holland

Sold and distributed in the U.S.A., Canada, and Mexico
by D. Reidel Publishing Company, Inc.
Lincoln Building, 160 Old Derby Street, Hingham,
Mass. 02043, U.S.A.

All Rights Reserved
Copyright © 1976 by D. Reidel Publishing Company, Dordrecht, Holland
Softcover reprint of the hardcover 1st edition 1976
No part of the material protected by this copyright notice may be reproduced or
utilized in any form by or any means, electronic or mechanical,
including photocopying, recording or by any informational storage and
retrieval system, without written permission from the copyright owner

To Catalina

TABLE OF CONTENTS

PREFACE

The local justification of beliefs and hypotheses has recently become a major concern for epistemologists and philosophers of induction. As such, the problem of local justification is not entirely new. Most pragmatists had addressed themselves to it, and so did, to some extent, many classical inductivists in the Bacon-Whewell-Mill tradition. In the last few decades, however, the use of logic and semantics, probability calculus, statistical methods, and decision-theoretic concepts in the reconstruction of inductive inference has revealed some important technical respects in which inductive justification can be local: the choice of a language, with its syntactic and semantic features, the relativity of probabilistic evaluations to an initial body of evidence or background knowledge and to an agent's utilities and preferences, etc. Some paradoxes and difficulties encountered by purely formal accounts of inductive justification, the erosion of the once dominant empiricist position, which most approaches to induction took for granted, and the increasing challenge of noninductivist epistemolgies have underscored the need of accounting for the methodological problems of applying inductive logic to real life contexts, particularly in science. As a result, in the late fifties and sixties, several related developments pointed to a new, local approach to inductive justification. Among them: The Rudner-Levi-Jeffrey debate on scientists' value judgments and the subsequent discussions of rules and criteria of acceptance, Carnap's shift to a normative, decision-theoretic treatment of inductive logic, Hempel's approach to epistemic utilities, the increasing influence of Popper's conception of the informational content of scientific hypotheses, Hintikka's reconceptualization of logical probability, information, and decision-theoretic rules of acceptance, etc. In his 1967 book *Gambling with Truth*, which has since become a classic of local induction, Isaac Levi summed up the trend: "In science, justification of belief is demanded only when the need for such justification arises in the context of specific inquiries". The ensuing controversies notwithstanding, the increasing influence of this trend is undeniable. The present volume attempts to offer a fair image of the present stage of research on local induction, characterized among other things by alternative approaches, reformulations of earlier positions, conceptual innovations, more attention paid to epistemological issues, and of course internal criticisms of specific proposals as well as external criticisms of the whole enterprise.

The volume opens with Isaac Levi's essay on 'Acceptance Revisited'. It is both a critical retrospect of his *Gambling with Truth*, and in several respects a step forward. An important epistemological development is Levi's treatment of the growth of knowledge as improvement or change. Thus, a large part of the paper is devoted to an examination of the conditions and criteria according to which the basic modifications in a corpus of knowledge, i.e. expansion or contraction, are contextually and yet rationally justified. In the process some problems dealt with in *Gambling with Truth* such as the criterion of maximizing expected utility, the multiplicity of epistemic desiderata, finite versus infinite partitions, interval versus point estimation, the bookkeeping problem, etc., are reexamined, and many objections are answered. The next two papers are directly concerned with Levi's view of local induction. Roger Rosenkrantz analyzes the cognitive decision-theoretic approach to inductive justification, particularly Levi's. According to Rosenkrantz, the major shortcomings of such an approach are the subjectivity of the criteria of acceptance and evaluation, and the undesirable multiplicity of epistemic utilities taken as formally independent objectives. These shortcomings can be avoided by adopting a bayesian theory of evidential support, based on the concepts of likelihood and sample coverage, and capable of handling any significant epistemic utility such as content, simplicity, accuracy, etc., as ingredients of support. William Goosens' paper takes the critique of epistemic utilities into the field of inductive evaluations based on generating-evidence experiments. Goosens maintains that there is "an integral connection between evaluation and experimentation" which poses a major challenge to an epistemic utility approach such as Levi's. In particular, he proves that the latter allows for a perfect experiment having lower expected utility than an imperfect experiment, thus leading to unreasonable preferences.

The papers by Keith Lehrer and James Fetzer indicate that the decision-theoretic and epistemic utility reconstruction of local induction is far from being monolithic. Keith Lehrer holds that local inductive methods can satisfy the basic objective of maxiverificity, i.e. acceptance of all and only those statements of a language which are true, and promote interpersonal agreement or consensus. The quest for maxiverificity in a certain context of inquiry is shown to be relative to an epistemic field in which it is reasonable to accept a maximally consistent set of statements which is more probable than any of its competitors. The rule of acceptance is set up in terms of utilities, *qua* costs and benefits of acceptance, and probabilities, *qua* consensual probabilities derivable from personal assignments made by members of a group. According to Lehrer, this approach can also account for major shifts in what a scientific group accepts as evidence and hypotheses. Like Lehrer, James Fetzer relates his theory of epistemic utility to a definite context of knowledge characterized as a minimally rational set of beliefs accepted by an individual at a given time in a certain language. The general inductive strategy of maximizing estimated epistemic utility is accounted for on a two-level basis. First, the notion of estimated epistemic utility of a hypothesis relative to a knowledge context is explicated as the product of the epistemic probability of that hypothesis on the relevant data and its epistemic utility. Second, the epistemic probability itself is explicated as a product of the degree of confidence in the given data and the likelihood of the hypothesis on that data; while the epistemic utility is explicated as the product of the logical scope of the hypothesis and its systematic power.

Two more papers, by Kl. Szaniawski, and G. Menges and E. Kofler, respectively, can also be included in the above category, although their treatment of cognitive decisions and utilities is closer to the working statistician's viewpoint. Starting from A. Wald's definition of sequential inference, Szaniawski discusses two rules of inference based on likelihoods and epistemic utilities, respectively. Then he introduces a third approach which interprets sequential inference as a search for information. The three construals are compared, and their merits and ranges of application are assessed. The problem of cognitive decisions under partial information is the subject of the paper by Menges and Kofler. The authors' view is that the notion of partial or local information should be analyzed relative to Cournot's concept of *a priori* probability (and not Laplace's), and also relative to the distinction between prior and posterior

probabilities. The formal framework of such an analysis is briefly formulated and explored.

The following two papers consider local induction relative to global
approaches and their philosophical motivation. The paper by Henry E.
Kyburg, Jr. is a detailed comparative analysis of several alternative
treatments of local induction with respect to the use and interpretation
of probability. Kyburg's own account of the latter, as epistemological
probability, is proposed as being better equipped to handle some specific
problems of local induction. It is argued, however, that no matter what
technical improvements are introduced, local induction still faces a general challenge, namely the adjudication of disagreements caused by initially
different data or bodies of background knowledge. Kyburg claims that
local induction cannot meet this challenge unless the very contexts of
justification it relies upon are suitably broadened. But such a move will
inevitably require a global theory of justification. The global inductivism
my paper is concerned with is that of Hume. Local induction is viewed
as an epistemologically motivated attempt to avoid the basic failure of
Hume's approach, namely his atheoretical view of knowledge and induction. I try to find the logical and epistemological sources of this conception in the context of its opposition to rationalism. I then argue that induction should be relativized to a theoretical context of knowledge, and
explicated as a relation of relevance between evidence and the ontological
claims made by a theory.

The remaining papers offer alternative frameworks for local induction
and suggest specific solutions to some local inductive problems. L. Jonathan Cohen's essay is a synthesis of his neo-classical and nonprobabilistic
theory of induction. The basic issues Cohen deals with are the nature of
inductive evidence, the syntax of inductive reliability, the justifiability of
induction, inductive reasoning versus probability, and the epistemology
of neo-classical induction. Unlike most approaches to local induction,
Cohen's is more concerned with scientific experimentation. The inductive conclusions reached in the latter's context are accounted for by his
method of relevant variables and evaluated within a modal, nonprobabilistic, formalism. Ilkka Niiniluoto's paper is an attempt to explicate the
theory of local inductive acceptance in terms of the question-making
and answer-providing strategies used in scientific inquiries. The types of
questions relevant to such inquiries are regarded as requests for informa-

tion. The formal framework of questions, problems, and inquiries is articulated with respect to various types and stages of scientific research. In this context, Levi's rule of inductive acceptance is criticized as being oversensitive to the choice of an ultimate partition and relative to closed questions only. Håkan Törnebohm examines the process of piecemeal knowledge-formation as a sequence of acts leading to the solution of a cognitive problem. His formal reconstruction of this process deals especially with the problem of positive and negative evidential support, the content and degree of truth of a hypothesis, its testing as well as the criteria of confirmation and elimination. Raimo Tuomela proposes a solution of the transitivity paradoxes of qualitative confirmation by relating the notion of confirmation to that of nomological explanation, and not to mere deducibility as it is usually done. In this way, the general principles of confirmation are restricted to contexts of explanation, and made relative to a certain scientific theory.

The papers included in this volume represent but a small sample of approaches to local induction. The bibliography which appears at the end of the volume will provide the reader with more information about contemporary research on local induction and related problems.

The editorial history of this volume is related to a sometimes difficult transition to a new world during which I was surrounded by an impressive human and professional solidarity. Many people in many places have supported me in more ways than I can here account for. To all of them I am and will remain deeply grateful.

As in so many other occasions, the first steps of this project, undertaken while still in Romania, were backed by a wonderful friend and matchless promoter of intellectual ventures, Dr Mircea Ioanid. Professor Jaakko Hintikka was from its very beginning a generous source of help, encouragment, and advice. D. Reidel Publishing Company and the contributors themselves will find here my warm thanks for their patience and confidence in the final outcome. I am grateful to Professor Keith Lehrer for his constant interest in my work, and to Professors Raimo Tuomela and Roger Rosenkrantz for their helpful suggestions. I owe special gratitude to Mrs Rivka Gisis and her wonderful staff, and to my old friend, Dr Ion Stamatescu, who did everything to make my work possible. Last but not least, I am grateful to Professor Henry E. Kyburg,

Jr., Stephen Pink, the faculty members and the graduate students of the
University of Rochester for their warm and stimulating companionship
during the spring semester spent here.

   I am afraid no words can express what I owe to my father. Nor can
they convey what my wife Catalina means to me. I dedicate this volume
to her love, luminosity, and courage.

Rochester, June 1975   RADU J. BOGDAN

ACCEPTANCE REVISITED

In 1967, I completed the perpetration of what Imre Lakatos once aptly called a 'book act'. Ideally the process ought to have served as a catharsis. Book acts, however, are public performances. They invite criticism both from the agents who perform them and from others. Unless one can follow the sage advice of Queen Victoria ('Never explain!'), the temptation to return to the scene of the crime becomes overwhelming. I fear I am subject to the temptation.

In this essay, I shall elaborate upon, defend, and modify various positions taken in my *Gambling with Truth*. The first two sections contain an outline of the epistemological position within the framework of which the question about 'acceptance' of hypotheses discussed in that essay becomes significant. In the sections which follow, I shall indicate some of the respects in which the proposals made in *Gambling with Truth* should, in my opinion, be modified, how the scope of application of these proposals might be extended, and shall offer responses to some of the objections made by critics.

I

Human knowledge is subject to change. In scientific inquiry, men seek to change their knowledge for the better. The central problem of epistemology ought to be, therefore, to provide a systematic account of criteria for the improvement of knowledge. Alternatively stated, the problem is to offer a systematic characterization of conditions under which alterations in a corpus of knowledge are legitimate or are justified.

Demands for justification of this sort arise in many contexts. Three kinds are worth mentioning here:

(i) At time t, X (who may be a person or group of persons such as a scientific community) has a corpus of knowledge $K_{X,t}$. The problem facing X is whether he should alter his corpus in some way or another. He must justify making any proposed modification *to himself*.

(ii) Instead of attempting to justify a given modification of his corpus

R. J. Bogdan (ed.), Local Induction, 1–71. *All rights reserved.*
Copyright © 1976 by D. Reidel Publishing Company, Dordrecht-Holland

to himself, X might be concerned to justify *his* making the shift to some other agent or group of agents Y.

(iii) At t, X has h in his corpus and Y does not. (This is one of many ways in which X and Y might disagree. I pick this way for illustrative purposes only.) X may be concerned to justify to Y why Y should modify his corpus so that it agrees with X's corpus.

In undertaking the second kind of justification, X may or may not be concerned to persuade Y of the views he adopts after he has made the shift from his initial corpus. His primary concern, however, is to indicate to Y the grounds on which he (i.e., X) modified his corpus. His justification will be valid if and only if it is a legitimate justification of the first type. That is to say, the same standards control the legitimacy of X's shifting from K to some other corpus K' whether X is addressing the question of legitimacy to himself or defending his making the shift to Y.

In this sense, I am presupposing that the legitimacy of X's modification of his corpus is subject to 'objective' or 'inter-subjective' appraisal. In taking this position, I do not mean to deny that whether X's modification of his corpus is legitimate or not depends on his circumstances or 'situation' and is, to this extent, 'subjective' or idiosyncratic. Whether the modification is appropriate depends not only on X's initial corpus but on other features of the context in which he makes or contemplates making the change. My assumption is that a systematic account can be given of what the relevant 'contextual parameters' are and how 'values' of these parameters determine whether a modification is admissible or not.

Of course, investigation may reveal that no system of contextual parameters can be identified such that, given specific values for these parameters, the legitimacy of X's modification of his corpus is determined according to adequate objective standards. In that case, such modifications of bodies of knowledge would be subjective or context dependent in a sense which put them beyond critical control. However, we would be obstructing the course of inquiry to assume that this is so at the outset.

Thus, I disagree with those who, starting with the correct observation that changes in knowledge are context dependent, conclude that, except for relatively minimal requirements of consistency, no general criteria applicable to all rational agents under a wide variety of circumstances can be devised which regulate the modifications they institute in their corpora of knowledge. The legitimacy of modifications is undoubtedly context dependent and 'pragmatic'. For this reason, it seems to me, we should undertake closer inquiries into the pragmatics of the matter. Otherwise we tacitly concede that the most important questions concerning the 'growth' or improvement of knowledge are to be psychologized, sociologized or historicized but not rationalized.

Once this is understood, even the third kind of justification can be assimilated, at least formally, to the first; for what X is attempting to do in case (iii) is to furnish Y with good reasons for Y to change his own corpus. From Y's point of view, X will have succeeded if and only if Y can use X's argument legitimately to justify Y's making the change to himself.

Of course, in case (iii), X is *not*, from his own point of view, attempting to justify his having h in his corpus *to himself*. X already accepts h in his corpus. He takes for granted that h is true. As far as he is concerned, it is quite certain that h is true. He assigns h a subjective or personal probability of 1. Moreover, from X's point of view, it is quite impossible that h is false. As far as he is concerned, there is no need for him to justify his 'acceptance' of h.

By saying that X is completely certain of h and considers it impossible that h is false, I do not mean that X's subjectively felt intensity of conviction is at a maximum. Clearly, when X is a group of agents, it is doubtful whether we should talk of subjectively felt convictions. Even in the case of an individual agent X, what X accepts in his corpus does not, I suspect, correlate well with such intensities.

When h is in X's corpus at t, X is prepared or committed to use h as a premiss in deliberations aimed at realizing whatever moral, political, economic or other practical objectives he is concerned to promote at that moment. He is also prepared to take the truth of h for granted in scientific inquiries. My claim that X is certain of the truth of h and regards it as impossible that h is false concerns X's *use* of h as a premiss in deliberation and inquiry. It does not concern his subjectively felt convictions or the origins or grounds of his having acquired h in his corpus in the first place.

One of the ways in which premisses, evidence or background in a corpus of knowledge are used in inquiry and deliberation is to distinguish between those logical possibilities which are 'serious' and ought to be taken into account in evaluating risks and those logical possibilities which may be ignored. (Of course, the evaluation of risks depends not only on what is seriously possible and what is not but upon assessments of probability and the agent's goals and values. When a logical possibility is discounted as nonserious by X, X assigns that logical possibility personal probability 0. Moreover, no matter what gains or losses are attributed by X to that logical eventuality, he is committed to ignoring them.)

Suppose, for example, that X is offered a gamble on the outcome of a toss of a coin near the earth and under fair conditions. X may take seriously the logical possibility that the coin will land on its edge and wish to take into account the benefits and losses should that eventuality obtain. But, if he is like most of us, he will not and should not take seriously in this way the logical possibility that the coin when tossed will fly straight to Alpha Centauri or that the earth will explode. If he did take such logical possibilities (especially the latter) seriously, he would never accept an otherwise attractive gamble on the outcome of the toss (e.g., one where he receives $1,000 whether the coin lands heads or tails). Even if these logical possibilities are regarded as extremely improbable, as long as they are attributed some positive probability of happening the risks entailed would be too great to contemplate accepting the gamble. Most of us would agree, I take it, that anyone who refused a gamble on the toss of a coin with a payoff of $1,000 whether the coin lands heads or tails because he takes seriously the logical possibility that the earth will explode is neurotic and that his practice ought not to be taken as a model of rationality.

Suppose an agricultural experimenter is interested in determining whether there is any significant difference between the size of corn of two different varieties. He designs an experiment involving the planting of the two different kinds of corn on a suitably prepared field. He may take seriously the logical possibility that a cow will break the fences protecting the field and munch his way through. But he will not and should not worry about the logical possibility that one of both of the varieties of corn will metamorphosize into gold. That logical possibility is not serious. In the sense of relevance to scientific inquiry and practical deliberation, it is impossible, as far as the experimenter is concerned, that the corn will turn into gold.

In ruling out the logical possibility that the coin will fly to Alpha Centauri or that the corn will turn into gold, X typically assumes not only the 'practical' impossibility of statements like 'the coin will fly to Alpha Centauri' and 'the corn will turn into gold' but assumes also that it is practically impossible that various theories, laws and statistical claims are false. X's corpus will contain not only singular statements or their negations but laws, theories, and statistical claims as well.

It may be suggested perhaps that instead of saying that h is in X's corpus of knowledge at t, I should say that h is in X's corpus of 'full' beliefs at t. In my opinion, there is no relevant difference, from X's point of view at t, between what he knows and what he fully believes. From his point of view, all items in his corpus are true. Moreover, he is certain of their truth in the sense that he is committed to assigning them probability 1 in evaluating risks. Finally, it is impossible that such items be false in the sense of 'possible' relevant to the use of such items in the conduct of inquiry and deliberation. They are, in this sense, not only certain but infallibly true. What other qualifications should items in X's corpus have in order to acquire, from his point of view, the honorific status of knowledge?

I do not, of course, mean to suggest that the distinction between knowledge and belief is totally useless. Thus, X may very well regard items which once were in his corpus but no longer are as past beliefs of his which were not known by him to be true. When Y's corpus contradicts X's, items in Y's corpus are, from X's point of view, falsely believed.

When rightly understood, even the distinction between true belief and justified belief has some value. Prior to adding h to his corpus, X considers h (where neither h nor $-h$ is in X's corpus already) possibly true and possibly false. Hence, adding h to his corpus could, from X's point of view, breed error or yield true information. Moreover, X may or may not be justified in adding h to his corpus. Hence prior to adding h to his corpus, X can acknowledge a distinction between adding h when h is true (coming to believe truly that h) and being justified in adding h to his corpus (being justified in coming to believe that h).

Thus, the distinction between coming to believe truly and coming to believe justifiably is significant. From X's point of view, however, once he has added h to his corpus and committed himself to its truth, it no longer matters to him how he came to add h to his corpus and whether he was justified in doing so. Why should he continue to be concerned with the pedigree of his conviction? Perhaps, because he is in doubt as to the truth of h. If he is in doubt, however, h is not in his corpus.

Of course, circumstances can arise where X will be concerned to decide whether he should retain h in his corpus (continue to believe that h) or remove h from his corpus (cease to believe that h). But whatever considerations are relevant to determining when X is justified in retaining h or removing it from his corpus, the justification must be relative to what is already in X's corpus – including h. Consequently, the considerations relevant to justifying adding h to X's corpus (coming to believe that h) ought not to be confused with the considerations relevant to justifying retaining h (continuing to believe that h). In principle, the considerations which did or did not justify X in adding h to his corpus initially need not be germane to determining whether he is justified in retaining h in his corpus when the need for such justification arises.

To be sure, when X is concerned to justify to Y adding h to Y's corpus (where X has already added h to his corpus), X may indicate the grounds on which he (X) added h to his corpus. (This need not, however, be the case.) But this does not mean that, from X's point of view, X's acceptance of h is in any way in need of support by reference to the grounds on which he added h to his corpus. From X's point of view, h is true, X is quite certain that this is so and regards it as impossible that h be false. There is no distinction for X between his believing truly that h and his being justified in believing that h. There is no useful meaning to be attributed to justifying belief unless it is justifying coming to believe or justifying retaining belief. Once X has h in his corpus and as long as no occasion arises where the issue of retaining or surrendering his belief arises, there is no relevant need for justification.

In a similar vein, if Y agrees with X that h is true but thinks that X was not justified in adding h to his corpus in the first place, this does not mean that Y should withhold the status of knowledge from X's belief that h. For Y to do this would be tantamount to his recommending to X that X should remove h from his corpus and start over again. But this is patently absurd. From Y's point of view, X should retain h in his corpus. Y's complaint that X was not justified in adding h to his corpus initially given the corpus with which X began is a criticism of the modification X originally made in his corpus in coming to believe (or know) that h. But once X has made the modification and it has turned out right (from Y's point of view), there is no need for further emendation. Where it doesn't itch, don't scratch![1]

As I stated, X is committed to treating all items in the corpus of knowledge he adopts at t as infallibly true in the sense that the logical possibility that one of these items is false is not, as far as he is concerned, a serious one.

Observe that this infallibility concerns the way items in a corpus of knowledge are to be used by the agent who adopts the corpus as a resource for subsequent inquiry and deliberation. Infallibility in this sense should be carefully distinguished from infallibility of sources of information such as the empirical senses or various would be authorities such as the Pope or Stalin. Even X who is committed to treating items in his corpus as infallibly true can, nonetheless, consistently acknowledge that Y need not treat items in X's corpus in this way and that even from his own point of view, some of his beliefs (e.g., many of those he adopted before endorsing the corpus he currently adopts) have been false.

Not only is the infallibility of knowledge in its use in inquiry and deliberation to be distinguished from alleged infallibility of sources of information but from claims that justifications of belief are in some sense infallible.

Recall that three different sorts of justifications have been distinguished:

(i)   Alleged justifications for having h in a corpus at a time when h is already admitted into that corpus.

(ii)   Alleged justifications for adding h to a corpus to which it did not belong initially.

(iii)   Alleged justifications for removing h from a corpus to which it did belong initially.

I shall call philosophers who demand of rational X that all the items in X's corpus of knowledge at time t be justified 'justificationists'. I am not sure that this is precisely what Popper has in mind by justificationism. But he surely must have meant to include justificationists in this sense in his sweeping condemnation of that point of view. In any case, I agree with Popper in rejecting this version of justificationism.

It is worth noticing, however, that one possible source of discomfort with my contention that items in a corpus of knowledge are to be treated by the agent who adopts that corpus as infallible in the conduct of inquiry and deliberation derives from a commitment to justificationism.

If X is obliged to justify having h in his corpus at the time t when he has h in his corpus, he cannot, without circularity, satisfy this requirement by appealing to premisses which contain h or logically imply h. Relative to the premisses he uses to justify having h in his corpus, h must be possibly false. This possibility must be taken seriously as long as the justification does not furnish a guarantee against the possibility that h is false. If such justification is fallible in the sense that it cannot furnish such a guarantee, X cannot be justified in refusing to take seriously the possibility that h is false.

Thus, a commitment to justificationism together with acknowledgement that the justifications available are fallible in the sense that they furnish no guarantee of the truth of conclusions implies that X should not treat items in his corpus as infallibly true even in their use as resources for inquiry and deliberation.

From such a justificationist point of view, the only items to be used in this way are the premisses which are the foundation of justifications of the sort justificationists demand. Normally, such premisses will include logical truths, mathematical truths, and other allegedly analytic, a priori or necessary truth. Sometimes suitably restricted testimony of the senses is allowed this status. Theories, laws, statistical claims and, for many justificationists, observation reports are not allowed to be treated as infallibly true.

I shall not rehearse the familiar roster of problems which plague justificationist epistemologies. As I have indicated, I reject justificationism. As a consequence, objections to the thesis of the infallibility of knowledge as a resource for inquiry and deliberation based on justificationist assumptions do not seem to me to be compelling.

Although I reject the requirement of justificationism which demands of a rational agent that all items in his corpus at t be justifiably accepted by him at t, I do also maintain that revisions of corpora of knowledge often require justification. One sort of context where such justification arises is where X considers adding h to a corpus to which it did not originally belong and whose members did not entail it. The premisses to be used to determine whether X is justified in adding h consists of the items in the initial corpus. Relative to these premisses, the logical possibility that h is false is a serious possibility. From X's point of view prior to adding h, such a move involves genuine risk of error.

In this sense, any justification for adding *h* to the corpus is a 'fallible' one. Nonetheless, I contend that *X* can be justified in this fallible sense in adding *h* to his corpus and, having done so, in ceasing to take seriously the possibility that *h* is false. *X* is justified, in short, in a fallible way in coming to treat *h* infallibly in subsequent inquiry and deliberation.

To add *h* to his corpus and to treat *h* subsequently as infallible is to incur a risk from *X*'s point of view prior to expanding his corpus. But the risk may be worthwhile provided that the information added to the corpus by the expansion is important enough. Sometimes it may not be. Then there is no justification. In other cases it will be. The crucial point is that the fallibility of the justification for adding *h* to a corpus does not prevent *X* as a rational agent from treating *h* as infallible once it is added to his corpus and is to be used as a resource for inquiry and deliberation.

One way in which *X* will often expand his corpus of knowledge is by appeal to his senses or the testimony of some witness. If prior to expansion *X* takes for granted that observations to be made are to be made under conditions under which the testimony of the senses is reliable or if the witnesses relied on are assumed to be expert and honest, expansion in this way may be legitimate. This is so even though *X* does not suppose that his senses (or the witnesses) are perfectly reliable – i.e., infallible. Once more, the crucial point is that the fallibility of a source of information used in adding new items to a corpus does not and should not prevent *X* from treating such items as infallibly true in subsequent inquiry and deliberation.

The other context in which justification is wanted concerns situations where *X* is contemplating removal of *h* from his corpus. I contend that conditions can and do arise where *X* is justified in contracting his corpus in this way. In short, the items in a corpus of knowledge are liable to revision, removal or correction. They are, in this sense, corrigible.

I think the most serious case which can be made against my contention that *X* should treat items in his corpus of knowledge as infallibly true is based on an appeal to the corrigibility of human knowledge. In my opinion, Popper's rejection of infallibilism is based on such an appeal. Indeed, for Popper as for Peirce before him, there is scarcely any difference between fallibilism and corrigibilism. Both authors typically appeal to considerations arguing in favor of the corrigibility of human knowledge in defending what they call fallibilism.

Even those who reject my infallibilism and would replace certainty with near certainty are often animated by a commitment to corrigibilism and not justificationism in developing their position. If h is counted as infallibly and certainly true and, hence, is assigned probability 1, authors like R. C. Jeffrey seem to hold that there is no legitimate way in which that probability can justifiably be revised downward at some subsequent time.[2]

Needless to say, I believe that corrigibilism can be consistently maintained along with the thesis that rational men can and should treat theories, laws, statistical claims in a corpus of knowledge as infallibly and certainly true when using that corpus as a resource for inquiry and deliberation. I shall develop my view on this point in the next section. If I am right, the major philosophical motivations for objecting to accepting statements as true, as infallibly true and as certainly true will have been undermined. The elaborate efforts of followers of Popper, on the one hand, and various 'bayesians' on the other to deny the prima facie obvious fact that in scientific inquiry and practical deliberation men do use knowledge in the manner I have suggested (and cannot avoid doing so) will no longer seem so pressing. Instead, attention may be focussed on what, in my opinion, is the main problem of epistemology – to wit, the systematic study of criteria for evaluating modifications of corpora of knowledge.

Strictly speaking, there is more to this 'main problem' than I have been suggesting. From X's point of view, every item in his corpus is infallibly and certainly true. But he is far from certain of everything. And in the conduct of inquiry and deliberation, he will not only be required to determine which possibilities are serious and which are not but also to distinguish between serious possibilities with respect to grades of probability. I shall call X's system of evaluations of the (subjective or personal) probabilities of hypotheses at t his 'credal state at t'. According to X's credal state, all items in his corpus receive probability 1 (although the converse need not hold). I do not suppose, as most bayesians do, that X's credal state can be characterized by a numerically determinate probability function. However, this difference between me and the bayesians can be ignored for the present.

An adequate account of the revision of knowledge ought, in my opinion, to be supplemented by an account of the revision of credal states. Indeed, a change in X's corpus of knowledge will entail some sort of change in his credal state as well.

Most advocates of bayesian approaches to rational choice, preference and credence would agree with the need for furnishing some sort of account of the revision of credal states. The most common view of the matter relies heavily on the so called principle of conditionalization. As is well known, application of this principle for revising credal states depends on X changing his corpus of knowledge. Hence, bayesians who rely on conditionalization do concede that rational agents do accept items into a corpus of knowledge and, moreover, use it as infallible in subsequent inquiries.

However, this circumstance has appeared unsatisfactory to some bayesians. One response (this was Carnap's) is to restrict the items on which revisions of credal states could be based via conditionalization to observation reports but to reject the idea that a corpus of knowledge could be revised other than via observation.[3] In particular, Carnap explicitly opposed the view that a corpus could justifiably be modified by adding items to it via 'induction' from data.[4] The prima facie fact that scientists do accept theories, laws, and statistical claims into a corpus was to be explained away by somehow replacing certainty by near certainty.[5]

Some bayesians have gone still further. R. C. Jeffrey has abandoned conditionalization as a criterion for the revision of credal states in favor of a principle which does not rely on modifications of a corpus of knowledge in my sense.[6] If a project like Jeffrey's could be developed successfully (needless to say I do not think so),[7] one would have an account of the revision of credal states which did not require of X that his corpus contain anything to be treated as infallibly true and certain other than logical, mathematical and other such 'necessary' or 'analytical' truth. One of the advantages of this view from the viewpoint of those who endorse it is that whereas if h is assigned probability 1, conditionalization cannot allow that probability to be revised downward, if it is not assigned probability 1, there is no obstacle to revising its probability. In this way, bayesians appeal to the thesis of corrigibility to oppose my thesis of infallibility.

It should be obvious that this sort of appeal to corrigibility is far from compelling. In order to account for the revision of credal states without using conditionalization, Jeffrey has introduced a principle which does not require appeal to items counted as certain. To be sure, Jeffrey has formulated his principle in such a way that if X were to shift the probability of h to 1, the reverberations through his credal state would be as conditionalization requires. But once conditionalization is abandoned, it remains obscure to me why one should opt for a principle for the revision of credal states which has this property. It is obvious that if items assigned probability 1 are to be considered corrigible, conditionalization cannot be regarded as universally legislative as a principle for revising credal states. Hence, those who invoke conditionalization to show that corrigibility implies fallibility simply beg the issue.

There is a more interesting argument for the view that corrigibility implies fallibility. It is at least implicit in the views of Peirce and Popper. I shall return to it in the next section.

In any case, bayesians like Carnap and Jeffrey do attempt to explain away the facts by insisting that when h is in a corpus of knowledge it is not treated as infallibly and certainly true in subsequent deliberation and inquiry but only as nearly or almost certain. (This is often called 'practical' certainty; but, of course, 'practical' is used in a different sense than I would use it in calling items in a corpus practically certain.) Their position does have one virtue. In denying my view of the way that knowledge is used as a resource for inquiry and deliberation, they offer another alternative view. If h is assigned probability $1 - \varepsilon$ for small ε, the possibility that h is false is serious; but the risk of error involved in acting on h is identifiable, and bayesians can offer a systematic account of how risks of this sort are to be taken into consideration in making decisions.

It seems clear that in daily life and scientific inquiry, we discount utterly all sorts of logical possibilities. We do not assign them small positive probabilities of being true. If we did proceed in this fashion, the probability that at least one of the many items so treated is true would have to be very substantial. Yet, we mean to discount not only the possibility that any given one of them is true but also the possibility that at least one of them is true.

Thus, neither Carnap nor Jeffrey apply their theories to gambles such as gambles on the outcome of a toss of a coin in such a manner as to take seriously the possibility that the coin will fly to Alpha Centauri. In practice, they assign probability 0 to that logical possibility. Is this a mere oversight? So be it. Strictly speaking, probability ε is to be assigned. Unfortunately, there are noncountably many other oversights which have to be taken into consideration in this way. The probability that at least one of these logical possibilities is true must itself be kept very small. Clearly, the practical and technical difficulties standing in the way of eliminating these mere oversights are substantial.[8] Whether they are insuperable in principle is an open question. It is incumbent on those who think that they are to show this. They have thus far failed to do so.

But there is yet more required of those who would substitute near certainty for infallibility. They have to provide an account of the revision of credal states which in no way involves reference to X's having a corpus of knowledge substantially stronger than one that consists of logical truth supplemented by whatever else is accorded similar status.

In particular, they have to explain why we are sometimes justified in modifying our credal states by shifting the probability of a theory or law which intitially has probability 0 or near 0 to one which has probability almost but not quite equal to 1. Harold Jeffreys and Abner Shimony have attempted to meet this challenge.[9] However, their approaches do take for granted that a corpus of knowledge is accepted substantially stronger than one restricted to items accorded the privileges of logical truth; and, in any case, their views are open to other objections which I shall consider subsequently.

On the view I advocate, this sort of difficulty does not arise. A theory which has a very low initial probability may sometimes be added to a corpus of knowledge (and its probability shifted to 1) even though the risk of error is very high; for the information obtained may be worth the risk. Of course, on some occasions the trade off between risk of error and information obtained favors refusing to add the theory to the corpus. We do not add items to a corpus of knowledge solely because it provides important information and without regard to the risk of error involved. Notice, however, that if, instead of adding a theory to a corpus and assign-ning it probability 1 because the possibility of error was worth risking due to the information to be gained, the probability was shifted to $1 - \varepsilon$, there would be no risk of error connected with the possibility that the theory is false. Hence, the rationale I propose is not available to those who would substitute near certainty for infallibility.

There may be ways in which these and other difficulties might be over-come. If they can (and that remains to be seen), the merits of an account of the revision of credal states which substitutes near certainty for infallibility would have to be compared with approaches such as mine which do not make such a replacement. At present, no theory of the required sort has been developed.

In spite of these reservations concerning views which replace infallibility with near certainty, they do have one attractive feature. The doctrine of infallibilism I advocate concerns the way in which agents should use their corpora of knowledge in the conduct of deliberations and inquiry. Those bayesians who deny that rational agents should accept items in a corpus in this sense do, nonetheless, offer an alternative view which continues to be concerned with the way in which agents should use what appear to be items in a corpus of knowledge in deliberation and inquiry. The dispute is a substantive one: Is a rational agent ever entitled to refuse to take seriously the logical possibility that h is false where h is neither a logical truth nor some other sort of statement bearing epistemological status similar to that accorded logical truth? Bayesians offer a negative answer to this question. However, they supplement this negative answer with a positive account of how nearly certain hypotheses are to be used in the conduct of inquiry and deliberation. This account is derivative from a general account of how X's credal state cooperates with his values and preferences to determine how he should evaluate options in the context of a given decision problem.

Other advocates of a fallibilistic attitude towards human knowledge are not so admirably clear. This is especially true of Karl Popper who like Carnap and Jeffrey would also reject my thesis of the infallibility of human knowledge as a resource for inquiry and deliberation.

Popper holds that the ultimate aim of inquiry is to obtain *the* truth by which he means apparently an error free and maximally consistent story of the world.[10] He claims that the best way to undertake this endeavor is to devise very strong hypotheses eminently vulnerable to falsification and to subject them to as severe test and criticism as is feasible.[11] He nowhere explains why this is the best way to proceed in attempting to reach the truth. He insists quite rightly that no proof or demonstration can be given that we will attain the truth in this way or even approach it.[12] But he owes us some reasons for supposing that his procedure is better than alternative ones for attaining that objective. To my knowledge, nothing he has offered in the way of such a reason is more than a circumlocuitous reaffirmation of the contention that it is.

Be that as it may, in any specific context of inquiry, the immediate or proximate aim is to devise testworthy hypotheses and to subject them to test. Progress is measured by the testworthiness of the hypotheses we have devised, the severity of the tests to which we have put them and the extent to which some such hypotheses have stood up to test.[13] At no point in this process does it become an important question as to whether some of these hypotheses should gain admittance into a corpus of knowledge and be counted as infallibly true.

My view is different. In the context of a specific inquiry, the immediate or proximate aim is to revise a corpus of knowledge. It is a common feature of all such aims to avoid error or to obtain error free information. Thus, avoidance of error is a critical desideratum of any proximate aim of inquiry. For Popper it is not. How could it be? Insofar as all that we are concerned to do is to devise testworthy hypotheses and to test them, considerations of error do not enter into the picture at all. A hypothesis' testworthiness is, in general, quite independent of its truth value. False hypotheses, as Popper emphasizes, can be eminently testworthy. They can, indeed, successfully pass very rigorous tests. If these are the sole characteristics of hypotheses which matter when they are evaluated in order to ascertain the 'progress' of science, truth does not enter into the picture at all – except, perhaps, in some unexplained way in connection with promoting the long run ultimate aim of coming nearer the truth. At best, truth is placed on a pedestal so remote from the immediate concerns of inquiry as to play no role in directing the conduct of such inquiry.[14]

I do not, of course, deny that it is important in inquiry to devise testworthy hypotheses and to test them. My contention is, however, that it is desirable to do so in order to best promote an effort to obtain error free information pertinent to answering the specific question or questions under investigation at the given occasion. We devise and test hypotheses in order to improve our body of knowledge so that it will better serve us in subsequent scientific inquiries and in the conduct of practical affairs. On the basis of inquiries, including testing hypotheses, we sometimes reach a point where we conclude that the trade offs between risk of error and informational benefits are such as to warrant adding some hypothesis to the corpus and so to convert its status from mere hypothesis to settled, established and infallible truth (where being settled, and established is only for the time being and not necessarily forever). Indeed,

I contend that there is little point in testing hypotheses unless on some occasions we can hope to reach such a conclusion. To devise testworthy hypotheses and severely test them solely for the sake of realizing the long run goal of coming close to the truth seems, in contrast, to be an attempt at intellectual masturbation – and a futile one at that.

Furthermore, as I have indicated previously, no hypothesis can be regarded as testworthy, no tests can be devised, no hypothesis can be considered to have passed a test unless substantially more is incorporated into a corpus of knowledge than what is incorrigibly secure. I have contended that what are used as the premisses or background of a test are, when they are so used, not themselves being subjected to criticism and test. They are taken for granted. They determine what are to count as serious possibilities and what are not. They are, in short, taken to be infallibly true.

Now Popper styles himself a fallibilist. Moreover, he acknowledges and, indeed, insists that 'background knowledge' is required in order to determine the testworthiness of hypotheses and their success or failure in passing tests. But he is not troubled by the difficulty which this poses for his own doctrine of fallibilism. He writes as follows:

> The fact that, as a rule, we are at any given moment taking a vast amount of traditional knowledge for granted (for almost all our knowledge is traditional) creates no difficulty for the falsificationist or fallibilist. For he does not *accept* this background knowledge; neither as established nor as fairly certain, nor yet as probable. He knows that even its tentative acceptance is risky, and stresses that every bit of it is open to criticism, even though only in a piecemeal way. We can never be certain that we shall challenge the right bit; but since our quest is not for certainty, this does not matter.[15]

In what sense is acceptance of background knowledge 'tentative'? To accept h tentatively is not to accept h as background to be used as premisses in testing a hypothesis but to regard its acceptance as contingent on the outcome of further investigations. In so doing, X may be willing, as we say, to accept h "for the sake of the argument" in order to explore the consequences that would ensue were he actually to add h to his corpus. Such tentative acceptances is not the same as taking h "for granted". Nor is tentative acceptance frought with risk of error as Popper suggests. What would be risky would be to add h to X's corpus. From X's point of view prior to adding h, to add h could possibly lead to error. But to tentatively add h is not to add h at all and incurs no risk of error.

Of course, Popper may mean to suggest that acceptance of background knowledge is 'tentative' only in the sense that even though it is accepted into the background it may, nonetheless, be removed from that status subsequently and be subject to criticism. I have no quarrel with this reading of his statement.

But Popper seems to mean more than this. He denies that we accept background knowledge as 'established', as 'fairly certain' or even as 'fairly probable'.[16] He seems to imply at the same time that we often continue to consider what we accept as background as improbable and frought with risk and never as certain or established. The risk may be worth taking but it is a risk nonetheless.

This contention is not to be confused with my claim that when X is concerned to add h to his corpus of knowledge, h may be extremely improbable and adding h to his corpus fraught with risk which is, nonetheless, worth taking. Popper is not talking about when it is legitimate to add items to a corpus of knowledge or to background knowledge but about what X's posture towards h should be at the time when it is already a part of his corpus of knowledge. Even while it is a part of his corpus, X should continue to regard it as highly improbable and his endorsement of it as highly risky. Clearly Popper's fallibilism endorses the thesis of the fallibility of human knowledge which I reject.

But in what sense could accepting h be judged risky? Relative to the corpus of background knowledge itself, h is not merely highly probable but bears probability 1. Relative to that corpus, there is no risk in accepting h. Popper must, therefore, mean to deny that rational men should assess risks relative to their background knowledge. Nor should they determine possibilities of error relative to such knowledge whether in practical decision making or in conducting tests.

Relative to what premisses then is the acceptance of h a risky affair? Presumably it is relative to a system of assumptions consisting of logical truths and other first premisses. And, indeed, if we evaluate risk relative to logical truth and whatever else passes for first premisses, accepting h in a corpus is, indeed, a risky matter even after it is so accepted.

Observe, however, that Popper has already conceded the necessity of using background knowledge in devising testworthy hypotheses and in testing them. It should be obvious that such background is needed precisely to determine what are serious possibilities and to assess risks. If this is so, the risks that are relevant to the conduct of inquiry including the testing of hypotheses are such that the acceptance of what is already in a corpus of knowledge carries no risk of error in the relevant sense.

If Popper thinks otherwise, this must mean that he considers the relevant risks to be taken into account in conducting tests are those relative to first premisses. But he cannot mean this without betraying his opposition to justificationism – an opposition I share with him and with proponents of the pragmatist tradition of Peirce and Dewey.[17]

It is to Popper's credit that he refuses to follow Carnap and Jeffrey in construing items belonging in background knowledge as being nearly certain. On the other hand, by denying the infallibility of such knowledge, he is left with no way of characterizing how such knowledge is to be used in deliberation and inquiry. He cannot say that it is almost certain and, hence, is to be used to assess risks in the way hypotheses assigned probability near 1 but not quite equal to it are used. Nor can he treat it as infallible and perfectly certain and so assign it probability 1. He is, therefore, bereft of any way of characterizing how background knowledge is used in devising testworthy hypotheses or conducting tests, or, more generally, how it is to be used not only in scientific inquiry (even as Popper characterizes it) but in the conduct of practical affairs as well. His notion of tentative acceptance of background knowledge is utterly obscure and he offers no clear account of why anyone would want to accept background knowledge in his tentative sense.

Popper's multifaceted point of view suggests a response to this dilemma. It may be granted that in the context when X uses a given corpus as background knowledge, the items in that corpus are treated as infallibly and certainly true in my sense. But this sense of infallibility and certainty is 'subjective' in that it depends on the agent's circumstances. When we consider the matter from an 'objective' vantage point which ignores the idiosyncratic features of the circumstances in which agents find themselves, background knowledge is possibly false, is fallible and uncertain.[18]

The sort of progress in science with which Popper is concerned is objective progress. That progress is to be measured, according to Popper, by our success in devising testworthy hypotheses, testing them severely and having some of these hypotheses, nonetheless, pass the tests. I suspect that Popper might concede (I cannot be sure of this) that in conducting any specific test, the background knowledge is counted as certainly and infallibly true. But the objective appraisal of the progress achieved through conducting the test will take the form of saying that relative to background knowledge *b*, hypothesis *h* passed (or failed) such and such a test. This claim can be made without considering the background knowledge *b* as certainly and infallibly true. And it might (dubiously I think) be contended that this claim is not dependent on context for its correctness and is, therefore, 'objective'.

By distinguishing between objective and subjective certainty (or infallibility), Popper can grant me my contention that background knowledge is certain and infallibly true from the point of view of the agent. But he will then qualify this concession by stating that the certainty and infallibility is subjective.

I have contended that the main problem of epistemology ought to be to provide a systematic account of the improvement of knowledge which is used to distinguish between serious and nonserious possibilities in the conduct of inquiry and deliberation. According to Popper, such revisions depend on context and are, in this sense, subjective. Because they are context dependent they are in some sense idiosyncratic. We are absolved (so it seems) from bringing such revisions under critical control. What we should do, instead, is to focus on the 'objective' progress of science.

Now I do not quite understand why progress in the sense in which Popper characterizes it is objective. But it seems to me that claims like "relative to background knowledge *b*, hypothesis *h* passed (or failed) such and such a test" are in themselves extremely unexciting. They become interesting when they are combined with the commitment to use *b* as an infallible and certain resource for inquiry and deliberation. I suspect that if we restrict ourselves to what is objective in Popper's sense, we will end up trivializing our concern with the growth of knowledge. What is clear is that Popper himself has neglected the problem of revising background knowledge in spite of his voluminous writings on the growth of knowledge. Save for a few casual comments, he has next to nothing to say on this matter.

I myself do not consider it an obstacle to bringing the problem of the revision of knowledge under critical control that revisions of knowledge are heavily context dependent. That they are context dependent means that a systematic account of the revision of knowledge will have to seek the relevant 'contextual parameters' which determine the legitimacy of revisions and attempt to explain how they operate in determining that legitimacy. To suppose that this cannot be done is, in my opinion, to endorse the antirationalist view that revisions of corpora of knowledge are not subject to critical control or adopt the obscurantist thesis which says that although men can acquire the skill to revise their knowledge in legitimate ways, no criteria can be offered for evaluating such legitimacy.

Thus, to the extent that Popper seriously does intend to dismiss the problem of revising corpora of knowledge as a 'subjective' or 'pragmatic' matter, he has trivialized the problem of the growth of knowledge. What is worse, he has tacitly conceded the field to those who insist that the growth of human knowledge is to be psychologized, sociologized or historicized – but not rationalized. Popper's lust for objectivity has led him to reduce the domain subject to objective criticism – not to enlarge it. I wish to proceed in a different direction.

II

In seeking to improve his corpus of knowledge, X can modify his corpus in one of two ways: by adding new items to it and by removing items from it. The former sort of modification I shall call an 'expansion'. The latter I shall call a 'contraction'.

In *Gambling with Truth*, I focussed attention on a certain kind of expansion and contended that no matter what the particular immediate problem which occasioned the concern to expand the investigator X was concerned to obtain error free information.

If this view or some variant of it can be generalized as a thesis concerning the invariant features of the proximate aims in terms of which all sorts of modifications of a corpus of knowledge ought to be evaluated, it appears to pose a difficulty for my rejection of the equivalence of fallibilism and corrigibilism.

On what grounds would X be justified in removing h from his corpus once it had become a member of it? Once h is in X's corpus, from X's point of view it is impossible that h is false. From his point of view, it is certainly true – i.e., bears personal probability 1. To remove h from his corpus would be tantamount to giving up information which X is certain is true.

Worse yet, in science and everyday life, we not only remove items from a corpus but also replace them with other claims incompatible with them. Grand examples are all too familiar. Kepler's laws were replaced by Newton's corrections of them. Phenomenological thermodynamics was abandoned in favor of statistical mechanics. Newtonian mechanics was superseded by quantum mechanics. But if X is concerned to obtain error free information, it appears as though X could not justify replacing a theory T_1 by a theory T_2 inconsistent with it. From his vantage point when T_1 is in his corpus, T_2 is certainly and necessarily false. For him to replace T_1 by T_2 would be tantamount to exchanging truth for falsehood.

Thus, it seems as though I must give up either my view of the characteristic aims of scientific inquiry or my rejection of the equation of corrigibilism with fallibilism.

I deny that I must do either of these things.

As a first step towards coming to grips with the objections just raised, I shall characterize somewhat more systematically what is to count as a change in X's state of knowledge. I propose to represent X's state of knowledge at t – insofar as that state is expressible in a suitably regimented language L – by a set $K_{X,t}$ (X's corpus of knowledge at t) of sentences in L satisfying the following conditions:

(i) $K_{X,t}$ is deductively closed – i.e., contains all deductive consequences in L of sets of sentences in $K_{X,t}$.

(ii) $K_{X,t}$ contains all items in the deductively closed set UK (the 'urcorpus' for L) consisting of all logical truths, set theoretical truths, mathematical truths and whatever else counts as 'incorrigible' claims expressible in L.

X's state of knowledge at t cannot be completely characterized by such a deductively closed set $K_{X,t}$ of sentences in L. For example, if h is in $K_{X,t}$, 'h is true in L' is not in $K_{X,t}$, that sentence is not in L. If we have a metalanguage L_1 containing L in which truth conditions for sentences in L can be specified in a Tarskian manner, we can consider the corpus $K_{X,t}^1$ of sentences in L_1. This corpus is deductively closed and contains the urcorpus UK^1. Moreover, there are several relations obtaining between $K_{X,t}$ and $K_{X,t}^1$. If $h \in L$, $h \in K_{X,t}^1$ if and only if $h \in K_{X,t}$ which in turn holds if and only if 'h is true in L' $\in K_{X,t}^1$ which in turn holds if and only if '$h \in K_{X,t}$' $\in K_{X,t}^1$.

We must, of course, countenance a still more inclusive metalanguage

L_2 with a corresponding corpus $K^2_{X,t}$ related to $K^1_{X,t}$ as the latter is related to $K_{X,t}$.[19] I ignore the technical details. The crucial point is that changes in the corpus $K_{X,t}$ will reverberate through the hierarchy of corpora $K^1_{X,t}$, $K^2_{X,t}$, ... in a definite manner.[20] For most purposes, we can restrict our attention to modifications of $K_{X,t}$ and ignore the rest.

As I have construed it, $K_{X,t}$ consists of all those sentences, expressible in L, X counts as true in L, which he is certain are at t true in L (i.e., to which he assigns probability 1), and whose falsity he regards at t to be impossible in the practically relevant sense. It would seem, in the light of this, absurd to require of a rational agent X that his corpus $K_{X,t}$ be deductively closed or that it contain all logical, set theoretical and mathematical truths. No one, not even Kurt Gödel, could come close to satisfying this requirement.

The objection is well taken. Instead, however, of weakening the condition that corpora of knowledge be deductively closed sets, I suggest construing the proposed representation of X's state of knowledge at t as a representation in terms of the set of sentences (in L) X is *committed* to counting as true, necessarily true and certainly true. X may be committed to holding h to be true without being explicitly aware of his commitment.

I do not know how to explicate the notion of commitment used here and suspect that the best way to do so is to indicate the conditions on commitment (such as deductive closure) and develop an account of how commitments ought to be modified over time. X's commitment is partially determined by the sentences he explicitly identifies as contained in his corpus but I know of no adequate way to explain the determination.

By interpreting $K_{X,t}$ as the set of sentences in L, X is committed to counting as true, being certain of and taking to be infallibly true, we may concede that X can retain his status as a rational agent even though he fails to conform to the deductive closure requirement in his deliberations – provided that he has an excuse for such failure such as breakdown of memory, computational facility or lack of imagination. If, however, X explicitly identifies h as true and, in addition, recognizes g as a deductive consequence of h and, yet, refuses to accord g the status of an infallible and certain truth, X has deviated from the requirements of rationality. He has reneged on his own commitments.

The same point applies to the requirement that all elements in the ur-corpus be members of X's corpus no matter what other items are. To suppose that someone who can barely add should explicitly regard all mathematical truths as necessarily and certainly true is blatantly absurd. I claim only that X is committed to the certainty and infallibility of such items.

Why are these requirements to be imposed on X's commitments? I take for granted that if X rules out the logical possibility that h is false, he is thereby committed to ruling out the logical possibility that any deductive consequence of h is false. Similarly, if X rules out the logical possibility that any one of a set of sentences (in L) is false, he is committed to precluding the logical possibility that any deductive consequence of that set is false.[21]

The deductive closure requirement guarantees that logical truths will be members of the urcorpus UK. Since some fragments of set theory are required to supply a system of truth conditions on Tarskian lines in L_1 and some of these set theoretical claims may be expressible in L, I have simply and extravagantly included set theory in L rather than a restricted fragment thereof. My extravagance extends to mathematics. The basic principle I have used in assigning claims to UK is that all such sentences are not subject to revision according to the criteria I propose to develop. In other words, by assigning items to UK, I impose a restriction of the scope of applicability of an account of the improvement of knowledge I wish to construct. Perhaps, these restrictions can be removed in some other theory. I do not, after all, insist that my own account or fragment of an account is itself immune to revision; but relative to the account I propose to develop, the items in UK are to be considered immune to revision.

Once this restriction in the scope of my proposed account of the improvement of knowledge is understood, certain other objections to my requirements on corpora of knowledge are at least mitigated. On my view, changes in knowledge or, more accurately, changes in commitments as to what is true, certainly true and infallibly true, are shifts from one deductively closed set to another.

As a consequence of this, if X should discover that g is a deductive consequence of his corpus $K_{X,t}$, that 'discovery' does not, on my view, constitute a change in X's state of knowledge. Moreover, there can be no change in X's corpus which represents a change in knowledge of logic, set theory or mathematics.

Given that X's state of knowledge as I construe it is a system of commitments of the kind specified above, this consequence does not imply
any denigration of the importance of logical or mathematical knowledge
or of changes in such knowledge. What it does imply is that my account
of changes in knowledge is limited in its scope of application. It cannot
deal with changes in logical or mathematical knowledge. This is a limitation which I do not know how to mitigate in any adequate way.

Notice that on my view the items in UK are distinguished from other
items in $K_{X,t}$ by their incorrigibility. They are *not* distinguishable with
respect to their truth, necessary truth or certainty. From X's point of
view at t, every theoretical assumption, statistical claim, universal generalization and observation report in his corpus at t is as certainly and
necessarily true as any truth of logic – at least as far as the conduct of
practical deliberations and scientific inquiry are concerned. Those who
think otherwise are guilty of confusing infallibility with incorrigibility.[22]

Every change in X's knowledge state (expressible in L) is a shift from
one deductively closed set K_1 containing UK to another K_2 also containing
UK. Given any pair of such sets K and K', it is possible to move from one
to the other by a sequence of shifts of the following two types:

(a)   *Expansions* where one shifts from K_1 to a K_2 containing K_1 as
        a proper subset.
(b)   *Contractions* where one shifts from K_1 to a K_2 which is a
        proper subset of K_1.

Every shift from K to K' is either an expansion, contraction or decomposable into a sequence of expansions and contractions.[23] Thus, if we
can find an adequate account of conditions under which expansions and
contractions are justified, we should be able to furnish a satisfactory view
of the conditions under which shifts in knowledge are, in general, justified.

Expansions fall into two broad categories:

(i)   expansions via routine, habit or conditioned response.
(ii)   expansions via deliberate decision.

When X makes some sort of observation, this often issues in his expanding his corpus of knowledge. I say 'often' and not 'invariably' because even though X may respond to sensory stimulation by acquiring an
intense feeling of conviction that h or a disposition to vigorously affirm

that *h*, he might – after the first blush of conviction is over – refuse to frame subsequent decisions on the assumption that *h* is certainly and necessarily true. We (both as individuals and as communities) may have little control over our immediate responses to sensory stimulation – but we often have considerable control over our considered responses. In science, this is what matters.

Nonetheless, it is true that one way in which a corpus of knowledge is expanded simulates direct response to sensory stimulation. *X* is committed to adding to his corpus sentences, which he is disposed to affirm to be true in response to sensory stimulation, provided that these responses are made under conditions which yield reliable responses. These conditions are specified by the corpus of knowledge with which *X* begins and are themselves subject to revision. Scientists have often revised and refined their judgments as to what kinds of conditions yield reliable observations. Though empiricism requires that some observation routines be counted reliable, it does not imply that all are or that such judgements as to which routines are reliable are immune to any revision during the course of inquiry itself.

Expansion via observation is by no means the only sort of routine expansion. *X* may follow the practice of adding information furnished by *Y* on a given topic to his own corpus on the grounds that *Y* is an expert. Expanding via the testimony of others is as much a form of routine expansion as is expanding via the testimony of the senses. In both cases, expansion is the outcome of following some procedure, practice or routine which may be applied over and over again and which has on any given application a constant statistical probability or chance of yielding error free information. Following a reliable routine (one with a high statistical probability or chance of avoiding error) over and over again can with considerable plausibility (which, nonetheless, requires explanation) be expected to inject error into *X*'s corpus only rarely. Of course, such routines do sometimes lead to erroneous expansions – indeed, to expansions where the result is an inconsistent corpus. (Thus, observations made according to reliable procedures can lead to contradictions with theories and other assumptions already in *X*'s corpus. *Y*'s testimony can contradict assumptions already in *X*'s corpus.) This is a regrettable feature of such routines; for expansion into error is inimical to the aim of obtaining error free information. However, following a routine which sometimes leads to such error may be compensated for by the fact that far more frequently the routine breeds error free information.

Because routines used for expansion either via observation or taking the testimony of other agents are rarely if ever perfectly reliable, epistemologists often conclude that knowledge acquired in this way is fallible.

It is, indeed, true that the routines which are used are fallible in the sense that they are less than perfectly reliable. And prior to using a routine to obtain new information, X will recognize the possibility that the result of such use will be an error. But once X has used a routine and has added the results to his corpus, the information obtained is used in subsequent deliberations and inquiry to rule out as impossible, in the practical sense relevant to such inquiries and deliberations, any claims inconsistent with these results. Relative to X's expanded corpus, the testimony obtained from the senses or from others is as certain and necessary as anything else in the corpus. In this sense, knowledge obtained via observation is infallible even though the routines used to obtain it are fallible.

The crucial point to emphasize here is that the fallibility of the routine employed (and of the source of information used in applying the routine) does not, from the point of view of X who has used the routine, infect the claims he incorporates into his body of knowledge once they are added. X uses a fallible routine in order to add to his corpus of items counted as certainly and infallibly true new information in subsequent inquiry. There would be no purpose in his employing such a routine if he were to remain uncertain of the testimony obtained via the routine after he has followed it. At any rate, the routine would no longer be a routine for expansion of a corpus but a routine of some other sort. Those who think that knowledge is fallible in the sense relevant to the conduct of inquiry and deliberation because such knowledge is obtained from fallible sources via fallible routines must either think the acquisition of such knowledge to be illegitimate because the way it is to be used requires counting it as infallible or have some view of the services rendered by knowledge in inquiry different from that proposed here. They must adopt a view of the services rendered by knowledge and belief in deliberation and inquiry which correlates the kinds of services rendered with the pedigree of the items of knowledge and belief under consideration. I reject this approach. According to my view, once an item is incorporated into a corpus, its pedigree becomes irrelevant to its uses in subsequent inquiries and deliberations. (This applies, by the way, not only to knowledge obtained via observation but to knowledge obtained via induction through deliberate expansion as well.)

Notice that as long as X assumes that a given routine for expansion is reliable and has the skill to apply it, there can be no question of justifying the result of expansion via that routine. The result itself requires no justification. It is, as we say, a routine affair. A problem of justification arises only when the question is whether the routine is reliable. But then what we are concerned with is changes in X's corpus concerning the reliability of a routine and not with changes in X's corpus due to the application of a routine assumed to be reliable.

By way of contrast, expansion via deliberate decision does require justification. In cases where such expansion is at issue, X has some question or problem he is considering, has somehow identified a set of relevant answers to that question (or, alternatively and equivalently, a set of potential expansion strategies) and is concerned to choose one of these in the hope of choosing an answer which at one and the same time will be true and will satisfactorily and relevantly answer the question being raised. In such expansion, one expansion strategy is compared with others to see which of them is best. In routine expansion, there is no comparison of alternative expansion strategies in some specific case but an invocation of a routine which singles out a definite strategy in response to the circumstances of application and without regard to the alternatives. The expansion strategy is not justified over its rivals as in expansion via deliberate decision.

Expansion via deliberate decision might also be called expansion via inference or, perhaps, via inductive inference if we are allowed to equate 'inductive' with 'nondeductive'. In any case, I think that expansion of this sort is appropriate when deciding which of rival theories to add to a corpus, in cases of inductive generalization, and in various sorts of 'inverse' statistical inference.

There is, of course, considerable controversy concerning expansion via deliberate decision. Authors like Peirce, Neyman, and Pearson seem to deny or have denied that such expansion is ever appropriate. They insist that all expansion (which Peirce would recognize as inductive inference and Neyman and Pearson call 'inductive behavior' on one interpretation) is expansion according to some principle, habit or routine whose merits depend on the 'long run' characteristics of its applications (its 'operating characteristics' in Neyman-Pearson jargon). I think this view is misguided; but the points under dispute are complex and cannot be considered here. My own position is that both sorts of expansion play an important role in the improvement of knowledge.

One important difference between routine expansion and deliberate expansion is that the rational agent X can expand into an inconsistent corpus via a routine expansion but not according to a deliberate one. Anyone who is interested in obtaining error free information will not regard a deliberate expansion strategy which will for certain yield error, as superior to one which does not. If X is deliberating as to which of alternative expansion strategies to adopt, he should never choose expansion into a contradiction.

Routine expansion, whether it be via observation or through reliance on the testimony of others, does not involve comparing alternative expansion strategies on the available evidence. Rather it is like choosing an expansion strategy on the basis of the outcome of a random experiment on some chance device such as a roulette wheel. Once X adopts such a stochastic routine, which expansion strategy is selected depends entirely on the outcome of the random experiment. Similarly, once X has committed himself to adding whatever Y says on a given subject to his corpus or whatever sentence he feels disposed to accept in response to sensory stimulation under conditions C, the expansion strategy he adopts depends entirely on the outcome of the random experiment (as long, of course, as X does not renege on the routine). Clearly, relying on such a routine over and over again could lead to adding items into X's corpus contradicting assumptions already in it.

Some empiricists seem to think that the fact that observation reports often contradict assumptions already in a corpus is a virtue of expansion via observation. In my opinion, exactly the opposite is true. It is a defect of such expansion (and expansion through reliance on the testimony of others) that it leads to error and contradiction. Unfortunately, the evil must be tolerated; for routine expansion furnishes types of information which cannot be gained legitimately in any other way – information which is often critical to the use of knowledge in the guidance of practical conduct.

The usefulness of empirical and human sources of information is not to be found in some alleged epistemological priority. Items added to a corpus via observation are no more certain or necessary than any other items in the corpus. Nor are they less immune to revision. The point is that such sources are often the most efficient (and sometimes the only) sources of information about specific events and situations relevant to practical conduct or to the testing of scientific hypotheses.

This is an unfortunate fact. I am an empiricist (in the sense that I believe that some routines for expansion via observation are legitimate) and rely on the testimony of others (in a similar sense) out of necessity. I also think Plato was quite right to disparage these sources of knowledge because they lead to error. I think he was wrong to suppose that we could transcend them.

Expansion via deliberate choice is the sort of modification of human knowledge on which I focused attention in *Gambling with Truth*. I shall return to this topic shortly.

For the present, however, I wish to consider contraction. It is contraction which poses one of the major difficulties to my contention that corrigibility is not equivalent to fallibility. In contracting his corpus, X must remove items from the corpus which from his point of view are certainly and necessarily true. In so doing, he suffers a loss of information. Surely to follow such a practice is inimical to his aim to obtain error free information.

I disagree. There are two sorts of occasions where X might justifiably contract his corpus:

(i) X detects a contradiction in his corpus – a contradiction which might perhaps have been generated by a routine expansion. When that is the case, X will contract in order to eliminate error. To be sure, he will lose information as a consequence. However, he will attempt to contract in such a way as to minimize the loss of information subject to the constraints that contradiction be eliminated and that where such contradiction can be eliminated in several ways by removing one of several conflicting claims each of which is more or less equally attractive from an informational point of view, each of these should be removed and regarded as a hypothesis. Once that is done, X is in a position to conduct further inquiries without begging questions in order to decide which of these hypotheses to add to his corpus subsequently in a new expansion step.

(ii) X may sometimes have good reason to contract his corpus even though he detects no contradiction in it. Suppose, for example, that X accepts in his corpus the truth of phenomenological thermodynamics. Assuming that statistical mechanics is, indeed, inconsistent with phenomenological thermodynamics, X would be committed to regarding statistical mechanics as certainly false. Even so, X might recognize that statistical mechanics systematizes a given domain more effectively than thermodynamics does. As such, it is informationally more attractive than phenomenological thermodynamics. (It must be kept in mind that the informational value of a hypothesis is independent (save in certain special cases) of its truth value. False claims may be informationally more attractive than true ones.) For this reason, X might have good reason for giving statistical mechanics a hearing. To do this, however, he will have to contract his corpus by removing phenomenological thermodynamics. If he does contract, thermodynamics will cease to be 'knowledge' or 'evidence' or 'background knowledge' and will become a 'hypothesis'. Statistical mechanics will cease being certainly false and will become a possibly true hypothesis. To proceed in this way is in keeping with the aim of obtaining error free information. In contracting, X does not import error – i.e., false assumptions. That can only be done via expansion – from X's point of view. Contraction does indeed lead to loss of information. That loss is compensated for by the fact that the hypothesis which prior to contraction was counted as certainly and necessarily false and after contraction is considered possibly true is informationally more attractive than the hypothesis removed from X's corpus in the contraction step and for this reason merits a hearing.

Notice that the fact that X contracts to give some hypothesis a hearing does not mean that X will subsequently be justified in adding that hypothesis to his corpus in an expansion step. Sometimes the old hypothesis will be reinstated. Sometimes the new one will. And sometimes neither will gain entry into his corpus.

Of course, on some occasions X will shift from a corpus K_1 containing h via contraction to a corpus K_2 which no longer contains h and relative to which both h and a contrary h' are rival hypotheses. He might subsequently shift via expansion to K_3 by adding h' to K_2. If we ignore the intermediate corpus K_2, this looks like a shift from a corpus K_1 containing h to another K_3 containing h' which is contrary to h. From X's point of view initially, he would be foolish to replace h by h'. He would be deliberately replacing certain truth by certain error.

The intermediate shifts should not, however, be ignored. The replacement is to be analyzed as a contraction followed by an expansion. X's aim at the contraction stage was to contract in order to give h' a hearing (presumably because of its explanatory or other information features). Once, however, this modification has been made, his point of view changes. His corpus is different. Moreover, his immediate project has altered. He previously was concerned to contract in order to give h' a hearing. Now he is concerned to expand in order to obtain new error free information. At each stage in the process, X has a definite proximate aim or goal. These distinct goals share certain features in common – they are both instances of aiming at acquiring error free information. But they are different instances. Moreover, they are proximate aims of the operations to be undertaken.

From the point of view of authors like Peirce and Popper this response seems absurd. According to both these authors, obtaining a true complete story of the universe is an ultimate and long run aim of inquiry. When X begins his replacement with K_1, he should, if he takes the long view, regard the sequence of shifts from K_1 to K_2 and then to K_3 as fruitless relative to this long run goal precisely because error is thereby imported into X's corpus.

In my opinion, it is because Peirce and Popper regard seeking the truth as the ultimate and long run aim of inquiry that they cannot endorse an account of how replacement of one theory by another is justified of the sort I have just outlined here. As a consequence, they insist that there could be no infallible knowledge unless that knowledge was also immune to revision.

But it is their position and not mine which seems untenable. In different ways, both Peirce and Popper come to the conclusion that avoidance of error is not a desideratum of the proximate aims of specific inquiries. For Popper, the proximate aim of a specific inquiry is to devise and test hypotheses. Testworthy hypotheses, however, need not be true ones or even probably true ones. Indeed, a good case can be made (and Popper makes it) that improbable hypotheses may be better than probable ones if the aim is to falsify them. Peirce, on the other hand, regards relief from doubt as the proximate aim of any specific inquiry. Both of them have considerable difficulty in showing how the pursuit of proximate aims of special inquiries of the sorts they consider help promote the ultimate aim of inquiry as they pursue it. As I suggested earlier, Peirce and Popper succeed in placing truth on a pedestal so remote from the immediate concerns of science as to make it irrelevant to scientific inquiry.

I wish to proceed differently. There is no single objective which is the ultimate aim of inquiry. Moreover, there are as many distinct proximate aims as there are distinct demands for the modification of bodies of knowledge. I contend that these distinct proximate aims share certain features in common. They are all cases of seeking to obtain an informative and error free corpus.

It is important, in this context, to be clear as to the sense in which X is concerned to avoid error in modifying his corpus. From his point of view at t when $K_{X,t}$ is his corpus, no item in $K_{X,t}$ is false. When he assesses the possibility of error in considering a modification of his corpus, $K_{X,t}$, he does so on this assumption.

Furthermore, I contend that avoidance of error is a proximate desideratum of his effort at modification. By this I mean that X is concerned to avoid error as assessed relative to $K_{X,t}$ only in shifting from $K_{X,t}$ to another corpus via contraction or expansion. He is not concerned with assessing the prospects of incorporating error into his corpus on some subsequent shift. To be sure, once he has shifted from $K_{X,t}$ to $K_{X,t'}$, he will be concerned to avoid error on a subsequent shift; but possibility of error is assessed relative to $K_{X,t'}$ and not $K_{X,t}$. Peirce and Popper seem to hold that X should consider the possibility of error in shifting from $K_{X,t'}$ to some new corpus even prior to X's shifting to $K_{X,t'}$. This, I think, is part of the force of their view that truth is an ultimate aim of inquiry. It is this contention that I deny.

Once this is understood, there seems to be no difficulty in rationalizing the contraction from K_1 to K_2 and the subsequent expansion from K_2 to K_3. From X's point of view at the time that his corpus was K_1 it would indeed be irrational for him to replace h by h'. But the problem before him at that time was whether he should remove h from his corpus or not in order to give h' a hearing. Even if he anticipates at that time that relative to K_2 he would be justified in adding h' and expanding to K_3 relative to the proximate aim he would have in deciding whether to expand at that time, that would not matter to him; for his concern is not with what the other consequences of using K_2 in inquiry and deliberation will be but only whether in shifting to K_2 he will be giving h' a hearing. As long as both h and h' are hypotheses relative to K_2 in a manner which begs no questions that condition will be met.

Thus, I concede that my rejection of the equation of corrigibility and fallibility does prove untenable if one endorses the view that avoidance of error is an ultimate and long run desideratum of inquiry. I reject that thesis about the aims of inquiry. But I do not abandon the view that avoidance of error is a desideratum in inquiry. This is one case where myopia is a virtue. Our concern with avoidance of error in considering any specific modification of a corpus is the shortest of short runs concerns. It applies only to the immediate revision under consideration and not to subsequent ones. If this thesis is granted, the objection to denying the equivalence of corrigibility and fallibility collapses.

III

The rules for inductive acceptance proposed in *Gambling with Truth* were devised as part of an account of expansion of a corpus of knowledge by deliberate choice. At t, X has some corpus $K_{X,t}$. He is concerned to answer some specific question. The question is a demand for new information of some sort.

Part of X's task is to identify potential answers to that question. This aspect of X's task corresponds roughly to what Peirce called 'abductive' inference. It is, perhaps, misleading to call such abduction 'inference'; for the 'conclusion' of an abductive inference is not the incorporation of new items into X's corpus of knowledge but the identification of a system of alternative expansion strategies or potential answers to the question under consideration.

At the abductive stage, X is concerned, in a sense, to clarify the question he is considering. Abduction so construed may be said, in a suitably circumscribed sense, to have its own 'logic'. Abductive logic, as I understand it, is a system of norms prescribing necessary conditions which a system of potential answers to any legitimate question should satisfy. Such an abductive logic should not be expected to uniquely determine, in any given context, what the complete set of potential answers are or should be. X's demands for information are relevant factors. So is his capacity for identifying potential answers.

In *Gambling with Truth*, I indicated how, given a list of strongest consistent potential answers represented by what I called an 'ultimate partition', a complete list of potential answers or expansion strategies could be generated.[24] The procedure for doing this is regulated by principles of abductive logic.

Of course, there are other principles to consider. Potential answers may be expected to be susceptible to testing in some appropriately specified way.[25] In contexts where X is seeking an explanation of some phenomenon, potential answers should satisfy whatever pass for necessary conditions for potential explanations. In *Gambling with Truth*, I did not consider matters such as these. Nor shall I do so here. However, one of the useful byproducts of the scheme proposed in *Gambling with Truth* or of improved versions of that scheme is that it can be used to check proposed models of potential explanation so that philosophers of science do not have to rely so strongly on presystematic observations concerning the uses of the verb 'to explain'. A proposed system of necessary conditions for potential explanations can be combined with the criteria for expansion I propose to ascertain which among rival potential explanations will be favored under various hypothetical circumstances.

In addition to identifying potential answers, X may wish to engage in collateral efforts to expand his corpus in order to answer subsidiary questions pertinent to his problem. Such expansion may be routine expansion via observation or appealing to the testimony of others. Or it may involve other expansions via deliberate choice, i.e., via induction.

At any given stage in his inquiry, however, X may consider the following question: Given the corpus of knowledge which he has at that stage together with the potential answers he has identified, which of the potential answers or expansion strategies would be the best to adopt were he to make a decision at that point?

The criteria for acceptance I proposed in *Gambling with Truth* were designed to determine which of the potential answers should be identified as the one to favor under such circumstances. However, I explicitly denied that identifying such a potential answer was sufficient warrant for adding it to one's corpus of knowledge.[26] To do so would be tantamount to supposing that X was entitled at that stage of inquiry to terminate deliberation and choose an expansion strategy. There are several reasons why X should on some occasions refrain from doing so. I cited some of these in *Gambling with Truth*.

One of the main obstacles to terminating inquiry might be that X recognizes more than one question as requiring an answer where the ultimate partitions associated with these two questions together with X's corpus of knowledge warrant expansions which yield an inconsistent corpus. When such a situation arises, X should not expand his corpus relative to either question. Rather he should reconsider his demands for information and return to the abductive stage so that he can formulate a question which will accommodate suitably modified versions of the two original questions. Alternatively, he may seek more data (expand his corpus via observation and the testimony of other agents) so that answers to the two questions can be given which do not clash. Or perhaps, further efforts at data collection should be undertaken.

Because I wished to distinguish between acceptance in the sense in which my rules licensed acceptance on the often counterfactual assumption that no obstacles exist to choosing an expansion strategy and acceptance when that assumption is factual, I distinguished between mere acceptance or acceptance as true and acceptance as evidence. h is accepted as evidence when it is admitted into X's corpus of knowledge via expansion and is subsequently counted as necessarily and certainly true. If h is merely accepted, then it remains, for X, less than certain and possibly false.

Many critics, including Ian Hacking[27] and Risto Hilpinen,[28] have complained about the sensitivity of my rules of acceptance to the choice of an ultimate partition. This sensitivity would be objectionable were acceptability relative to an ultimate partition itself a sufficient warrant for acceptance as evidence. But this was clearly denied in *Gambling with Truth*. Mere acceptance is relative to the choice of an ultimate partition; for it is relative to a specific question and the potential answers identified for that question. This seems to me to be precisely as it should be.

Acceptance as evidence via inductive expansion, on the other hand, is relative to X's problem situation – i.e., the set of questions he recognizes to be serious at the time and the systems of potential answers he has identified for those problems. I assume (and I think that Peirce would agree) that not every question which with logical possibility might be raised at a given time is a serious one. And it is logically possible for X to have identified potential answers to any question serious for him even though he does not do so. But expansion no more takes into account the questions it is logically possible to take seriously than it does the logical possibility that some sentence in his corpus prior to expansion is false.

It is important to keep in mind that X's problem situation is open to revision just as his corpus of knowledge is. A major topic in epistemology should be to provide a systematic account of considerations which might rationally lead to such revision or, at any rate, to determining the extent to which such revision is subject to critical control. I did not consider this topic in *Gambling with Truth* nor shall I do so here. However, since I did not set out to discuss the problem of revising problem situations in *Gambling with Truth*, it is not an objection to the proposals made there that I failed to do so. On the other hand, because my criteria for acceptance are sensitive to the problem situation, the account of the improvement of knowledge offered in *Gambling with Truth* is incomplete. I readily acknowledge this.

One of the central tenets of *Gambling with Truth* is that criteria for rational acceptance ought to be such that expansion strategies favored by such criteria are optimal strategies for promoting the investigator's effort to obtain information pertinent to his problem while avoiding error. Moreover, the criteria for optimality to be employed are the same criteria which are employed in assessing options in any 'practical' decision problem. The difference between 'practical' decision problems and cognitive ones is to found in the aims generated by such problems and the options pertinent to realizing these aims – not in the general criteria for rational goal attainment.[29]

To develop such an approach requires adopting a system of criteria for rational goal attainment. In *Gambling with Truth*, I adopted the criterion of maximizing expected utility. In my opinion, this is one of the major defects of the theory developed in that essay. What is objectionable here is not that maximizing expected utility is recommended as rational in those cases where X has the subjective probability and utility judgments necessary to apply that principle. I think that in such cases, maximizing expected utility remains an adequate criterion. Rather it is a fact that in real life men often lack the subjective probability and utility judgments required in order to apply the principle. This is so whether the problem is a practical or a cognitive decision problem. What is needed is a more generally applicable criterion for rational choice which can treat the "bayesian" case when the principle of maximizing expected utility does apply as a special 'limiting case'. I have recently published an outline of just such a theory of rational choice and hope eventually to bring it to bear on the question of acceptance rules.[30] Space does not permit an elaboration of these proposals here. I shall, therefore, continue to operate with the fiction that relative to his corpus of knowledge X has a 'credal state' represented by a probability function assigning to all sentences in L a numerical probability consistent with the requirement that all items in his corpus bear probability 1. I shall also assume that he adopts a 'confirmation function' which specifies for each 'potential corpus' to which he might shift (i.e., to each deductively closed expansion of the urcorpus UK) an associated credal state. If $c(h; e)$ is that confirmation function and K is the expansion of UK obtained by adding e and forming the deductive closure, X's credal state is represented by a function $Q(h; g)$ such that $Q(h; g) = c(h; e \& g)$. (g is consistent with $K_{X, t}$ and e with UK.)

All of this is excessively unrealistic as is the parallel assumption that in the context of a given decision problem X's goals and values can be represented by a utility function defined over the possible consequences of his options unique up to a linear transformation. Nonetheless, some light can be shed on the more realistic situations of daily life and scientific inquiry by considering these idealisations.

In *Gambling with Truth* and a subsequent article, 'Information and Inference', I proposed a utility representation (unique up to a linear transformation) of the invariant features of a certain class of cognitive decision problems concerned with expansion. I regarded such epistemic utility functions as themselves a function of two conflicting desiderata: an effort to avoid error regardless of the information afforded and an effort to obtain new information regardless of whether error is avoided or not. For each of these desiderata, I introduced a separate utility function and then represented the goal of obtaining error free information as a weighted average of these utility functions subject to the constraint that no potential error should be accorded higher utility than any potential correct answer.[31]

The technical details of these proposals and my reasons for proceeding in this way were outlined in the two essays just cited. When the resulting epistemic utility function is combined with X's credal state to determine an expected utility for every expansion strategy available and a suitable rule breaking ties for optimality with respect to expected utility is introduced, the resulting acceptance rule is the following:

> *Rule A*: Reject every element h of the ultimate partition whose subjective probability $Q(h)$ relative to $K_{X,t}$ is less than $qM(h)$.

The index q can take positive values less than or equal to 1. It represents the way in which the utility for avoiding error and the utility for information are weighted. q increases with an increase in the degree of 'boldness' exercised or a decrease in the degree of caution.[32] The function $M(h)$ is a probability measure defined over the ultimate partition which is such that $1 - M(h)$ is the informational utility or value of adding h to X's corpus (and forming the deductive closure) when considerations of truth value are ignored.

It is important to keep the two probability functions mentioned in rule A distinct. If $K_{X,t}$ is the deductive closure of e and UK, $Q(h) = c(h; e)$. The Q function is used to compute expected utilities both in cognitive and practical decision problems relative to $K_{X,t}$. $M(h)$ is neither equal to $c(h; e)$ nor to $c(h; t)$ (where t is in UK). If it were required to be equal to $c(h; e)$, rule A would preclude rejecting any element of the ultimate partition and no expansion of $K_{X,t}$ could ever take place. If $M(h)$ were equal to $c(h; t)$, we would obtain a result equally as unattractive. Suppose that h and h' are equivalent given $K_{X,t}$ but not equivalent given UK. It could be the case that $c(h; t)$ does not equal $c(h'; t)$ although, of course, $c(h; e) = c(h'; e)$. Now given $K_{X,t}$, any expansion strategy involving rejection of h involves rejection of h' as well. The same increment of information is obtained by rejecting the one as rejecting the other. Yet, if $M(h)$ is equated with $c(h; t)$ and $M(h')$ with $c(h'; t)$, this could not be the case. Moreover, the application of rule A would lead to the untenable result that relative to the same question the same expansion strategy could be both recommended and rejected.

In a paper published in 1963, I proposed equating $M(h)$ with $c(h; t)$ without realizing its objectionable character.[33] By the time I wrote

Gambling with Truth, I had already abandoned it on the grounds just indicated. I did not explain my reasons at the time since it seemed to me that to do so would be to engage in a private soliloquy. Unbeknownst to me at the time, Hintikka and Pietarinen were proposing the same idea[34] – an idea which has subsequently been defended by Hilpinen. In 'Information and Inference', therefore, I did explain my objection.[35] Since Hilpinen has, nonetheless, insisted on defending the equation of $M(h)$ with $c(h; t)$, the objection bears repeating.[36]

The main point to keep in mind is that the Q-function used in rule A has a different function in inquiry than the c-function and that both of these probability measures have different functions from the M-function. The latter is used to represent informational utility. The c-function together with X's corpus determines his credal state – i.e., his Q-function. And, X's Q-function determines expected utilities in both cognitive and practical decision problems. The idea that all of these probability measures must be related in a definite way derives historically from uncritically accepted assumptions to be found in the pioneer writings on information of Popper, Carnap and Bar Hillel. There is no reason in the world why we must embrace them.

In *Gambling with Truth*, I restricted the scope of application of my proposals to cases where the information determining probability measure $M(h)$ assigns equal value to all elements of the ultimate partition. Because of this, I did not claim that these proposals could be employed in situations where some elements of an ultimate partition are theoretical hypotheses. I suggested that for those cases we should regard the cognitive goal of expansion as including other desiderata besides avoidance of error and the acquisition of information. I suggested that we might introduce utility functions representing the simplicity, explanatory power, and other desirable features of theories and that a weighted average of these utility functions, the utility of truth, and the utility of information might be constructed to represent the composite utility function relative to which expected values were to be computed.[37]

In 'Information and Inference', I changed my position.[38] I suggested that in all cases of expansion via induction, the epistemic utility function could be construed as a weighted average of the utility of truth and the utility of information. At the same time, I abandoned the assumption that all items of the ultimate partition be assigned equal M-value. In many contexts, a uniform distribution of M-values over elements of the ultimate partition will prove appropriate. This will be so in many cases of simple inductive generalization and statistical estimation. On the other hand, when it comes to choosing between theories, such uniform distributions will typically be inappropriate.

Suppose that X is seeking a theory to systematize some subject matter and has identified two theories T_1 and T_2 as potential solutions to his problem. His ultimate partition will then contain three elements: T_1, T_2 and a residual hypothesis R asserting that T_1 and T_2 are both false. (Typically, X's corpus of knowledge $K_{X,t}$ will allow for the possibility that R is true.) Given the demand for information involved, the informational value of R will be less than that of the other elements of the ultimate partition. Adding R to his corpus will yield a relatively unsatisfactory answer to his question. Moreover, X might regard T_1 as simpler in some sense or explanatorily more satisfactory according to some criterion than T_2. These discriminations will mean that he will rate T_1 informationally more valuable than T_2 which in turn is informationally more valuable than R. This implies that $M(R)$ will be greater than $M(T_2)$ which will be greater than $M(T_1)$. Thus, in applying rule A, R will be prone to rejection even though its Q-value is very high whereas the Q-values of T_1 and T_2 might need to be considerably lower before these hypotheses are rejected.

On this approach, considerations of simplicity, explanatory power, etc. no longer function as desiderata additional to informational value and truth which are to be averaged in to the epistemic utility function. Rather they serve as factors which determine the informational values of rival hypotheses.

This approach has the advantage over the method I originally suggested of treating all cases of inductive expansion in a unitary way. Rule A applies across the board. To achieve this result, it is necessary to abandon the assumption that all elements of the ultimate partitition must be counted as equally informative. But as Richard Jeffrey has rightly pointed out, my arguments for imposing this requirement were extremely weak.[39] I now take the position that the appropriate M-function to adopt depends upon the demand for information occasioned by the question under consideration. Arguments concerning the adoption of one M-function rather than another are to be evaluated (insofar as there is a right and a wrong to the matter) according to principles of abductive logic just as is the eligibility of hypotheses to be elements of an ultimate partition for a given question.

Another deficiency of the proposals made in *Gambling with Truth* is that the proposals were restricted to cases where the ultimate partition is finite. The difficulties concerning cases where the ultimate partition contains countable and even non-countable elements (as when estimating the value of some real valued magnitude) derive from technical problems concerning the application of the principle of maximizing expected utility and the introduction of expectation determining probability measures and utility functions over infinite sets of elements.

I know of no unitary and satisfactory way of treating cases where infinity rears its perplexing head. This is true, however, not only in the context of cognitive decision problems but in cases of practical decision making as well. Nonetheless, there are ways of handling specific categories of situations which are fairly noncontroversial and satisfactory extensions of the bayesian approach to decision theory. If a given option in a decision problem has a continuum of consequences indexible by real numbers in a given finite interval, a continuous expectation determining probability distribution can be defined over that system of hypotheses and a continuous utility function can be defined over the same domain, the expected utility of the option becomes $\int f(x) u(x) \, dx$ where the integral is taken over the interval of reals, $f(x)$ is the expectation determining density function and $u(x)$ a corresponding utility determining density function. A different approach may be required when the alternatives are countably infinite or where the real variable is unbounded at one or both ends (especially where the ultimate partition consists of specifications of n-tuples of real values). Whether the problem is a practical or a cognitive decision problem, the technical problems are substantially similar and to the best of my knowledge are currently handled with the aid of a strong dose of common sense and a clear understanding of the problem under consideration.

Suppose that X uses an expectation determining density function $f(\alpha)$ where $\underline{\alpha} \leqslant \alpha \leqslant \bar{\alpha}$. The Q-value assigned to the hypothesis that the true value of α falls in some subinterval of $[\underline{\alpha}, \bar{\alpha}]$ (or, more generally, in some measurable subset) is obtained by evaluating the appropriate definite integral of the function $f(\alpha)$. Similarly, X can use an information determining density function to determine the information determining probability of such a hypothesis.

Let h_A assert that the true value of α falls in some measurable set A. The expected utility of accepting h_A as strongest is equal to $\int_A f(\alpha)\,d\alpha - q\int_A m(\alpha)\,d\alpha = \int_A (f(\alpha) - qm(\alpha))\,d\alpha$. The maximum value this integral will take for any measurable set A of values of α will belong to any such set A^* such that (i) every value of α such that $f(\alpha) - qm(\alpha)$ is positive is in A^* and (ii) no value of α such that $f(\alpha) - qm(\alpha)$ is negative. The rule of ties recommended in *Gambling with Truth* favors picking the set A^* such that conditions (i) and (ii) obtain and, in addition, all values of α such that $f(\alpha) - qm(\alpha) = 0$ belong to the set.

The upshot is that in cases of estimation of the sort under consideration, X should reject all and only elements of the ultimate partition such that $f(\alpha)$ is less than $qm(\alpha)$. This, in effect, is rule A applied to cases of estimating the value of a parameter taking all real values in some finite interval and where both expectation determining and information determining probabilities are characterized by densities. Familiar examples of this are estimation of binomial parameters and of n-tuples of multinomial parameters.

Special difficulties of other sorts arise when the number of elements in the ultimate partition is countably infinite or where these elements are all the real values from $-\infty$ to $+\infty$ or from some finite value to either $-\infty$ or to $+\infty$. In particular, if one wishes to assign equal informational value to all elements of the ultimate partition or equal expectation determining probabilities to all these elements, it seems that one cannot do so without violating the requirement of countable additivity. Other strategems are available for handling such cases. I shall not discuss these matters here.

Suppose that X has a corpus K_1 and a question under consideration. Rule A (or one of its variations) recommends a given conclusion at an appropriately high level of caution and X's problem situation does not contain other questions for which rule A recommends conflicting conclusions. X is in a position to accept as evidence the conclusion which K_1 and rule A warrant his merely accepting. In that case, X shifts to the corpus K_2 which is the result of such expansion. He can then proceed to consider whether relative to K_2 rule A licenses adding still more information pertinent to the question X is considering. X can ask the same question over and over again until he has exhausted the resources of using rule A in this reiterated way.

In *Gambling with Truth*, I called this process 'bookkeeping'.[40] Hilpinen has objected to it without, as far as I can make out, offering any reason. He appeals to a condition of adequacy which states, in effect, that bookkeeping is illegal. But that is scarcely an argument. It is an expression of Hilpinen's disagreement.

Once what is merely acceptable satisfies the other conditions for acceptance as evidence, it beomes as certainly and necessarily true as other items already in X's corpus. It has as much claim to be used as premisses or evidence in subsequent investigation (such as considering the old question relative to the new knowledge) as any other item in X's corpus. That is the sense in which I understand acceptance as evidence. Hilpinen may wish to follow Jeffrey and eliminate all acceptance as evidence (save, perhaps, for logical and other such truths) although he does not seem to want to endorse such a position. If he does, however, reject the legitimacy of ever accepting any statement as evidence, in what sense does he think we accept hypotheses? In the sense of mere acceptance? My understanding of that notion suggests that there would be no interesting use for it or for rule A unless acceptance as evidence via induction were sometimes legitimate. And if acceptance as evidence is sometimes legitimate in this way, I can see no basis for rejecting bookkeeping which is a straightforward consequence of such legitimacy.

Once the legitimacy of bookkeeping is recognized, investigator X can consider what the results of bookkeeping would be relative to his question and what he knows were he justified (perhaps counter to fact) in converting what is merely accepted to what is accepted as evidence. Such calculations may have relevance to determining whether, given his problem situation, conflicts between questions will arise and so prevent acceptance as evidence and may also suggest directions in which X should undertake new investigations.

In *Gambling with Truth*, I did not appreciate the usefulness of the bookkeeping device in imposing constraints on degrees of caution. In that essay, I thought that the only restrictions to be imposed on the value of q is that it should be greater than 0 and less than or equal to 1. In a highly qualified way, I continue to endorse that view; but the qualifications are important.

Let the ultimate partition be finite and $q=1$. Let $Z(h)=Q(h)/M(h)$. It is demonstrable that by reiterating the process of bookkeeping a finite number of times all elements of the ultimate partition except the one bearing the maximum Z-value will be rejected. When all elements of the ultimate partition bear equal M-values, this means that the unrejected element will be the one bearing maximum Q-value.

As the value of the q-index is decreased, other elements of the ultimate partition bearing lower Z-values (or, where all elements of the ultimate partition have equal M-values, lower Q-values) will tend to escape rejection. Indeed, when Q-values are very low, bookkeeping will lead to only marginally stronger conclusions than those allowed by a single application of rule A.

Where X is concerned to estimate the value of a parameter taking real values in some finite interval, the process of bookkeeping will not necessarily terminate in a finite number of steps, but it will lead asymptotically to the conclusion that all elements of the ultimate partition but those bearing the maximum value of the ratio $z(\alpha)=f(\alpha)/m(\alpha)$ are to be rejected. Of course, where the information determining density is uniform, the asymptotically unrejected point values are those for which the expectation determining density is a maximum.

Keith Lehrer has persistently advocated a rule of acceptance which recommends rejecting all elements of an ultimate partition except those bearing maximum (expectation determining) probability.[41] Rule A and bookkeeping lead to the same result when all elements of the ultimate partition are equally informative and $q=1$. I do not think this is a happy result. Lehrer's proposal leads to conclusions which conflict with presystematic judgments on one of those rare occasions where presystematic judgment is virtually noncontroversial.

If I toss a coin known to be unbiassed ten times, I am not prepared to predict that it will land heads exactly 5 times. If I toss it 100 times, I am not prepared to predict that it will land heads exactly 50 times. My reluctance is not based on any doubts as to the truth of the claim that the chance of heads is 0.5. I would not make the predictions even if I was certain that 0.5 is, indeed, the chance of heads. The point is that I would refuse to rule out hypotheses asserting that the relative frequency of heads will differ from heads by some small amount. Yet, Lehrer's rule leads to the contrary result. So does bookkeeping with my rule when $q=1$ and all elements of the ultimate partition are equally informative.

I do not think that this constitutes an objection either to rule A, to bookkeeping, or to regarding elements of the ultimate partition as equally informative. Rather it points to the dubiety of exercising the minimum degree of caution as indexed by $q=1$.

Nonetheless, there is a strong case for refusing to prohibit using the index $q=1$ under all conditions.

In estimating the value of a real parameter by the procedures described when q is less than 1, the best one can do via rule A and bookkeeping is to obtain an interval estimate of the value of the parameter α. Obtaining additional data may lead to narrowing the interval but it cannot lead to a point estimate. When the interval does become suitably small, the investigator X might quite reasonably conclude that acquisition of further data will yield very small dividends and cut off such inquiry.

Sometimes such a small interval estimate will be sufficiently satisfactory to warrant termination of the inquiry. But this will by no means always be the case. Three examples may serve to illustrate the point.

(1) In estimating the value p of the chance of obtaining heads on a toss of a coin, imagine that X observes a large number of tosses and they all land heads. I shall analyse this kind of example in greater detail in the next section. It will become clear that under suitably specified conditions, X will be able to conclude that p falls in an interval from $1 - \varepsilon$ to 1 when the value of q is less than 1 but will not be able to conclude that p equals 1. Yet, on some occasions, it may be appropriate to conclude that the coin will always land heads on tosses and $p=1$.

(2) Abner Shimony has considered the case where it is assumed that the electrostatic force on an object is directly proportional to $r-\alpha$ where r is the distance between the object and some other electrically charged object. The problem is to estimate the value of α. It is typically concluded that α equals 2 exactly and not merely that it falls within some small interval around 2.[42]

(3) Suppose that a population of organisms which are all hybrid with respect to some pair of alleles are randomly mated and data as to the percentage of offspring which are purebred dominant, purebred recessive and hybrid collected. On the basis of the data, one might conclude that the chance of obtaining a purebred dominant offspring from hybrid

parents is approxmately $\frac{1}{4}$, of obtaining purebred recessive offspring is also approximately $\frac{1}{4}$ and of obtaining hybrid offspring is approximately $\frac{1}{2}$. However, there is considerable pressure to conclude that the values for these chances are exactly $\frac{1}{4}$, $\frac{1}{4}$, $\frac{1}{2}$ respectively.

These three cases share certain important features in common. The available knowledge warrants accepting a rather strong interval estimate of the value of a parameter. (In the third example, the estimate is of the values of a triple of multinomial parameters.) Further data might yield narrower interval estimates but not precise point estimates according to my procedures when q is less than 1.

In all three cases, however, there are under these circumstances pressures to obtain a point estimate of the parameter in question. Moreover, the pressures derive from similar considerations. In case 1, the value $p=1$ is distinguished from values near it by the fact that when $p=1$ one can conclude via deduction what the percentage of heads will be on an arbitrary number of tosses of the coin. This is not true for any of the values of p arbitrarily close to 1. As Shimony points out, in the second example, the value $\alpha=2$ implies that the flux across the surface of a sphere with the point source at its center is independent of the radius of the sphere. This is not true for values of α arbitrarily close to 2. Appealing to a generalization of this claim, Shimony notes that values for α other than 2 would yield inverse α laws not readily embeddable in available comprehensive formulations of the principles of electricity.[43] In the third example, the precise values $\frac{1}{4}$, $\frac{1}{4}$, $\frac{1}{2}$ could be readily embedded in simple versions of Mendelian genetics in a manner in which arbitrarily close but different point estimates could not.

Thus, in all three cases, one point estimate is distinguished from all the others allowed as possibly true according to the interval estimate obtained via rule A and bookkeeping with a low degree of caution. Their distinguished status derives from the problem the investigator is considering and his demands for information. These factors might conceivably lead to his regarding these distinguished estimates as informationally more valuable than their rivals. But this need not and, I suspect, will not, in general, be the case. Typically one point estimate will count as equally informative as any other. What will happen, however, is that when X has concluded that the true value of the parameter lies in some small interval including the distinguished estimate, he will shift his attention from the question as to which point value is the correct one to the question as to whether the distinguished point value is the correct one or not. That is to say, he will change his ultimate partition so that instead of having an ultimate partition containing all point values in the interval, he will consider only the two hypotheses which assert that the true value is the distinguished one and the point value is not the distinguished one. The distinction of the distinguished value will occasion an interest in settling the question as to whether it is the correct value.

But now we face a new technical problem generated by considerations of infinity. The informational value of the point estimate (the hypothesis h_α) will equal 1; for its informational determining probability $M(h_\alpha)$ equals 0. Similarly, its expectation determining probability $Q(h_\alpha)=0$. Moreover, the corresponding M and Q values for the residual hypothesis $-h_\alpha$ are both 1. Applying rule A will lead to failure to decide between these two hypotheses at any q-value.

Remember, however, that h_α has a density $f(h_\alpha)$ (which is conditional on the assumption that the true value of α falls within the interval estimate now accepted as evidence). Moreover, this density is greater than $qm(h_\alpha)$ where the m-value is the corresponding information determining density and the q-index is the one originally used to obtain the interval estimate.

Let h_α^δ assert that the true value falls within a narrow interval around the distinguished point α of width δ and consider the application of rule A to the partition consisting of this hypothesis and its negation. (δ is narrower than the width of the original interval estimate.) Suppose that for arbitrarily small δ, h_α^δ is accepted – i.e., its negation is rejected. It seems plausible to regard this as sufficient warrant for accepting the point estimate expressed by h_α when it and its negation are the two elements of the ultimate partition.

Observe, however, that if the q-index of caution is less than 1, the residual hypothesis $-h_\alpha$ cannot be rejected by using this asymptotic version of rule A. Nor, for that matter will h_α be rejected if the q value used is the one used to obtain the original interval estimate. However, if q is then set equal to 1, the results of using the asymptotic version of rule A run as follows: if $f(h_\alpha)>m(h_\alpha)$, then the negation is rejected. If $f(h_\alpha)<m(h_\alpha)$, then the point estimate is rejected.

These considerations suggest that we treat those situations where exercise of a low degree of caution yields an interval estimate containing one distinguished value and where the distinction derives from the demands of the problem of a sort which make it desirable to render a verdict as to whether that value is the correct one or not by shifting to a two-fold ultimate partition (relative to the assumption that the interval estimate is accepted as evidence via rule A, bookkeeping, and a low value for the q-index) and by 'throwing caution to the winds' through shifting the value of the q-index up to 1.

Thus, it seems to me that occasions can arise where an investigator has reached a stage of inquiry where his demands for information may very well induce him to modify his ultimate partition and change his degree of caution. This seems entirely legitimate in just those cases where a decisive verdict concerning some hypothesis is wanted and where the prospects for obtaining it are dim through continuing inquiry through data collection or in some other way.

This position is in substantial agreement with some of the themes developed by Abner Shimony in his excellent monograph 'Scientific Inference'. It emphasizes, as he does, the role of the investigator's problem situation in determining how he evaluates hypotheses and it concedes, as Shimony does, that "seriously proposed" hypotheses be allowed a hearing with the prospect of receiving a favorable verdict.

There is, however, one fundamental point of principle with respect to which Shimony's approach and mine differ. According to Shimony's 'tempered personalism', identifying a serious hypothesis is itself sufficient warrant for the investigator to modify his expectation determining probability judgments. His view seems to be that in situations like those under consideration, distinguished point estimates should be accorded positive Q-values even though prior to their estimates becoming distinguished in this way, such assignments need not have been made.[44]

I do not object to the fact that Shimony allows for changes in credal states of confirmation functions without their being any change in X's corpus of knowledge. As I argue myself in 'On Indeterminate Probability', revision of confirmation functions (or confirmational commitments) is an important topic in epistemology.[45] Moreover, I agree that X's demands for information is a factor relevant to justifying such changes. However, adoption of a system of probability judgments to assess risks of error seems to carry with it some sort of judgment about the way the world is just as accepting an item into a corpus of knowledge does. Although it may be inappropriate to baptise confirmation functions as true or false in the sense in which items in a corpus of knowledge are, there does seem to be something analogous to the possibility of error involved in changing probability judgments. And that factor (however it is to be spelled out) ought to play a role along with X's demands for information in modifying expectation determining probabilities.

Shimony's tempered personalism allows no place for such considerations. On his view, it is enough that a hypothesis has acquired the status of being seriously proposed to juggle probabilities so as to accord it positive probability. This is so even though the considerations he himself appeals to in defending the consideration of a hypothesis as seriously proposed refer to features of such a hypothesis which have much to do with its informational value but little to do with whether it is true or false. In this sense, Shimony's approach seems to endorse a form of wishful thinking which, in my opinion, is unacceptable.

The approach I favor locates the privileged status of seriously proposed hypotheses where it belongs – namely as a reflection of X's demands for information. I, like Shimony, agree that as the course of inquiry develops the problem situation might require making modifications. But the modifications are to be made either (i) in the assessments of informational value, (ii) the choice of an ultimate partition or (iii) in the choice of a degree of caution. None of these kinds of alterations demand a change in X's judgments concerning probabilities which determine risk of error.

I am not sure that the particular technical stratagems I have proposed here to handle the question of point estimation are the best that can be devised for this purpose. However, they do conform to the philosophical requirements I have just insisted upon in opposition to Shimony. The pedantic details require further scrutiny; but these proposals do illustrate that one does not have to endorse tempered personalism in the form apparently implied by Shimony's proposals in order to confront the important questions which he raises.

The considerations which might warrant throwing caution to the winds in the context of estimation of the value of a real valued parameter may also be relevant in other contexts as well. Suppose that X is concerned to decide between two theories T_1 and T_2 and some residual hypothesis R

asserting that the two theories under consideration are both false. In the course of investigation, X might devise some new theory by repartitioning the residual hypothesis. But whether he does so or not, he can well recognize that save for limitations of time and his own imagination, he might have done so. Be that as it may, the residual hypothesis is not a theory. Given his demands for information, its informational value will be substantially less than those of either of the two theories. As a consequence, its M-value will be substantially higher.

It may be argued that $M(R)$ should equal 1 on the grounds that it is equivalent to a 'disjunction' of an infinite number of unspecified rival hypotheses all of which are as informative or nearly as informative as either T_1 or T_2. Perhaps, a similar argument might be invoked to assign $Q(R)=1$. I do not know how to adjudicate this matter. But if the argument seems convincing, then I submit that it would be fair to regard the problem as analogous to the problem of estimating the value of a real valued parameter – at least in a qualitative manner. Doubts about the propriety of the analogy, it should be noticed, raise doubts about the original argument to the conclusion that $M(R)$ and $Q(R)$ both equal 1. The solution of the problem stands or falls with the legitimacy of the problem in the first place.

Suppose, however, that both T_1 and T_2 are assigned positive but low M-values. $M(R)$ would then be high but not equal to 1. We may suppose that Q-values are also distributed so that initially T_1 and T_2 have low but positive values.

If the q-index of caution is set at a low value, then as long as $Q(R)$ is greater than $qM(R)$, the residual hypothesis will not be rejected by rule A unless $Q(R)$ is very low. Ex hypothesi, it is not very low and may not become very low even after data has been obtained. What may happen after data has been collected is that one of the two theories will be rejected leaving the other theory and R as the two survivors.

Here, as in the case of estimating a real valued parameter, the investigator may demand a more decisive verdict as to the truth value of the surviving theory. It is desirable to have the matter settled; for the investigator is looking for a theory to use in subsequent investigations as part of his background knowledge or evidence. If the theory in question will suit, his demand will be satisfied. If it does not, he should go back to his drawing board and attempt to carve a new theory out of the residual hypothesis which does survive. But a decision depends on a verdict concerning the surviving theory and the residual hypothesis.

I suggest that, under the circumstances as described, the investigator should throw caution to the winds and shift to $q=1$. He should do so, it should be emphasized, only after he has rejected all specific rival theories using a higher degree of caution. X should be very cautious in rejecting rival theories. He may be less so when deciding between a theory and a residual hypothesis and where the focal demand becomes to render a verdict one way or the other.

Suppose that X does proceed in this fashion and ends up accepting T_1 as evidence. Subsequently, an informationally attractive theory T_3 is brought up for consideration which conflicts with T_1. T_3 presumably is carved out of the discredited residual hypothesis R. From X's current point of view T_3 is false. Yet, if the theory is sufficiently attractive, he may contract by removing T_1 from his corpus and suspending judgment between T_1 and T_3. I mention this point to emphasize that the approach I am advocating here does not prevent reconsideration of conclusions reached at a previous stage. It does, however, insist that such reconsideration is legitimate only when the rival theory proposed is itself informationally attractive. Open mindedness ought not to degenerate into frivolous reconsideration of assumptions no matter what damned fool idea is proposed.

After X has contracted so that he now has T_1 and T_3 to consider, he need not consider a residual hypothesis as well. After all, in contraction, he should take care to minimize the loss of information. In my opinion, the importance of considering residual hypotheses looms large in those situations where the initial background knowledge contains little in the way of settled theoretical information concerning the problem in question. In a context where there is such a background and X contracts in order to give some novel hypothesis a hearing, there is no need for seriously entertaining a residual hypotheses.

In deciding between T_1 and T_3, therefore, the pressures which lead to throwing caution to the winds no longer exist. X may proceed to examine the two hypotheses exercising a very high degree of caution.

To sum up this discussion of the adoption of a q-value indexing degree of caution, I suggest that the q-value ought to be substantially less than 1 unless there is some excusing circumstance along the lines described above. I have no firm conviction as to what an appropriate numerical value for q ought to be which will satisfy "all but the virtually sceptical" but I suspect that many sceptics will remain dissatisfied unless it is substantially less than 0.5.

Before terminating this section, it may be useful to mention that instead of dealing directly with the problem of selecting a suitable q-index, one can adopt a variant of the notion of degree of confidence of rejection which I explicated in *Gambling with Truth*. Let $d(h)$ – the degree of confidence with which h is rejected – equal $1 - D(H)$ where $D(H)$ is the maximum value for q relative to which h fails to be rejected by bookkeeping with rule A. If h fails to be rejected even when $q=1$, $D(h)=1$ and $d(h)=0$. This version of the measure of degree of confidence is slightly different from the version proposed in *Gambling with Truth*.[46] However, it has the same formal properties – which are those attributed by G. L. S. Shackle to his measures of potential surprise.[47]

Using this concept, instead of selecting a definite degree of caution for use in employing rule A, we can require that a hypothesis be rejected only if it is rejected with a specified degree of confidence. The two formulations are equivalent. Furthermore, if we take the degree of confidence of acceptance $b(h)$ to be equal to $d(-h)$, as I did in *Gambling with Truth*, we can reformulate the requirement as stating that h is acceptable only if it is acceptable with a sufficiently high degree of confidence. This formulation may help ease the discomfort of those who think that a necessary condition for accepting an item into evidence is that prior to such acceptance it be 'believed' to a very high degree.

IV

To illustrate the workings of rule A in numerical detail, consider once more the problem of estimating the value of a binomial parameter p as in the case of estimating the chance of obtaining heads on a single toss of a given coin.

Prior to making observations, suppose that X's corpus K_1 and credal state B_1 is such that he assigns a uniform expectation determining probability distribution to the values of p between 0 and 1. Moreover, his problem situation is such that he is concerned to identify the true value of the parameter p so that he does assign equal informational value to each point estimate. His expectation determining density function $f(p)=1$ for all values of p. His information determining density function $m(p)$ also equals 1 for all values of p.

Suppose that X observes n stochastically independent tosses of the coin and expands via observation by adding e_r asserting that r of the tosses land heads and $n-r$ land tails to his corpus. As a result, he obtains corpus K_2. Assuming that his credal state is changed according to conditionalization, his expectation determining density function $f(p; e_r)$ is equal to

$$\frac{(n+1)!}{r!(n-r)!} p^r (1-p)^{n-r}.$$

This distribution is unimodal with a maximum value at $p=r/n$. The mean of the distribution is $(r+1)/(n+2)$ and its variance is $(r+1)$ $(r+2)/((n+2)^2 (n+3))$. As n increases, it can be approximated by the normal distribution.

If $q=0.1$, applying rule A requires rejecting all point estimates of p for which $f(p; e_r)$ is less than 0.1. If bookkeeping is invoked, then the result can be computed in the following way. (This method will work in this example and many variants of it but not necessarily in all cases.) Let $Z(p)$ equal $Q(h_p; e_r))/M(h_p)$ where h_p asserts that the true values of the parameter is some value whose density is greater than or equal to $f(p; e_r)$. $Q(h_p; e_r)$ is the Q-value of h_p conditional on e_r which is obtainable by computing the integral $\int h_p f(p; e_r)\, dp$ where h_p is now taken to be the set of values of the parameter meeting the condition just specified. $M(h_p)$ is equal to $\int h_p m(p)\, dp = \int h_p\, dp$ which is the length of the interval in which all and only members of h_p fall.

In the special case under consideration, there will be at least one and at most two values of the parameter meeting the following condition: $f(p; e_r)=Z(p)$. It is demonstrable that bookkeeping will lead to rejecting all and only values of p whose densities are less than the densities of the points satisfying this condition.

The estimate of the true value of p obtained in this manner will assert that p falls in an interval approximately centered around the observed relative frequency of heads r/n. The larger the number of tosses for fixed degree of caution, the narrower the interval is and the more symmetrical the interval is around r/n. How narrow the interval estimate will be depends, of course, on the index of caution.

These results conform as well qualitatively with presystematic judgment as anyone could expect. Of course, most rival approaches can claim much the same in the binomial example. Differences between approaches are exhibited in differences in the lengths of the interval estimates for given sample size and in the rationales and epistemologies on which these approaches are based. Presystematic judgment does not speak clearly at all concerning precise numerical values of end points of interval estimates. A fortiori the points under philosophical dispute are not to be settled by reference to judgments of this sort. l claim only that my approach can be applied as effectively as alternative approaches to problems of interval estimation. This, in my opinion, holds true not only for simple binomial problems but for other situations where presystematic judgment speaks fairly clearly. I shall not, however, enter into other cases here.

There is, to be sure, one deficiency in the procedures described here. I have presupposed that the prior expectation determining density for $q(p)$ is uniform. This assumption is questionable in two respects. First it is not irrational for X before sampling to employ a nonuniform density. Second, X might initially have a credal state which is not representable by a numerical density.

The first of these objections is partially mitigated by the fact that no matter what prior density X employs, for sufficiently large n, the qualitative results described above will take hold. This is only partial mitigation; for when the number of observations is small, the results can diverge in striking ways. Indeed, if X's prior is heavily biassed – say in favor of values of p around 0.01, rule A and bookkeeping could lead to accepting an interval estimate even prior to sampling and even when X is very cautious.

This is scarcely an objection to rule A. If X does have such a prior credal state, then I see no reason why X should refuse to reach conclusions without making observations. Such conclusions are scarcely a priori (whatever that is supposed to mean) and objectionable as Hilpinen makes them out to be. [48] If X's corpus of knowledge and credal state are initially biassed in favor of the value 0.01, X may have very good reason to expand his corpus without the benefit of observation.

Why should X expand via observation first in a case such as this? Some might say: in order to test X's bias in favor of 0.01. But if X's corpus and credal state are such as to yield this bias, from his point of view there is no need to test the bias. To test the bias would be to question it. But it would then be incoherent for him to adopt that bias (which is being subject to test) in computing a posterior density.

But perhaps it is not the bias which is being tested. Rather it is the battery of hypotheses about the true value of p. But why should we test these hypotheses? Popperians seem to believe that such testing is virtuous either for its own sake or because somehow (the Popperians never explain this clearly) it helps promote the acquisition of truth in the long run. I have already rejected these views. In my opinion, testing hypotheses via observation is desirable when that is the optimal way to obtain a corpus of knowledge on the basis of which expansion via induction can legitimately be undertaken in order to answer the question under consideration. But when the prior corpus and credal state are heavily biassed in favor of the value 0.01, observation is, perhaps, not needed for this purpose. Given that making observations often incurs economic, moral, political or other practical costs and that, in general, observation routines are rarely perfectly reliable and, hence, incur risks of error of their own, there may be very good reason to expand relative to the prior corpus and credal state.

Of course, the more serious objection is the second one. When we think of cases of estimating the value of parameter p presystematically, we normally consider situations where X's credal state and corpus are such as to render him substantially 'ignorant' of the value of p. Bayesians assume often that such ignorance amounts not only to suspension of judgment as to what that true value is but to lack of bias in X's probability judgments in favor of one value rather than another. Hence, for them, such states of ignorance are adequately represented by a uniform prior. My own view is that adopting a uniform prior is an expression of lack of bias. But that is far from ignorance in the appropriate sense. If X is not 'ignorant' when he thinks that probabilities are biassed in favor of 0.01, he is not ignorant when he uses an unbiassed probability distribution. Both credal states are quite as definite as one another. They both lead to as decisive recommendations concerning conduct. 'Ignorance' of the required sort, in my opinion, obtains when X's probability judgments are numerically indeterminate and cannot be represented by a numerical probability distribution.

As I stated earlier, I do not intend to discuss in detail an account of indeterminate probability judgment here. Speaking very roughly, however, I propose to characterize X's credal state by a convex set of probability distributions. For example, X's prior credal state for the value of p might be all weighted averages of beta distributions of the form $\Gamma(x+y)/\Gamma(x)\,\Gamma(y)\,p^{x-1}\,(1-p)^{y-1}$ where x and y are both positive and $x+y=1$.

Again speaking roughly, the rule of acceptance appropriate relative to such a credal state or to a credal state obtainable via conditionalizing on observations of the outcome of tosses (which credal state would be a convex set of beta distributions as well) involves first rejecting all values of p which are rejected according to rule A when any density in the set is used and then reiterating the procedure according to bookkeeping. In the example I have just given, prior to observation, X should refuse to reject any value of p. He would have good reason to conduct experiments. If he observes only a small number of tosses, the conclusions reachable via rule A remain very uninformative so that he should look for more data. As n increases, he will be able to reach stronger conclusions via induction – conclusions which agree qualitatively with those sustained by pre-systematic judgment on the assumption that X is initially ignorant.

Thus, the deficiencies in my initial account of estimation of the value of p derive not from defects in rule A but from the dubious assumption that when X is 'ignorant' his probability judgments are numerical precise. This deficiency can, in my opinion, be remedied in keeping with the epistemological approach which argues in favor of rule A in the first place. Moreover, the results of proceeding in this way seem to remain in excellent qualitative agreement with presystematic judgment.

In spite of the reservations I have just expressed concerning the bayesian use of uniform priors to represent ignorance, there are, as the example under discussion illustrates, situations where such uniform priors do simulate the qualitative results obtainable with the use of more realistic indeterminate priors and computations using uniform priors may often serve to provide helpful approximations to more realistic results (although this claim deserves much more careful critical scrutiny than I am in a position to offer here). For the present, I shall continue to use them.

Let us now consider a numerical example of estimating the binomial parameter p when the prior density is uniform. Suppose that X observes

20 tosses of the coin and 10 of them are observed to land heads. His posterior beta distribution $q(p; e_{10})$ for the parameter p equals $21!/10!^2 p^{10}(1-p)^{10}$.

If X's degree of caution is indexed by the value $q=0.1$, rule A recommends accepting the interval estimate $0.23 < p < 0.77$. Bookkeeping strengthens the conclusion to $0.257 < p < 0.743$.

The error probability customarily computed by bayesian statisticians for this result is 1.5%. It should be emphasized, however, that my rejection rule is not based on consideration of error probabilities.

In order for rule A to warrant rejecting the hypothesis that $p=0.5$, X would have had to observe 4 or less tosses landing heads or 4 or less landing tails in the 20 tosses. If bookkeeping is legitimate, X would have had to observe 5 or less tosses landing heads or 5 or less tosses landing tails. If the procedure is construed as a significance test for the null hypothesis that $p=0.5$, the significance level (corresponding to bookkeeping) is 1.18%.

As I pointed out before, the recommendations yielded by rule A are not automatically to be admitted into X's corpus through expansion. His problem situation must not present obstacles due to conflicting demands for information. Moreover, the degree of caution used must be such as to satisfy all but the virtually sceptical.

I am assuming that in this example, no such obstacles stand in the way of expansion. There is, however, a further consideration which might appear to stand in the way. Suppose that X has the opportunity to toss the coin 180 more times. Given his demands for information, he clearly should do so provided that the practical costs of doing so are not excessive and the risk of error in expanding via observation of 200 tosses is not too great. No matter what the outcome of such observation might be, X will anticipate being able to expand via induction based on such observations to a much narrower interval estimate than before. If, for example, he should observe 90 heads and 90 tails in the new observations and use bookkeeping, he would obtain an interval estimate that $0.41 < p < 0.69$.

The question which might be raised here is whether X should, prior to making the observations of the 180 tosses, expand relative to the observations of the first 20 tosses or, instead, refrain from expansion pending the outcome of the next 180 tosses? In other words, does X's intention of acquiring more data constitute an obstacle to his expansion relative to the data he now has?

In my opinion, it should not do so. Yet, there is a consideration which might seem to argue in favor of the contrary view. Given what X knows after having observed that 10 out of 20 tosses land heads and prior to expanding via induction relative to that data, it is possible that all 180 tosses will land heads so that 190 out of 200 tosses will have landed heads with a relative frequency of 0.95. If X were to expand relative to the initial 20 tosses, he would have ruled out the hypothesis that $p=0.95$. When he subsequently observes the additional 180 tosses which land heads, unless he contracts his corpus by removing the first interval estimate, he will be precluded from estimating that the true value of p falls within an interval around 0.95. And should he be justified in contracting (I shall not consider this matter here), it seems as though his initial expansion would be pointless. This does not appear to be an attractive predicament.

This objection is not as serious as it seems. Given X's credal state after observing the twenty tosses, the probability of his being confounded in this manner must be quite small. This need not be true if X uses a high value for the degree of caution. But when the degree of caution exercised is considerable, the prospect of being confounded will also be minimized. Whatever degree of caution X uses, that represents his view of what is a fair trade off between risk of error and informational benefits. That judgment is also an index indirectly of his anxiety concerning being confounded by information to be obtained subsequently. Consequently, anyone who is reasonably cautious should not, given his commitments, be anxious about the prospect of being confounded.

Of course, X might be confounded by his subsequent observations. In that event, he faces a problem as to whether he should contract his corpus and if so how he should do it. This is a question which arises not only in cases of estimation of the sort under consideration but in other and often more important contexts of scientific inquiry. It is not, however, the topic under consideration here.

We can make the binomial estimation problem do double service for us. In his earlier formulations of confirmation theory,[49] Carnap considered simple languages containing only one place predicates as extralogical predicates and where every such predicate is equivalent to a disjunction of so-called 'Q-predicates'. In case the universe of discourse consists of n distinct individuals (for which there are n distinct individual constants in the language), one could consider the possible frequency distributions of Q-predicates among the n-individuals (structure descriptions). Each such frequency distribution corresponds formally to a possible outcome of n stochastically independent trials of an k-faced die where k is the number of Q-predicates and possible outcomes are described in terms of the frequency with which each face of the die lands up. Such a process generates a multinomial distribution with k parameters $p_1, p_2, ..., p_k$ where the parameters sum to 1. If the values of these k-parameters are unknown, we can consider a probability density over the possible $(k-1)$-tuples of independent parameters and, in this way, induce a definite probability distribution over the structure descriptions. Furthermore, if n is allowed to go to infinity, we can consider infinitely large structure descriptions as being characterized by a $(k-1)$-tuple of real numbers representing limits of relative frequency of the corresponding Q-predicates. The probability distribution induced over these k-tuples will be identical with the probability distribution over the k-tuples of multinomial parameters. Thus, the problem of drawing conclusions concerning which structure description is the true one relative to information about some set of the individuals in the universe is formally the same as the problem of estimating the values of multinomial parameters. And the binomial estimation problem is the simplest case of the general problem of multinomial estimation. The discussion of the binomial problem previously applies with only small modifications to the general multinomial case.

All of the confirmation functions in Carnap's lambda system belong to a subclass (unduly restricted) of the family of beta distributions. In particular, his favored measure c^* in the binomial case corresponds to a uniform prior density over the values of p and in the general multinomial case to a uniform distribution over the $(k-1)$-tuples of multinomial parameters.

Setting aside qualms about the excessive simplicity of Carnap's models, they can be used to illustrate an important point about inductive inference (provided allowances are made for the excessive simplicity). It has often been objected that Carnap's confirmation measures assign 0 confirmation values to universal generalizations in infinite universes even when the

number of positive confirming instances is large and there are no dis-
confirming instances. This is taken to be a defect to Carnap's scheme and
some authors sympathetic to Carnap's program – most notably Hintikka
and his school – have undertaken to introduce confirmation measures
which do not have the defect.[50]

I think the anxiety is gratuitous. If one uses a uniform prior over the
values of p in the binomial case and obtains a large number of heads
and no tails in a sequence of tosses, then even when the q-index adopted
is very low, rule A and bookkeeping favor concluding that the true value
of p falls between 1 and $1-\varepsilon$. We have a situation where the narrow
interval estimate is justifiably accepted into evidence and where one point
in the interval – namely, $p=1$ – bears special theoretical interest. Moreover, further observation of experiments will not yield a point estimate
but only a narrower interval estimate than before. Given the urgency of
obtaining a verdict as to the truth value of $p=1$, X might be justified in
shifting to an ultimate partition consisting of the hypothesis that $p=1$
and its negation and throwing caution to the winds by setting $q=1$.
(Notice that in this example, throwing caution to the winds will yield
the same results whether X shifts his ultimate partition or not; for the
posterior density for p is at a maximum where $p=1$.)

Thus, inductive generalization can be rationalized on my approach
without assigning positive probabilities to all 'constituents' in Hintikka's
sense. Moreover, the results described for the binomial case extend in a
natural way to multinomial situations where there are more than two
Q-predicates. (In those cases, variety of instances can be recognized as a
relevant factor). Finally, the methods I propose allow for situations where
the data warrants accepting a statistical hypothesis as well as for situations where data justifies accepting a universal generalization. Normally
(as is the case in real life) the statistical hypothesis accepted will be
'composite' (e.g., an interval estimate of the value of p). But in those
situations where some 'simple' hypothesis is distinguished by the demands
of the question (as in the Mendelian example cited previously), data can
be such as to warrant the acceptance of the precise point estimate of the
statistical parameter. Hintikka's theory as it is now presented is quite
incapable of handling that situation or cases like those discussed by
Shimony.

I do not mean to suggest that Hintikka's measures are somehow incoherent and should not be adopted by rational agents. My view of confirmation theory as partially sketched in my recent 'On Indeterminate Probabilities' implies that rational agents might, on some occasions, adopt measures from the Hintikka family as their own. However, I do not think that it is a good reason to do so solely in order to be able to accept universal generalizations. Not only is that kind of reason objectionable due to the same kinds of considerations which argue against Shimony's 'tempered personalism', but it is unnecessary to assign positive probabilities to universal generalizations in order to accept them.

Of course, my alternative to the approach favored by Hintikka, Pietarinen and Hilpinen involves the use of rule A and bookkeeping. Hilpinen has objected both to rule A and to bookkeeping. Followers of the Finnish School hold fast to a high probability rule for acceptance and so, from their point of view, are pressured into assigning positive probabilities to constituents.

But aside from the objection I raised against Shimony's following a somewhat parallel approach, it is important to appreciate that unlike Shimony and Harold Jeffreys before him, members of the Finnish School adopt confirmation measures which allow them to assign positive probability to universal generalizations forgetting that on some occasions we seek point estimates of statistical or other parameters which, on their view, must continue to be assigned 0 probability. It is a virtue of Shimony's approach that he does not forget this fact. My rules for acceptance can be used to justify the appropriate recommendations in the cases of concern to the Finnish School while avoiding a commitment to assigning universal generalizations positive probability and assigning simple statistical hypotheses 0 probability.

The problems of binomial and multinomial estimation which are at the core of the discussion of this section do not, of course, begin to exhaust the problems of statistical inference in the light of which the merits of rule A can be evaluated. Space does not, however, allow further discussion of how rule A works in standard cases of estimating normal means with or without knowledge of the value of the variance, of estimating normal variance with or without knowledge of means, estimating differences or ratios of means, etc. I do intend to develop some of the more elementary applications (those within the scope of my own limited mathematical abilities) elsewhere.

V

For the present, however, I shall turn to another unrealistic example which, however, has some general theoretical interest for statisticians and which Ian Hacking[51] and later Hilpinen[52] have used as the basis of criticism of my proposals.

Suppose that X faces a deck of cards. He knows that the desk was obtained at random from an urn of decks of 52 cards, $\frac{1}{3}$ of which were 'normal' decks, $1/78$ of which contain 52 aces of spaces, $1/78$ of which contain 52 Kings of spades, and so on for the 52 possible abnormal decks.

There are two questions X might ask:

Question A: Which of the 53 possible kinds of deck is the deck drawn from the urn? Here X's ultimate partition consists of 53 hypotheses: h_n asserting that the deck is normal and 52 hypotheses specifying 52 different kinds of abnormal deck.

Question B: Is the deck normal or abnormal? Here X's ultimate partition consists of two hypotheses: h_n and its negation.

Suppose that X is concerned only with question A. Assuming that all elements of the ultimate partition bear equal M-values of $1/53$, then if X should adopt a value for q greater than $53/78$, he will reject all elements of the ultimate partition except h_n which he will accept. Should he accept h_n as evidence? Given that no question in his problem situation conflicts with question A, that presents no obstacle. But the degree of caution he exercises might well do so. It must be greater than $53/78$ which is greater than $\frac{2}{3}$.

Suppose that X is concerned only with question B. Assuming that both elements of the ultimate partition bear equal M-values of $\frac{1}{2}$, then as long as q is greater than $\frac{2}{3}$ X will reject h_n. Again X's problem situation presents no obstacle to acceptance as evidence. But, once more the degree of caution X exercises might do so.

Both Hacking and Hilpinen think it a defect of rule A that, relative to question A, h_n is accepted and, that relative to question B, it is rejected. They give no reason for this except some appeal to intuition. But presystematic judgment is scarcely a guide in these matters unless one is clear as to the exact nature of the situation. As I explicitly stated in *Gambling with Truth*, the conclusions warranted by rule A are conclusions warranted in order to gratify a specific demand for information. Relative to a different demand, it is to be expected that a conflicting conclusion is to be expected. It is a virtue – not a defect – of rule A that it mirrors this feature. Moreover, the feature is of considerable use in inquiry.

I suppose that were X actually confronted with the situation, both questions A and B would be serious questions relative to his problem situation. By examining how rule A yields conflicting answers to these two questions, X would have good reason for refusing to accept either of them into evidence. He should, under the circumstances, defer further expansion until he has had an opportunity to inquire further into the matter.

Of course, as I just pointed out, the conflicting results are obtained only if X uses an unduly high q-value. And this will also prove an obstacle to expansion. Hence, X has two good reasons to refuse to expand his corpus relative to what he knows at the outset. Thus far, my analysis conforms to common sense. I see no difficulty for rule A but only an illustration of how it works as an indicator of considerations which warrant further inquiry rather than expanding relative to the information already available.

Suppose now that X does draw a card at random from the deck and observes that it is an ace of spades. Relative to both questions, his ultimate partition now consists of h_n and the hypothesis h_1 asserting that the deck consists exclusively of aces of spades. The probability he should now assign h_n remains as it was – to wit $\frac{1}{3}$. The probability of h_1 shifts from 1/78 to $\frac{2}{3}$.

There is now no conflict between the two questions. No matter which q-value X adopts, rule A will yield the same answer for both questions. If q is greater than $\frac{2}{3}$, h_n will be rejected. Otherwise X should suspend judgment. But the excessively high q-value may very well continue to present an obstacle to expansion as evidence.

It may be thought that there is another obstacle to expansion. X can consider that if he draws another card from the deck, there is a $\frac{1}{3}$ probability that he will draw a card different from an ace of spades. Hence, if he should reject h_n relative to what he knows prior to such a draw, there is a significant chance that subsequent observation will conflict with his corpus and require him to contract. As I pointed out previously, however, if X is prepared to be so bold as to adopt a q-value of $\frac{2}{3}$, then he is prepared to risk a probability of error of $\frac{1}{3}$ in order to obtain the information promised by rejecting h_n. And if he is prepared to do that, he should also be prepared to risk being confounded by subsequent observations so that he will have to contract his corpus after the later observation in order to remove contradiction. Of course, setting $q = \frac{2}{3}$ is to exercise a degree of caution far in excess of what, in my view, is warranted.

Thus, suppose that instead of the 'prior' probability of h_n being $\frac{1}{3}$ it was 1/53. After observing the ace of spades, the probability would remain 1/53. If q is slightly greater than 2/53 – say 0.04 – h_n would be rejected on the basis of the observation of a single card. But notice that the probability of being confounded on observing the results of a second draw are very small – namely 1/53. I suggest that if X is prepared to risk the error entailed by adopting $q = 0.04$ and expanding relative to information about the first toss, the prospect of being confounded by the results of the second toss will not and should not deter him.

When the prior probability of h_n is 1/53, how should X proceed prior to observing the first card? Let $q = 0.04$. Relative to question A, he will reject no element of the ultimate partition. Relative to question B, he will reject h_n. If both questions are serious, X should refuse to expand his corpus and should wait for further information. But if question B alone were serious, no obstacle to expansion would present itself. He could expand by rejecting h_n.

Hilpinen seems to think a result like this is objectionable on the grounds that it implies that X is entitled to reach a conclusion without the benefit of observation.[53] But I see no reason why we must always make new observations before modifying our corpus. When should we make observations in the conduct of inquiry? There are three important considerations: (i) the observation routines available to use in making the observations are known to be highly reliable; (ii) it is expected that whatever the outcome of observation will be the information obtained via observation will lead to a corpus on the basis of which one can draw more informative inferences via induction relevant to the question than one could without the observational data; (iii) the economic, moral, political and other practical costs of observation are not too high.

Now, in our example, the information which X could obtain via induction is just as useful to him as what he can hope to obtain after observation. Moreover, the risk of error entailed is quite low – perhaps as low as that to be obtained by using the observation routine. The benefits of observation are marginal. It is far from clear that X should conduct experiments and make observations under the circumstances. The knowledge he already has suffices for him to obtain the information he is looking for with an acceptable risk of error.

If we were to take Hilpinen's objection seriously, we would never expand via induction. We would always feel obliged to continue making further observations. The corpus we adopt at t cannot be used as the basis for expanding via induction; for to do that is to expand relative to the background we have at that time. We would need further observations. When could we stop? Surely it is not merely a matter of the number of observations. It depends on the corpus we have and on the demands for information being made. But once that is acknowledged, then the force of Hilpinen's objection evaporates.

In discussing the card example, I have attributed a definite prior probability distribution to the hypotheses about the constitution of the deck. I have secured a modicum of realism for this by supposing that X knows that the deck was drawn at random from a population of decks in which decks of the 53 varieties occur in known proportions.

We might, however, consider the, perhaps more realistic, version of this example where X has no such information about the origins of the deck. I suspect his probability judgments concerning the deck prior to making observations would be quite indeterminate. This could prevent reaching any conclusion relative to question A or question B. Moreover, except for incorporating the deductive consequences of making a single observation, X could not reach any definite conclusions on the basis of observing the result of drawing one card. Observing the second card does yield a decisive result.

Thus, in what seems the most realistic version of this example, X would suspend judgment until he had seen two cards.

In discussing Hacking's example, I have considered various interpretations of the example in order to see what could be the foundation of an objection to applying rule A in substantially the way I already recommended applying it in *Gambling with Truth*. Relative to any specific characterization of the example which fills in the relevant details concerning the context I find no clash between rule A and presystematic judgment.

Of course, if one does not take care to specify the values of the relevant 'contextual parameters', rule A can be made to yield conflicting recommendations. But to criticize rule A for this is tantamount to demanding that criteria for acceptance be formulated which are independent of the features of the context of application which I hold are relevant to the legitimacy of an expansion. I do not know how to prove to a sceptic that the factors I hold to be relevant are, indeed, relevant. Yet, none of the factors that I do take to be relevant are clearly irrelevant as far as presystematic judgment, statistical practice or scientific inquiry are concerned. Indeed, the presumption seems the other way around. Among the factors I take to be relevant are (a) X's corpus of knowledge prior to expansion (including both 'background knowledge' and observational data), (b) X's credal state, (c) X's problem situation, (d) the particular question under investigation, (e) the system of potential answers X has identified for that question, (f) the informational value X has attributed to the potential answers (which includes in appropriate contexts matters of simplicity, explanatory power and the like) and (g) the degree of caution exercised by X. Factors very much like these have often been recognized to be relevant to the legitimacy of inferences. What is novel about my proposals is not my exploitation of these factors but the manner in which I exploit them to furnish an acceptance rule. My proposals may be faulted, perhaps, for failing to take these factors into account adequately, for neglecting other relevant contextual parameters, or for regarding as relevant factors which are irrelevant. But to object to my proposals merely because they furnish conflicting recommendations depending on how these contextual parameters are specified is to miss the point. Of course, conflicting recommendations can be obtained by altering contextual factors. That is to be expected when one proposes criteria which are heavily context dependent for their application. Far from being a defect of my proposed criteria, this context dependence is a virtue.

Some authors seem to think that insofar as the legitimacy of expansion (or the legitimacy of revisions of credal states or the status of arguments as potentially explanatory or whatever) is dependent on context, it becomes a 'subjective' or 'pragmatic' affair and that seeking criteria for evaluating such legitimacy is tilting at windmills. As a consequence, they will object to my criteria because they are too subjective or pragmatic and will demand more 'objective' criteria for expansion or, while agreeing with my emphasis on the context dependence of legitimate expansion, will question my attempt to furnish criteria for legitimate expansion.

Whatever the actual merits of my proposals may turn out to be, I contend that they constitute a prima facie counterinstance to all such views. My criteria are heavily context dependent. The legitimacy of an expansion depends on the circumstances of the investigator X and his circumstances. Yet, my criteria rule out certain moves as illegitimate and recognize others as legitimate when the relevant contextual factors are specified. In this respect, they inject an element of 'objectivity' into what are often taken to be 'subjective' or 'pragmatic' affairs.

It seems to me that, all too often, philosophers tend to stop analysis precisely at the point where it becomes important to pursue it further. So-called 'pragmatic' issues are shunted to one side, relegated to sociologists and psychologists, or taken to be questions of skill (like riding a bicycle) not to be sullied by systematic investigation. We are invited to focus attention, instead, on those criteria which regulate all expansions (credal states, potential explanations or whatever) for all agents at all times under all circumstances.

But precious little can be said to be obligatory on all agents at all times and under all circumstances. Whether in science or in practical conduct, the most important activities depend for their evaluation on circumstances which are subject to change. To refuse to consider criteria for the objective evaluation of such activities (including modifying a corpus of knowledge or altering a credal state) because they are heavily context dependent is to stop philosophical inquiry at precisely the point where it becomes both interesting and important.

Columbia University

NOTES

[1] At t, x has in his corpus the sentences h_1 ('Smith owns a Ford'), h_2 (Smith works in the office') and, as a consequence, g ('Someone who works in the office owns a Ford'). At t, a second agent Y has a corpus consisting of $-h_1$ together with f_1 ('Jones owns a Ford') and f_2 ('Jones works in the office'). Of course, Y also has g in his corpus.

From Y's viewpoint, X should remove h_1 from his corpus because it is false. According to Y, X does not know that h_1 but falsely believes it. But Y would surely not recommend that X remove g from his corpus. To the contrary, he would favor X's retaining g. Should or would Y say that X knows that g? Once one understands that, from Y's point of view, X should keep g in his corpus, whether Y should say that X knows that g or not ceases to be an interesting question.

[2] R. C. Jeffrey [10], p. 160.

[3] Carnap's position is not entirely clear to me. But he does seem by and large to restrict his own discussion to situations where the only alterations in a corpus of knowledge are through the addition of observation reports. See, for example, R. Carnap [1], pp. 309–310.

[4] Carnap [1], pp. 316–317. That is to say, Carnap thought that no 'rational reasons' could be offered for 'accepting' a 'new proposition' on the basis of data.

[5] This is not quite accurate. Only instances of laws which are presystematically alleged to be accepted are regarded as almost certain. See Carnap [3], pp. 571–575.

[6] R. C. Jeffrey [10], Ch. 11.

[7] I have criticized Jeffrey's proposal in [19] and [16]. Jeffrey's response is to be found in M. Swain [26], pp. 157–185.

[8] Among the mere oversights will be such possibilities as that the coin will explode with the force of a hydrogen bomb or, still, worse, that tossing the coin will cause the earth to explode. How small will ε have to be before we would be prepared to accept a gamble on a toss of the coin if such possibilities are serious?

[9] See, in particular, A. Shimony [25], pp. 97–121.

[10] K. Popper [23], pp. 228–231.

[11] K. Popper [23], pp. 228–231 and 234–235.

[12] K. Popper [23], p. 226, p. 234, and pp. 235–236.

[13] K. Popper [23], pp. 215–217.

[14] Popper is sensitive to the charge ([23], pp. 230–231). But nowhere in the many pages Popper devotes to the question can I find any argument which shows that devising highly falsifiable and testworthy hypotheses and severely testing them is a better strategy to adopt in order to come closer to the truth than persistently modifying some cherished theory by means of ad hoc moves so as to render it immune to severe testing. I am not asking for any guarantee that Popper's favored strategy will succeed if pushed far enough but only for some good reasons, given our 'situation' for favoring it. The claim that it is a better strategy is far from self evident.

[15] Popper [23], p. 238.

[16] Popper [23], p. 238.

[17] On the very page on which the passage quoted occurs, Popper denies that we must start 'from scratch'.

[18] I conjecture that this is part of Popper's meaning in 'Two Faces of Common Sense' [22], pp. 78–81.

[19] I do not claim that these conditions exhaust those needed to link $K^i{}_{x,t}$ with $K^{i+1}{}_{x,t}$.

[20] $K^{i+1}{}_{x,t}$ might be altered in certain ways without changing $K^i{}_{x,t}$. But changes in $K^i{}_{x,t}$ will, perforce, breed modifications in the upper level corpus.

[21] Thus, my defense of deductive closure rests on the assumption that a corpus is to be used to distinguish between those logical possibilities which are serious and those which are not. Hence, I presuppose that the distinction between claims whose truth is logically possible and those whose truth is logically impossible can be made.

Hence, if X adds h to his corpus and also adds g, he is committed to ruling out the logical possibility that h is false and also the logical possibility that g is false. To do that is, I submit, to rule out the logical possibility that h & g is false.

In [15], I pointed out that the conclusions X is entitled to reach on the basis of what already is in his a corpus depend on the context – in particular, on the question he raises and the potential answers he has identified for that question. (These do not exhaust all the relevant factors.) Given a complete set of values for relevant contextual parameters, X might be justified in reaching one conclusion h_1. Given another set of values, he might be justified in reaching another conclusion h_2. Yet, there may be no accessible system of values for the contextual parameters which warrants concluding h_1 & h_2. In this sense, and in this sense alone, do I concede to H. Kyburg that 'conjunctivitis' is a disease. I doubt, however, that this will satisfy him. See H. Kyburg [12], pp. 55–82. By the way, I do think that from X's point of view, everything in his corpus is true. (See Kyburg [12], p. 59.)

22 Thus, items in *UK* are 'incorrigible' relative to my theory of the improvement of knowledge. These claims correspond roughly to those philosophers call 'analytic' or 'a priori' or 'necessary'. I prefer avoiding the use of these categories. Analytic sentences are allegedly true 'in virtue of meaning'. I have no good understanding of this notion which is especially relevant to the question of the improvement of knowledge. Presumably, to say that items in *UK* are a priori is somehow to make a claim about the justification for their admittance into *UK* or the causes for their being admitted. I have nothing to say about this. Finally, if 'it is necessary that p' is equivalent to 'it is impossible that p is false' and possibilities are those logical possibilities that are serious in conducting inquiry, everything in X's corpus $K_{X,t}$, from his point of view at t, is necessary whether it is in *UK* or not.

23 That is to say, all changes are so decomposable for purposes of analysis. I claim that when an argument is offered for shifting from K to K', it should be decomposed into arguments for shifts from K to K_1 to K_2 to ... to K' where each step involves a contraction or expansion from the previous step. I do not claim that this analysis faithfully reflects the conscious or subconscious workings of X's mind.

24 I. Levi [15], pp. 32–38. My account was restricted to finite ultimate partitions.

25 The positivists explored necessary and sufficient conditions for sentences to have truth values. Popper's demarcation criterion was designed to discriminate between hypotheses with respect to testworthiness. I suggest that abductive logic should be concerned with necessary conditions for being a potential answer to question (belonging to some suitably specified category of questions). I would agree that any potential answer to a given question should (a) have a truth value and (b) be such that the outcome of further inquiry could, in principle, (in some explication of that obscure notion). lead to a verdict in favor of that answer or against it. I am not prepared, at present, to spell out a more adequate version of condition (b). However, such a surrogate for a 'verifiability' or a 'falsifiability' principle would not be either a criterion of 'meaningfulness' or 'bearing a truth value' or, alternatively, a demarcation criterion in Popper's sense.

26 Levi [15], pp. 28–29 and pp. 149–152.

27 I. Hacking [5], pp. 446–447.

28 R. Hilpinen [6], pp. 100–102.

29 Levi [15], p. 20.

30 Levi [18], pp. 409–418.

31 Levi [15], Chapters IV and V and [17], pp. 369–391.

32 R. C. Jeffrey proposed the felicitous phrase 'index of boldness' in [11], p. 316.

33 Levi [14], pp. 307–313.

34 J. Hintikka and J. Pietarinen [8], pp. 107–108.

35 Levi [17], p. 374 and Footnote 19.

36 R. Hilpinen [6], pp. 103–111.

37 Levi [15], pp. 104–108.

38 Levi [17], Footnote 34.

39 R. C. Jeffrey [11], p. 319.

[40] Levi [15], p. 151. I argued that the legitimacy of bookkeeping is "inevitable for any view that concedes that a sentence can legitimately be accepted as evidence from other evidence that does not entail it". Hilpinen [6], p. 103, says that my claim is false and cites a rule which avoids the need for bookkeeping. He is right. If acceptance rules are so formulated that either no element of an ultimate partition is rejected or all but one is, there is no scope for applying bookkeeping. By restricting the probability measures used (as Hilpinen, following Hintikka, does) and restricting the ultimate partition employed in suitable ways (as Hilpinen once more does), one can avoid the need for bookkeeping. Thus, Hilpinen restricts the elements of ultimate partitions to constituent structures and probability measures to those belonging to the family considered by Hintikka. In my opinion, these restrictions are excessive. We should have a theory to account for justifications of expansions where they do not obtain as well as when they do. It seems to me situations do arise (for example, in estimating the values of statistical parameters) where the data, in the first instance, do warrant ruling out some point estimates but do not warrant ruling out all but one point estimate. In such situations, there is sometimes scope for applying bookkeeping. If the restrictions proposed by Hilpinen are imposed, we cannot come to terms with these problems – as far as I can see.

[41] K. Lehrer [13].

[42] A. Shimony [25], pp. 112–113.

[43] *Loc. cit.*

[44] *Loc. cit.*

[45] Levi [18].

[46] Levi [15], Chapters VIII and IX.

[47] G. L. S. Shackle [24]. Jonathan Cohen [4] has proposed measures of 'support' sharing some but not all of the properties of Shackle's measures. Moreover, he apparently intends them to be applied in ways different from those proposed either by Shackle or me. However, I have not been able to clarify for myself precisely what use Cohen's measures are supposed to have. Cohen elaborates on the conditions under which hypotheses can be compared with respect to support and can be assigned grades of support. But, as far as I can make out, he offers no explanation of what a scientist is supposed to do with evaluations of hypotheses with respect to support once he has them.

[48] R. Hilpinen [6], p. 121.

[49] I assume familiarity with the terminology of Carnap's [2] and [3].

[50] Hintikka's theory is stated in [7].

[51] I. Hacking [5].

[52] R. Hilpinen [6], pp. 100–101.

[53] R. Hilpinen [6], p. 121.

BIBLIOGRAPHY

[1] R. Carnap, 'The Aim of Inductive Logic' in [20].

[2] R. Carnap *The Continuum of Inductive Methods*, University of Chicago Press, Chicago, 1952.

[3] R. Carnap, *Logical Foundations of Probability*, 2nd ed., University of Chicago Press, Chicago, 1962.

[4] L. J. Cohen, *The Implications of Induction*, Methuen, London, 1970.

[5] I. Hacking, Review of [15] in *Synthese* **17** (1967), 444–448.

[6] R. Hilpinen, *Rules of Acceptance and Inductive Logic*, North Holland, Amsterdam, 1968.

[7] J. Hintikka, 'A Two Dimensional Continuum of Inductive Methods', in [9], pp. 113–132.

[8] J. Hintikka and J. Pietarinen, 'Semantic Information and Inductive Logic', in [9], pp. 96–112.

[9] J. Hintikka and P. Suppes (eds.), *Aspects of Inductive Logic*, North Holland, Amsterdam, 1966.

[10] R. C. Jeffrey, *The Logic of Decision*, McGraw Hill, New York, 1965.

[11] R. C. Jeffrey, Review of [15], *Journal of Philosophy* **65** (1968), 313–322.

[12] H. Kyburg, 'Conjunctivitis' in [26].

[13] K. Lehrer, 'Induction: A Consistent Gamble', *Nous* **3** (1969), 285–297.

[14] I. Levi, 'Corroboration and Rules of Acceptance', *British Journal for the Philosophy of Science* **13** (1963), 307–313.

[15] I. Levi, *Gambling with Truth*, Knopf, New York, 1967. (Reprinted in paper by MIT Press, 1973.)

[16] I. Levi, 'If Jones only Knew More', *British Journal for the Philosophy of Science* **20** (1969), 153–159.

[17] I. Levi, 'Information and Inference', *Synthese* **17** (1967), 369–391.

[18] I. Levi, 'On Indeterminate Probabilities', *Journal of Philosophy*, **71** (1974), 391–418.

[19] I. Levi, 'Probability Kinematics', *British Journal for the Philosophy of Science* **18** (1967), 197–209.

[20] E. Nagel, P. Suppes, and A. Tarski (eds.), *Logic, Methodology and Philosophy of Science*, Stanford University Press, Stanford, 1962.

[21] K. R. Popper, *Conjectures and Refutations*, Basic Books, London, 1962.

[22] K. R. Popper, *Objective Knowledge*, The Clarendon Press, Oxford, 1972.

[23] K. R. Popper, 'Truth, Rationality and the Growth of Knowledge', in [21], pp. 215–250.

[24] G. L. S. Schackle, *Decision Order and Time*, Cambridge University Press, Cambridge, 1961.

[25] A. Shimony, 'Scientific Inference', in *The Nature and Function of Scientific Theories* (ed. by R. Colodny), Univ. of Pittsburgh Press, Pittsburgh, 1970.

[26] M. Swain (ed.), *Induction, Acceptance and Rational Belief*, Reidel, Dordrecht, 1970.

ROGER ROSENKRANTZ

COGNITIVE DECISION THEORY

1. Introduction

That scientists make decisions no one doubts. They allocate research funds, in effect deciding which problems are most worth attacking, they decide how large a sample to observe, which experimental methods to use, and even which outlying observations to reject or discount. These are all garden-variety decisions of a well understood sort, even though some of the utilities that enter might be deemed 'epistemic'. It is thought, however, that other decisions lurk back of all these which are far more germane to the scientific enterprise. I mean, of course, decisions to accept or reject hypotheses. These decisions are held by Levi (1967) to be purely cognitive in the sense that all the pertinent utilities are epistemic. Levi's program is to develop an appropriate cognitive decision theory which will enable scientists to compute and compare the expected epistemic utilities associated with the acceptance of different hypotheses.

Conceptions of science like the cognitive decision theoretic recur again and again (it is the ghost of Cardinal Bellarmine!), but in recent years this chronic illness has threatened to reach epidemic proportions, touching even many writers who would be disinclined to deploy their views under that banner. High time, then, that it came in for sustained criticism. The critique we offer here necessarily proceeds on many levels and is rather complex, and so it would be well to sketch the main outlines of the argument before proceeding.

Cognitive decision theory has two major liabilities. On the one hand, it appears to reduce theory appraisal to subjective whim. Even supposing that satisfactory objective measures of the various epistemic utilities were at hand, there would remain the problem of processing these several utilities into an overall appraisal. Levi is quite explicit on the need for assigning arbitrary weights to the different utilities (Levi, 1967, pp. 105 f.). Secondly, there is the vexatious matter of what counts as a *bona fide* epistemic utility. Does special virtue attach to a theory's ability to predict

R. J. Bogdan (ed.), Local Induction, 73–92. All rights reserved.
Copyright © 1976 by D. Reidel Publishing Company, Dordrecht-Holland

novel effects, as opposed to explaining effects already known? Is mathematical complexity an epistemic (or merely a practical) disutility? I am aware of no criterion by which cognitive decision theorists propose to adjudicate such questions.

These two areas of vulnerability must sooner or later occur to anyone who gives the matter serious thought. At its most extreme, cognitive decision theory justifies maintaining any theory in the face of any data on any grounds one likes. Amazingly, awareness of its potential for reducing theory evaluation to personal taste has been insufficient to dislodge its proponents. The clear need is for a clear and viable alternative. We shall be developing one such alternative conception according to which evidential support is all in all. One theory can be rationally preferred to another if and only if it is better supported. In local inductive contexts characterized by a partition of mutually exclusive and jointly exhaustive hypotheses, ordinary Bayesian measures of support (likelihood and average likelihood) will serve. But in non-local contexts, as when we wish to assess a theory without reference to alternatives or to compare non-exclusive alternatives, a non-trivial extension of Bayesian method is required. Such an extension will be developed in section 4 below and forms the heart of our positive contribution.

Given a satisfactory measure of support of the requisite scope, the matter of what counts as an epistemic utility can be handled as a matter of course. The only epistemic utilities properly so called are such by virtue of being reflected in a theory's evidential support. Readers familiar with the Bayesian literature know that the 'likelihood principle', according to which the entire import of an experiment (vis-à-vis a partition of hypotheses) is conveyed by the likelihood function[1] of the observed outcome, is a powerful criterion of relevance. It rules all of the following irrelevant: the experimenter's intentions when to stop sampling, whether the experimenter sought to confirm or refute a supported hypothesis, whether a supported hypothesis was predesignated or prompted by the experimental results, and much more. With an extended measure of support, applicable where there is no likelihood function or where it cannot be computed, we obtain a yet more powerful criterion of relevance. But it would be hasty to conclude that all purported epistemic utilities are ruled out. In fact, simplicity (or content) proves to be ingredient in support. But if this is so, there is no need to incorporate it separately into appraisals of a theory over and above what it contributes to support. The details of this argument, together with our explication of simplicity, are presented in Section 3.

Although criticisms of cognitive decision theory and of Levi's particular formulation of it will be given, our case must rest ultimately on the plausibility of our positive contention that evidential support is the measure of all things theoretical. If our thesis proves tenable, it may well result in one of the largest slum-clearance projects ever. Still, to wholly expunge the disease requires careful consideration of the reasons (I hesitate to say 'arguments') adduced on behalf of a cognitive decision theoretic conception, We cannot hope of course to examine every reason that has led anyone to travel that road, but the ones we take up are hopefully representative.

2. THE CASE FOR COGNITIVE DECISION THEORY

The cognitive decision problem, as Levi conceives it, is to decide which disjunction of elements of one's ultimate partition of hypotheses to *accept as strongest* in the sense of providing maximal relief from agnosticism at minimal risk of error given the decision maker's rate of exchange between the two. Interval estimation might be considered an exemplar (although Levi's treatment is confined to finite ultimate partitions). One needs to balance off the length of the interval held to contain the true parameter value against the risk that it does not contain it. Exemplars of this sort do not argue but perhaps invite a cognitive decision theoretic formulation.

But why should we single out a single interval? What purpose is served by such patently arbitrary choices? For a Bayesian, one's state of knowledge can be adequately represented only by the complete posterior probability distribution of the parameter, which determines, for each possible interval, the probability that it contains the true value. Likewise, relief from agnosticism, if we wish to speak in that way, is measured by the entropy of the posterior distribution. This is all ye know and all ye need to know for practical or cognitive purposes. I take it there is no argument on the practical side; let us consider the cognitive.

If the cognitive goal in estimating a theoretical parameter is to make a prediction, then one could not do better than weight the predictions based on each possible value of the parameter by its posterior probability. If one 'accepted' a particular interval of values as best or 'strongest' in Levi's sense, how would one then arrive at a best prediction? Given that his theory is purportedly based on relating cognitive decisions of acceptance and rejection to the aims of inquiry, Levi has precious little to say about this relation *in concreto*.

Suppose next that the cognitive goal is explanation. In order to explain an effect we need to assume that a theoretical parameter lies outside a highest density interval[2] whose probability of containing the true value is $1 - p$. We can aptly describe this posit as the incurring of an evidential debit measured by p: the smaller p, the larger the debit or strain of credence. *Ceteris paribus*, a scientist will be reluctant to assume such a debit, and his reluctance will grow (and should grow) with the size of the debit. But other things are almost never equal. It may be necessary to assume a debit in order to explain an effect or correct a theoretical prediction. Theorizing is very much a case of robbing Peter to pay Paul. This is illustrated in the following (somewhat dated) passage from Bondi *et al.* (1960) on the steady-state cosmology, pp. 17–18:

Bondi: Of course, it [the steady-state theory] deviates from ordinary physics in assuming the phenomenon of continual creation of matter which is, indeed, a major infringement of present formulations of physics. Dr Bonnor has argued that this process of continual creation violates the principle of conservation of energy which has withstood all the revolutions in physics in the last sixty years and which most physicists would be prepared to give up only if the most compelling reasons were presented; but this seems to me unsound. The principle of the conservation of mass and energy, like all physical principles, is based on observation. These observations, like all experiments, and observations, have a certain measure of inaccuracy in them. We do not know from the laboratory experiments that matter is absolutely conserved; we only know that it is conserved to within a very small margin. The simplest formulation of this experimental result seems to be to claim that matter must be absolutely conserved. But this is purely a mathematical abstraction from certain observational results that may contain, indeed are bound to contain, errors.

Now, in fact, the mean density of the universe is so low, and the time scale of the universe so large, by comparison with terrestrial circumstances, that the process of continual creation required by the steady-state theory predicts the creation of only one hydrogen atom in a space the size of an ordinary living-room once every few million years. It is quite clear that this process, therefore, is in no way in conflict with the experiments on which the principle of the conservation of matter and energy is based. It is only in conflict with what was thought to be the simplest formulation of these experimental results, namely that matter and energy were precisely conserved. The steady-state theory has shown, however, that much simplicity can be gained in cosmology by the alternative formulation of a small amount of continual creation, with conservation beyond that. This may, therefore, be the formulation with the greatest overall simplicity. There is thus no reason whatever, on the basis of any available evidence, to put the steady state theory out of court because it requires this process of continual creation.

This passage is beautifully illustrative of what I have been saying about theorizing; Lyttleton's contribution to the same volume is even more so, but is unfortunately too long to quote in full. As Bondi would have it, a small evidential debit (viz. the support lost by slightly complicating a law which fits the extant data) is more than compensated by a major simplification of theory, resulting in an overall gain of simplicity and support. The *form* of the argument is irreproachable; that steady-state cosmology no longer appears in accord with the facts is quite beside the point.

Certainly the conservation principle is as good a candidate for an 'accepted' law of physics as any; yet Bondi 'rejected' it. Levi would no doubt account for this by the high expected epistemic utility in accepting a small departure from the law given the cognitive aim of explaining the putative steady-state of the universe. And this cognitive aim has high epistemic utility because the steady-state theory is so much simpler than any theory which allows the state of the universe to vary in ways which are, at the outset, unspecified. We must weigh Levi's account against the account in terms of evidential debits and credits. In principle, at least, evidential support can be objectively assessed; but the relative importance assigned to simplicity qua epistemic utility vis-à-vis other such utilities seems patently subjective.

The cognitive decision theorists appeal to examples is buttressed by the claim that scientists do in fact accept and reject hypotheses, and that very fact seems to certify their conception, for as Rudner (1953) argued, the critical probability levels that determine when a scientist is warranted in accepting an hypothesis are functions of the importance he attaches to mistakes. In that sense, Rudner concluded, the scientist qua scientist makes value judgments. But he apparently meant no more by 'accept' than a disposition to act on an hypothesis. And of course we may 'act on' an hypothesis we accord a low probability, as when we buy insurance or gamble on a long-shot. Given Rudner's tepid sense of 'accept', Jeffrey (1956) found it easy to dismiss Rudner's claim, for the disutilities incurred by mistakes can be expected to vary from one practical decision making context to the next. Thus, the probability that a vaccine has dangerous side-effects may be sufficiently high to prevent its application to humans but not to pet monkeys. So either we find ourselves admitting that scientists both accept and reject the same hypothesis, or else we choose to regard the utilities as unspecified and the Rudnerian decision problem as correspondingly open-ended. At this moment of the dialectic, Levi (1960) entered the suggestion that acceptance of hypotheses is relative to a set of purely cognitive goals that are comparatively stable and impersonal. This possibility, at any rate, suggests a form of Rudner's conclusion not clearly at odds with the value-neutral character of science, Levi argued, for the critical probability levels required for acceptance may indeed be a function of the importance of mistakes, but the importance of mistakes in a given problem context may be determined by the goals of inquiry assumed to be shared by all scientists. As Levi put it (1960, Section IV), "the canons of inference might require of each scientist that he have the same attitudes, assign the same utilities, or take each mistake with the same degree of seriousness as every other scientist". But this implausible suggestion is quietly dropped in Levi (1967) without noting that its demise re-admits Rudner's strong claim that acceptance cannot be squared with the value-neutrality thesis. To be sure, the importance a scientist attaches to mistakes is conditioned, on Levi's account, by purely cognitive considerations, and not, for example, by such considerations as probable loss of reputation that would follow the publication of erroneous hypotheses. But the crucial point here is not whether the relevant utilities are practical or cognitive but whether the seriousness of mistakes is to be left to the personal assessment of the individual scientist. *Gambling with Truth*, in point of fact, candidly avers that the expected utility of accepting an hypothesis as strongest is a function of a caution index q whose value is set by the individual cognitive decision maker in ways that presumably reflect his personal rate of exchange between relief from agnosticism and risk of error. Finally, Levi's sense of 'accept' turns out to be no more exciting than Rudner's. It, too, amounts to 'acting on', but now in a cognitive, rather than a practical, decision making context. For his acceptance rule is relativised to a partition of hypotheses, and so it is clear that an hypothesis accepted in one local context (or relative to one set of cognitive aims) may be rejected in another. Above all, Levi's rule has no application to such tendentious claims as that General Relativity is currently accepted. Nor are even these claims philosophically interesting. Their cash value seems to be the claim that the given theory is supported by the known evidence and that contrary evidence is either lacking or of doubtful reliability. The only other construction I can place on such statements is a sociological one to the effect that the bulk of those in the relevant scientific specialty have a strong propensity to premiss the hypothesis or theory in question.

Similar remarks apply to the long-established but varying uses of 'accept' and 'reject' in the statistical literature. In the theory of hypothesis testing due to Neyman and Pearson, 'accept' is used primarily in the sense of 'acting on', Neyman's presumption being that every scientific problem is overtly or covertly a practical decision making problem. (That is why he eschews the term 'inductive inference', and prefers to speak of 'inductive behavior'.) In the older approach to significance tests due to R. A. Fisher (who heatedly disavowed Neyman's decision theoretic formulations), 'accept' and 'reject' connote no more than agreement and disagreement with the data as determined by a suitable criterion of fit. Current statistical usage seems to fluctuate between the Fisherian and Rudnerian senses. But neither usage suggests the existence of a clear all-or-nothing concept of acceptance of the sort cognitive decision theory seems to import. And, in any case, it is hard to see what features of scientific behavior such a concept would explain or what methodological prescriptions it would suggest. Whether we choose to denigrate hypotheses that are ill-supported by the available evidence as 'rejected', 'refuted', 'falsified' or what have you is immaterial, for such hypotheses, all connotations of finality notwithstanding, are never entirely banished from further consideration; they can be recalled from exile in the light of new evidence. The more descriptive and less misleading term 'ill-supported' would serve scientists at least as well.

The contentions of Levi (1970) that Bayesians need to accept evidence statements are no more persuasive. Strictly speaking, evidence statements are indeed less than certain, but that is no bar to conditionalizing on them as Jeffrey (1965) has shown. Similarly, the need to accept assumptions that a set of hypotheses collectively exhaust the possibilities is only apparent. Bayesian inference must be understood as providing a framework within which exclusive hypotheses can be compared; all Bayesian inferences are implicitly conditional on the assumption that the relevant partition of hypotheses is exhaustive. A lumpen alternative to this assumption always lurks in the background and can be thought of as having a non-zero probability. The data may single out some member of the lumpen alternative as particularly likely and the partition of hypotheses will then be expanded to include it. The point is that hypotheses which were not initially included explicitly in the partition are not thereby assigned zero probability.

In the last analysis, it is hard to see what is at issue here. Scientists premiss statements (sometimes merely to see what follows) and their activity of premissing is guided by probabilities or degrees of evidential support, at least when they are interested in obtaining correct results. Cognitive decision theorists insist that such rational goal-directed premissing is also guided by additional epistemic utilities, like simplicity, testability, explanatory power, and the like. My reply is that non-spurious epistemic utilities all get reflected in evidential support, and I think that is the real issue between us. But if that is so, the mere fact that scientists premiss hypotheses in some sense cannot by itself argue the case for the cognitive decision theoretic conception.

Let me now pass on to a third source of cognitive decision theory.[3] On the face of it, adherence to probability as a yardstick commits Bayesians to preferring hypotheses of lower content, for these are more probable. Given a partition of hypotheses, it seems, a Bayesian should prefer the disjunction of all of them as being most probable. Levi, Popper and others propose to avert this objection by giving due weight to content; science aims at *interesting* truth. "Without truth as a desideratum", Levi writes (1967, p. 120), "the risk of error is unimportant", and "without other desiderata ... no one would be justified in risking error".

The Bayesian reply to this objection has three prongs. Let me say at the outset that no Bayesian would controvert the emphasis on interesting or highly informative truth. As we will see in the next section, simpler theories are more confirm*able*; roughly speaking, the simpler of two theories in equally good agreement with the data will be better supported. Even if a simpler theory starts out less probable, it will quickly acquire higher posterior probability. (That is reason enough for testing simpler theories first.) Indeed, this fundamental product of Bayesian analysis shows *why* simplicity and content (which come to much the same thing on our account) are important desiderata.

But simpler theories are typically such by virtue of being special cases.

Circles are simpler than and special cases of ellipses, polynomials of lower degree are simpler than and special cases of polynomials of higher degree, and so on. Now it will quite properly be objected that the more complicated theory must, by the probability calculus alone, have higher probability than any of its more restrictive special cases.

This seemingly formidable objection is easily handled by taking logical differences. That is, to compare a model with a special case of itself, we compare their logical differences, thus obtaining the partition of exclusive hypotheses upon which Bayes' theorem can operate. This is even done tacitly by scientists in every case I can think of. E.g., the general polynomial of k-th degree is understood to have a non-vanishing leading coefficient (hence to be 'properly' of k-th degree). Note, though, by allowing the leading coefficient to approach zero, we enable the k-th degree polynomial to fit any observations fitted by the $(k-1)$st degree polynomial, so that effectively, if not logically, the latter remains a special case of the former. Once logical differences are assumed, there is no reason whatever to insist, as Popper so often has, that simpler hypotheses or theories are necessarily less probable. (Popper is wont to equate 'prior probability' with 'logical (sic. Laplacean) probability'.)

Still, it might be insisted, there are times when one wishes to compare an hypothesis with a disjunction of hypotheses containing it. This, of course, is the framework of Levi (1967). One *can* do just this by the methods of Section 4 without resorting to epistemic utilities not reflected in support, but I must confess I find Levi's formulation remarkably artificial and far-removed from scientific practice. A Bayesian will introduce a partition of hypotheses each one of which constitutes a 'relevant answer' to the question posed. (Thus, the emphasis on content merely comes in at a different stage, namely that of putting forward certain potentially interesting hypotheses for consideration.) From the fact that the elements of his ultimate partition are relevant answers it doesn't follow, however, that disjunctions of relevant answers are relevant. One may of course find the evidence equivocal, but in that case, one doesn't 'accept' some disjunction; one merely stores away the probabilities of the hypotheses of the partition. And this remains so when a number of ultimate hypotheses are frontrunners and the others virtual dead-letters. We are now in a better position to perceive the faintly question-begging character of the objection that Bayesians are committed to the most probable hypothesis willy-nilly, for it presupposes both Levi's questionable framework and that Bayesians accept hypotheses in something like Levi's sense. (In fact, Bayesians are only committed to the most probable element of the considered ultimate partition.) Perhaps enough has now been said to expose the completely inconclusive character of the considerations that have called forth Levi's formulation.

3. SIMPLICITY

Complicating a theory to improve its accuracy presents itself as an especially auspicious case for the proponents of cognitive decision theory. Accuracy and simplicity appear to function in such contexts as countervailing epistemic utilities: one can only get more of one by sacrificing some of the other. And what is to determine the rate of exchange between them?

This is a serious challenge, but it can be met. Although the argument is completely general, let us confine attention to complications which assume the form of adding an adjustable parameter. Auxiliary hypotheses usually take this form in any event. Thus, when we 'save' a predicted Mendelian ratio by pleading recessives are less viable than dominants, we introduce in effect a parameter of differential viability. Similarly, if we plead the existence of an unseen planet or star to account for the irregularities of a given orbit, we thereby introduce the orbital elements of the unseen body as new parameters.

On a strict Bayesian analysis support for a model with free parameters is measured by its average likelihood.[4] Consider now a proposed complication of the relevant type, say the introduction of a parameter of differential viability or a linkage parameter to save a particular Mendelian model. Should the scientist follow this course? Should he abandon the model? Should he abandon the particulate theory? Or should he attempt to explain away the exceptional ratio observed in his experiment? This would seem to be a cognitive decision problem if anything is, and the solution would be held to depend on the experimenter's personal rate of exchange between accuracy and simplicity. Bayesian analysis shows, however, that the rate of exchange is *not* arbitrary; it is determined by the average likelihood. When we add a new parameter, we increase the number of special cases of the model over which we average and so, *cet. par.*, reduce the average likelihood. On the other hand, we introduce new special cases, and some of these may be more accurate than the original model, and this improved accuracy will be reflected in higher maximum likelihood, and that will tend to increase the average likelihood. The improved accuracy compensates the loss of simplicity just in case the average likelihood is increased. In fact, given the approximate normality of the likelihood function, it proves possible to approximate the average likelihood by the maximum likelihood as follows:

$$(3.1) \qquad \int \ldots \int p_x(\theta_1, \ldots, \theta_k) \, d\theta_1 \ldots d\theta_k \doteq (2\pi)^{k/2} \, \mathbf{I}^{-1/2} p_x(\hat{\theta}_1, \ldots, \hat{\theta}_k),$$

where $\hat{\theta}_i$ is the maximum likelihood estimate of θ_i and $\mathbf{I}$ is the determinant of the information matrix (inverse to the covariance matrix) of the parameters θ_i. Using this approximation one can say by how much the maximum likelihood of the new model obtained by adding a parameter (i.e., the likelihood of its best fitting special case) must exceed the maximum likelihood of the original model in order for the average likelihood to increase. Using some of T. H. Morgan's data for the vinegar fly, I have shown in this way that the result of adding a linkage parameter to Mendelian theory radically augmented its support (cf. Rosenkrantz, 1975, Section 5).

The result has further application to polynomial regression, the case of the classical curve-fitting problem where it is assumed that the true curve is a polynomial. Here the parameters are the coefficients of the general k-th degree polynomial for $k = 1, 2, 3, \ldots$, and the information matrix has a known simple form (given in any standard text on regression). It is then an easy matter to determine the critical levels which the ratio of maximum likelihoods must exceed for the polynomial of next higher degree to be better supported. The maximum likelihoods are also easy to compute, being essentially $\exp(-\frac{1}{2}\sum e^2)$, where $\sum e^2$ is the residual sum of squares. Yet, many discussions of the curve-fitting problem leave the reader with the impression that the balancing of simplicity and accuracy must be performed at a wholly judgmental level.

I hasten to add that orthodox methods for regression and similar problems also determine a rate of exchange between simplicity and goodness-of-fit, though a different rate than that determined by the Bayes method. Whenever an F-test or a chi square test is used, complexity is penalized by the loss of a degree of freedom for each parameter that must be estimated from the data. Casual claims that the trade-offs between simplicity and accuracy required to make a theory more tenable are at the discretion of the individual carry, then, a heavy burden of proof.

While our argument to this point appeals to a particular widely shared intuition about simplicity (that special cases of a theory are simpler), little has been said about the explication of simplicity in general.

By the *sample coverage* of a theory with respect to an experiment I mean the probability that the outcome will fit the theory given that the universe is maximally disorderly. For present purposes we may identify a 'maximally disorderly' universe as one in which all outcomes of the experiment are equiprobable. Sample coverage then reduces to the proportion of outcomes which fit the theory, a criterion of fit being presupposed. I take sample coverage to measure simplicity: the smaller its sample coverage, the simpler a theory. (Needless to say, we can speak of sample coverage with respect to a set of experiments, as well as with respect to a single experiment.)

Now under any reasonable criterion of fit a model with parameters will fit any outcome fitted by any of its special cases and will therefore have larger sample coverage. Our explication thus captures the basic intuition about simplicity, but it also captures intuitions that seem on the surface unrelated. Hintikka's constituents are a case in point. Of k possible kinds, a constituent specifies which of them are exemplified in the population considered. Hintikka rightly supposes that constituents are simpler the fewer the number of kinds they affirm to be present in the population. Here a 'fitting' sample can only be one compatible with the constituent, hence one which exemplifies a subset of its designated kinds. Obviously, the smaller the number of kinds it designates as non-empty, the smaller the number of samples with which a constituent is compatible, so Hintikka's intuition, too, is preserved.

Of greater interest is the fact that, at least in a very broad spectrum of cases, the simpler of two theories in equally good agreement with the data will be better supported.[5] In particular, a model with free parameters is never better supported (i.e., never has higher average likelihood) than its best-fitting (maximally likely) special case, for the average of the likelihoods can never be larger than the maximum of the likelihoods. Indeed, as the sample size increases, the likelihood function becomes ever more concentrated in a small neighborhood of its maximum, so that, in the limit, the best fitting special case of a model becomes infinitely better supported. (Whenever a complication of a theory is support increasing, of course, the original theory is not the best-fitting special case of the new theory that includes the original theory as a special case.) Even where simpler theories of an ultimate partition of hypotheses start out with smaller prior probabilities – and they needn't – they will come to enjoy higher posterior probabilities, given consonant data. From a Bayesian point of view, then, there can be no reason to count simplicity (or content) as an additional epistemic utility to be considered apart from support or posterior probability, for it is already reflected in the latter. Content, of course, is also measured by sample coverage: the smaller the sample coverage, the higher the content. Hence, greater scope or generality, greater precision or specificity, and paucity of adjustable parameters, all make for greater content under our measure. In particular, a theory which explains more or determines more has greater content.

Consider now Levi's formula

$$(3.2) \qquad U^*(H, x) = P(H/x) - q \operatorname{Cont}(\sim H, x)$$

for the expected epistemic utility of accepting H as strongest in the light of the experimental outcome x. Here H may be any disjunction of members of one's ultimate partition of hypotheses, q is an index of caution reflecting the experimenter's rate of exchange between relief from agnosticism and risk of error, and Cont is a measure of content. I take it (3.2) is intended to measure the acceptability of H, though in an oversimplified way, since, in a full-blown treatment, Levi would presumably wish to build in other epistemic desiderata. But let us focus on this version.

If content is measured as we propose, (3.2) counts content twice, for it is already reflected in the posterior probability $P(H/x)$. But Levi elects to measure content in quite a different way by the Laplacean (or 'logical') improbability of an hypothesis, relativising to an ultimate partition rather than to an experiment. Thus if the ultimate partition comprises m hypotheses $H_1, ..., H_m$, $\operatorname{Cont}(\sim H_i) = 1/m$, $i = 1, ..., m$. His motivation is clear: the more alternative relevant answers an hypothesis excludes (among answers not already rejected or excluded by the background knowledge), the greater its content. The parenthetical qualifier is essential given the motivation, and this point cannot be overemphasized. Hence,

$\text{Cont}(\sim H, x)$ is the content of $\sim H$ (or the Laplacean probability of H) in the truncated partition induced by x, which omits elements of the original ultimate partition exluded by x. If x excludes n of the m elements, $\text{Cont}(\sim H_i, x) = 0$ or $1/(m-n)$ according as H_i is among those excluded by x or not. In particular, when x entails H_i, $\text{Cont}(\sim H_i, x) = 1$.

The hypothesis accepted as strongest should maximize $U^*(H, x)$. Happily this hypothesis is easily found, for U^* defined by (3.2) has the additive property that the expected utility of a disjunction is the sum of the expected utilities of the disjuncts. Hence the rejected H_i have negative U^*, i.e.,

$$(3.3) \qquad x \text{ rejects } H_i \quad \text{iff} \quad P(H_i/x) < q\,\text{Cont}(\sim H_i, x),$$

and the strongest accepted H is the disjunction of the unrejected H_i. Let us write H_x for the strongest hypothesis in question.

At first blush it may seem a matter of indifference whether or not we include in H_x those disjuncts for which $U^*(H_i, x) = 0$. Indeed, given Levi's motivation, it might seem best to exclude them, since we thereby obtain a stronger hypothesis with the same expected utility. But that would amount to rejecting H_i iff $P(H_i/x) \leqslant q\,\text{Cont}(\sim H_i, x)$. Suppose, though, that some of the H_i have prior probabilities less than the Laplacean benchmark, $1/m$, while others have priors larger than $1/m$. A *perfectly equivocal* outcome x would drive all posterior probabilities to $1/m$, in which case *every* H_i would be rejected by the modified rule. On the other hand, (3.3) itself runs afoul of *perfectly definitive* outcomes. For if x_i entails H_i, then both $\text{Cont}(\sim H_j, x_i) = 1$ or 0 and $P(H_j/x_i) = 1$ or 0 according as $j = i$ or not, whence $P(H_j/x_i) = q\,\text{Cont}(\sim H_j, x_i)$ for all j, and no H_j is rejected. Hence the vacuous disjunction of all H_i is accepted as strongest, and so Levi's rule recommends suspension of belief in the face of a perfectly definitive outcome.

The rule is easily repaired by having it reject elements of the partition logically excluded by the outcome. In that event, if x_i entails H_i, H_i is properly accepted as strongest. Still, not all is well, for consider a nearly definitive outcome x_i which accords H_i a probability close to one without entailing H_i or excluding any of the alternatives. Then $U^*(H_i, x_i) = 1 - \varepsilon - q/m$, while, in the case that x_i entails H_i, $U^*(H_i, x_i) = 1 - q$. Odd that a perfectly definitive result should lead to smaller expected utility than an imperfectly definitive result. The strangeness is thrown into sharper relief when we extend Levi's acceptance rule to a ranking of experiments. For, as Goosens shows elsewhere in this volume, a perfect experiment all of whose outcomes are perfectly definitive can have lower expected utility than an imperfect experiment. Goosens' example exploits the conditionalized character of Levi's content measure, but I have argued that feature is inescapable given Levi's underlying conception.[6]

Neither horn of the dilemma our discussion poses is very inviting. Either content is measured by the exclusion of alternative relevant answers *not already excluded,* in which case the Levi ranking of experiments has the strongly counterintuitive property Goosens points out, or else content is measured by sample coverage, in which case Levi's formula for $U^*(H, x)$ counts content twice.

Even if Levi could find a way out of these difficulties, my earlier reservations remain. Rules of acceptance are highly arbitrary at best. Standard decision theoretic formulations of testing familiar from statistics do not even allow for suspension of judgment, and Levi's rule is a step forward in that respect. But, even under Levi's rule, the probability difference between an accepted and a rejected hypothesis can be made as small as you like. And when one speaks of accepting disjunctions of relevant answers, artificiality is piled on artificiality.

4. GLOBAL COMPARISONS

There is a *prima facie* problem of how to combine evidence from disparate sources. Those who despair of doing so take the view that the variety of evidence is a factor in itself to be weighted apart from a theory's ability to predict the results of the experiments taken individually. In a similar vein, one may regard the definitiveness of the evidence and the sheer amount of evidence as two separable factors in acceptability. In this section I will argue that all of these allegedly separate factors are, like simplicity or content, ingredient in support. And I will also propose a way of amalgamating disparate evidence.

Consider the second sort of case first, where we have two bodies of evidence that may differ both in their definitiveness and their sheer amount. A case of this sort is discussed by Feller (1957), where interest focuses on the efficacy of a new vaccine. Three experiments result, respectively, in 0 infections among 10 vaccinated, 1 infection among 17 vaccinated, and 2 infections among 23 vaccinated. The percentages of uninfected subjects among the vaccinated are: 100%, 94.1% and 91.3%, respectively, so that the results are progressively less definitive. On the other hand, the samples are progressively larger. One could easily imagine a cognitive decision theorist arguing that an arbitrary assignment of weights to sample size and sample definitiveness is required to determine the bearing of these experiments on efficacy, or else maintaining that the experiments are incomparable.

In fact, such examples occasion little real difficulty in practice. Suppose the normal rate of infection for the disease is 25%, and that an 'effective' vaccine reduces the rate of infection to at most 5%. An average likelihood test is then in order. Feller, on the other hand, offers a significance test of the standard sort. It involves comparing the three experimental outcome probabilities conditional on the null hypothesis that a vaccinated animal's chance of infection is the same as an unvaccinated animal's chance (hence 25%). This test does not always give the same ranking of experiments as the average likelihood test, but there is a variant of the standard significance test that almost always will.

Rather than compute the improbability of the outcome conditional on the null hypothesis that the vaccine makes no difference, we might ask how well the outcome agrees with the alternative hypothesis of efficacy, and more pertinently still, how *improbably* well the outcome agrees with this hypothesis. Here we are able to measure agreement by probability, but the present method is more general and can be applied even where outcome probabilities on the alternative hypothesis cannot be computed. It requires only an intuitive rank ordering of possible outcomes as agreeing more or less well with the hypothesis of interest. And in practice such rank orderings are not hard to come by. In the present instance, agreement can be measured by the paucity of infections among vaccinated subjects.

Our present measure of support has an obvious connection with average likelihood and our earlier glosses on the relation between support and simplicity. The sample coverage of an hypothesis at an assigned criterion of fit, you recall, is the probability that an outcome will fit the hypothesis in a chance universe. The present concept is the special case where the outcome determines the criterion of fit, only those outcomes which agree as well being counted as 'fitting'. I call this special case of the sample coverage concept *observed sample coverage* (or **OSC** for short). In the general case, the **OSC** is the a priori probability of agreement with the hypothesis as good as that observed; in the special case where a 'chance universe' is one in which all outcomes are equiprobable, **OSC** reduces to the proportion of possible outcomes which agree as well as the observed outcome with the hypothesis of interest.

OSC, like average likelihood, is a joint function of simplicity and accuracy. Hence we should expect **OSC** and average likelihood to be inversely related: the higher the average likelihood at an outcome, the smaller the **OSC** at that outcome. Although this 'sample coverage rule', as I call it, has known exceptions, they are almost all of the practically unimportant kind where the outcome is in poor agreement with both hypotheses of a comparison. At the very least, then, the sample coverage rule appears to be a rather good rule of thumb. And, consequently, **OSC** can be viewed as an index of support which naturally extends average likelihood to contexts in which the likelihood function is unavailable.

Although Fisher frames the test of significance in terms of the probability on the null hypothesis of an outcome as deviant as that observed, this probability is typically just **OSC**. It is unfortunate that Fisher was wont to accentuate the negative, both here and elsewhere. Instead of reporting the **OSC** as a measure of the evidence in favor of efficacy, he would report the significance level at which the null hypothesis of no efficacy is rejected as a measure of the evidence against the null hypothesis. He even says (Fisher, 1935, p. 19): "every experiment may be said to exist only in order to give the facts a chance of disproving the null hypothesis". Fisher's 'negative way' is not only needlessly roundabout, but objectionable in not requiring that the data really do agree with the alternative hypothesis of interest. The general practice has been to equate the level of significance at which the null hypothesis is rejected with the strength of the evidence in favor of whatever alternative the experimenter happens to light upon as plausible. This common practice also fails to take the simplicity of the alternative hypothesis into account.

Fisher was surely right to insist that a suitable null hypothesis enters in the appraisal of any hypothesis or theory, but it enters, in our view, only as a fixed point of comparison against which to gauge the improbability of a theory's accuracy. In assessing the claims of a man who professes ability to find water with a hazel prong, we are not interested in his percentage of successes *per se*, but in whether he scores an appreciably higher percentage of successes than someone digging at random in the same area. Fisher came to view the null hypothesis as the real focus of a statistical test, but, in our view, the null hypothesis enters in determining the a priori distribution of experimental outcomes that fixes the limits beyond which a theory's accuracy becomes too great to be ascribable to chance.

I come at last to the primary concern of this section. We have seen that **OSC** serves as an index of support *for* an hypothesis without regard to alternatives. By computing their respective **OSC**'s, one can compare the strength of the evidence for any pair of hypotheses whatever, even of non-exclusive pairs drawn, perhaps, from disparate fields of science. What we have not yet suggested is a procedure for amalgamating data from several experiments.

To this end, we treat many experiments as a single composite experiment whose outcome space is the Cartesian product of the outcome spaces of the several component experiments. The **OSC** of the composite experiment is then defined to be the product of the **OSC**'s of its components. This allows a composite outcome to be counted in as good agreement with the theory as the observed composite outcome when one or more of its component outcomes is in worse agreement – provided that the other component outcomes are in better enough agreement for their product of **OSC**'s not to exceed that of the observed composite outcome.

Here, of course, I am assuming that the experiments are independent. Where dependencies occur, we must appeal to the general concept of **OSC** defined in terms of the possibly non-uniform probability distribution over outcomes. Consider the set of all composite outcomes whose agreement with the theory is no worse than the observed composite outcome in the sense just defined (i.e., the product of its component **OSC**'s is no larger than the product for the observed composite outcome). Then the sum of the probabilities of these composite outcomes is defined to be the **OSC** of the theory for the composite experiment.

Notice what happens in the simple case of two experiments when there is dependence. At one extreme, the experiments are perfectly correlated and the probability of a composite outcome is the probability of the first component outcome. At the other extreme, the experiments are independent and the probability of a composite outcome is the product of the probabilities of the component outcomes. In general, the latter is much smaller than the former, and so, other things being equal, the **OSC** will be much smaller (and the support much higher) for independent components than for strictly or highly correlated components. It is precisely this feature that accounts for the desirability of 'varying the instances of a law', sampling different strata of a population, or testing a theory under more than one set of experimental conditions. Like other genuine epistemic desiderata, this one is reflected in higher evidential support.

5. CONCLUSION

Kuhn (1957) describes a number of arguments Copernicus advanced on behalf of the heliocentric model as 'harmonies' of the system, thereby suggesting that their appeal was merely aesthetic. These arguments had no appeal to laymen, he says, "because they were unwilling to substitute minor celestial harmonies for major terrestrial discord" (p. 181). The most important of the arguments to which Kuhn refers is the heliocentric model's determination of the order of the inferior planets and the relative distances of all the planets from the sun, using only naked eye observations. There was no analogous Ptolemaic determination of the relative distances of the other bodies from the earth, so that these distances enter as free parameters in the geocentric model. Quite clearly such over-determination radically reduces the **OSC** of the streamlined version of the Copernican model without epicylces, so much so, in fact, that one suspects it is smaller than the **OSC** of any version of the Ptolemaic scheme. In any event, what Kuhn passes off as a mere 'harmony' or aesthetically pleasing feature of the new astronomy does in fact strongly augment the support it derives from whatever accuracy it possesses. Very few of the great scientists, I think, perceived simplicity as a merely aesthetic bonus, or even as an epistemic desiderata independent of support. It is more difficult to ascribe the accuracy of a simpler theory to chance, and, even at an intuitive level, that would tend to make one think a simpler theory in agreement with the data better supported.

Virginia Polytechnic Institute

NOTES

[1] The function which associates $P(x/H)$ with each hypothesis H of a partition is called the *likelihood function* at x. In this terminology, the 'likelier' hypothesis is that which accords higher probability to what was observed.

[2] A *highest density interval* (or region) is one for which the probability of each included point exceeds that of every excluded point.

[3] Correspondence with David Miller helped sharpen the objection.

[4] The *average likelihood* of a theory is the integral of the likelihood function with respect to the probability distribution of its parameters – i.e., the weighted average of the likelihoods of its special cases.

[5] Cf. Rosenkrantz (1975), Sections 4–5 for details.

[6] The relevant quantity for comparing two experiments for the same partition of hypotheses is $\Sigma\, U^*(H_x, x)\, P(x)$. *Goosens' example* (slightly modified). Tickets numbered $1, 2, ..., n, n+1$ are assorted at random into three urns, 1, 2, 3. Urns 1 and 2 are found to contain one ticket each. Let H_1 assert that ticket 1 is in urn 1, and let H_2 be $\sim H_1$. If x_i entails H_i, H_{x_i} is H_i and $U^*(H_i, x_i)=1-q$, since $\mathrm{Cont}(\sim H_i, x_i)=1$, as noted above. Hence $U^*(e^*)=1-q$ for the perfect experiment e^*, being the average of quantities each equal to $1-q$. Now consider the following imperfect experiment e. Outcome x entails that ticket 1 is in urn 2, whence H_x is H_2 and $U^*(H_x, x)=1-q$, x being definitive. Outcome y, assumed to have non-zero probability, entails that ticket 1 is not in urn 2. Hence $P(H_1/y)=1/n$, and so, for positive q, n can be taken so large that H_1 is rejected. For such n, than, H_y is also H_2, and $U^*(H_y, y)=(n-1)/n-(q/2)$, which, for large enough n, is larger than $1-q$. Hence $U^*(e)>1-q=U^*(e^*)$, as required.

BIBLIOGRAPHY

Bondi *et al.*, 1960, *Rival Theories of Cosmology*, Oxford.

Feller, W., 1957, *An Introduction to Probability Theory and Its Applications*, Vol. I, Wiley, New York.

Fisher, R. A., 1935, *Design of Experiments*, Oliver and Boyd, London.

Goosens, W. K., 1976, 'A Critique of Epistemic Utilities', *this volume*, p. 93.

Hilpinen, R., 1968, '*Rules of Acceptance and Inductive Logic*', Acta Philosophica Fennica **21**, North-Holland, Amsterdam.

Hintikka, K. J. J., 1967, 'Induction by Enumeration and Induction by Elimination', in I. Lakatos (ed.), *The Problem of Inductive Logic*, North-Holland, Amsterdam.

Jeffrey, R., 1956, 'Valuation and Acceptance of Scientific Hypotheses', *Philosophy of Science* **23**, 237–246.

Jeffrey, R., 1965, *The Logic of Decision*, McGraw-Hill, New York.

Kuhn, T. S., 1957, *The Copernican Revolution*, Harvard Univ. Press, Cambridge, Mass.

Levi, I., 1960, 'Must the Scientist Make Value Judgments?', *Journal of Philosophy* **57**, 345–357.

Levi, I., 1967, *Gambling with Truth*, A. Knopf, New York (new paperback edition, MIT Press, Cambridge, 1973).

Neyman, J., 1950, *First Course in Probability and Statistics*, Henry Holt, New York.

Popper, K., 1959, *The Logic of Scientific Discovery*, Hutchinson, London.

Rosenkrantz, R., 1975, 'Simplicity', in C. K. Hooker and W. Harper (eds.), *Proceedings of the International Workshop on Foundations of Probability and Statistics and Statistical Theories of Science*, D. Reidel, Dordrecht and Boston.

Swain, M. (ed.), 1970, *Induction, Acceptance, and Rational Belief*, D. Reidel, Dordrecht.

WILLIAM K. GOOSENS

A CRITIQUE OF EPISTEMIC UTILITIES

1. INTRODUCTION

The pioneering work of Isaac Levi (1963; 1967a; 1967b), building on the work of Hempel (1960; 1962) and extended by Hilpinen (1968), has shown how to apply decision theory in a fascinating way to problems of inductive logic. The approach has three major accomplishments.

1.1. *Decision Theory Can Be Applied to the Problem of Evaluating Hypotheses in a Manner Which Is Compelling*

The applicability of decision theory had been questioned on two fronts. First, it was held that decision theory applied to acts, not to establishing truth or falsity. Second, it was held that evaluation of hypotheses had nothing to do with making decisions. The support (or acceptability) of hypotheses is some function of evidence. Acceptability is a measure, no more decision-theoretic than length. The outcome of evaluation is the assignment of some degree of support (or acceptability). Evaluation is just not an act of acceptance or rejection.

Levi has shown how to view the problem of evaluation in a way that is compellingly decision-theoretic, in that no one can reasonably decline to accept or reject. Let *H* represent a set of relevant hypotheses, where each hypothesis is inconsistent with the others, and at least one must be true. Although one and only one element of *H* is true, the evidence might not justify picking out only one element of *H* as true. Each subset of *H* represents a possible choice as the most restricted set acceptable on the basis of the evidence. Some subset of *H* is the strongest acceptable on the evidence. This decision problem is compelling in that there must always be some subset which is acceptable – minimally *H* itself. Once these actions are seen to be the only compelling alternatives, it is obvious that they cannot be chosen from the viewpoint of minimizing the chance of error. For accepting *H* itself as strongest never has any risk! Obviously, acceptable utilities must go beyond an interest in truth alone.

R. J. Bogdan (ed.), Local Induction, 93–113. All rights reserved.
Copyright © 1976 by D. Reidel Publishing Company, Dordrecht-Holland

1.2. *Hypotheses Can Be Accepted in Such a Way That Closure Conditions Are Satisfied*

Once it is seen that decision theory can be used to evaluate hypotheses in a classificatory manner, the question naturally arises what sort of logical properties the classification has. Given that the intended applications include 'is acceptable' and 'is justified', then reasonable requirements would be that if *h* satisfies the condition, then so does anything entailed by *h*, and that if *h* and *k* separately satisfy the condition, then so does (*h* & *k*). Historically, the epistemic utility approach showed how to develop the mathematical theories in which such closure conditions could be proved.

1.3. *The Concept of an Epistemic Interest Can Be Clarified and Defended*

Persons held that the evaluation of hypotheses depended on values and the consequences of acceptance and belief, so that evaluation of hypotheses is not distinctively cognitive in a way that contrasts with practical consequences and ethical values. The approach of epistemic utilities takes this challenge head on by showing how to construct utilities for the evaluation of hypotheses that incorporate only such cognitive interests as truth and strength. The need for decision in science does not import or legitimize just any sort of utility. One can explicate the notion of epistemic interest in a way precise enough, for example, to demonstrate to statisticians that the utility functions they have usually considered in the statistical testing of hypotheses do not adequately reflect cognitive interests.

These accomplishments show that the epistemic utility approach is conceptually and philosophically powerful and fruitful. The purpose of this paper is to detail a major problem for the epistemic utility approach. Can the epistemic utilities be developed so as to result in reasonable preferences over possible evidence and experiments for generating evidence? As the next section argues, there must be an integral connection between evaluation and experimentation. The bulk of this paper shows how one extension of epistemic utilities leads to unreasonable preferences. What the future prospects are for the program is discussed briefly in the final section.

2. EVALUATION AND EXPERIMENTATION

A decision-theoretic approach to the evaluation of hypotheses is committed to the view that epistemic interests in evaluation can be represented in terms of decisions of acceptance and rejection. Let us call the information used in evaluation *evidence*, and the process by which this information is generated *experimentation*. There is an integral connection between evaluation and experimentation, which can be brought out most clearly by consideration of an idealization. Suppose that the only interest in experimentation is to serve evaluation. This sets aside (among other things) all consideration of the cost of experimentation. Experimentation is viewed solely from the viewpoint of the value of the information generated for evaluation. Once we specify how well off we are epistemically in evaluation in coming to have evidence, the expectation of this function over possible experimental outcomes defines a preference relation over experiments. This preference relation is not actual preference over experiments, but an idealized preference relation based solely on the epistemic impact of the outcomes on evaluation.

Let H stand for a fixed set of exclusive and exhaustive hypotheses under evaluation, and X stand for an experiment whose exclusive and exhaustive possible outcomes are $e_1, e_2, ..., e_n$. Let b stand for the fixed background information or evidence. Let $A_H(e, b)$ be the utility of *adding e to b* for evaluating H, and $I_H(d)$ be the utility of coming to have d as (total) evidence for evaluating H. Then the value of adding e would be the gain in having the augmented evidence:

(2.1.) $A_H(e, b) = I_H(e \ \& \ b) - I_H(b).$

Moreover, letting $P(u \mid z)$ be the probability of u conditional on z, the utility of doing experiment X for the evaluation of H would be

(2.2.) $U_H(X) = \sum P(e_i \mid b) \cdot A_H(e_i, b).$

Letting $U_H^*(X) = \sum P(e_i \mid b) \cdot I_H(b \ \& \ e_i)$ for all experiments X, then for experiment Y with outcomes d_i

(2.3.) $U_H(X)\{\gtreqless\} U_H(Y)$ if and only if $\sum P(e_i \mid b) \cdot U_H^*(X)\{\gtreqless\}$
$\sum P(d_i \mid b) \cdot U_H^*(Y)$

since $I_H(b)$ is a constant with respect to both X and Y.

The derived preference relation over experiments must then satisfy the following rationality conditions.

(R-1.) *The minimality of the sure experiment.* If C is an experiment such that one of its possible outcomes has an initial probability of 1 (for some outcome e, $P(e \mid b) = 1$) then, for every experiment Y, $U_H(Y) \geqslant U_H(C)$.

(R-2.) *The desirability of further experimentation.* Letting XY stand for the joint experiment of doing both experiments X and Y, $U_H(XY) \geqslant U_H(Y)$.

(R-3.) *The maximal preferability of the ideal experiment.* Let B be any experiment such that each one of the possible outcomes of B entails one (and only one) of the hypotheses in H. Then, for all experiments Y, $U_H(B) \geqslant U_H(Y)$.

By calling these conditions rationality conditions I mean to require that every adequate theory of hypothesis evaluation must have these conditions as provable theorems[1]. In Sections 6 and 7, a second group of rationality conditions will be developed, based on what I call the efficiency of experiments.

The three rationality conditions, as they stand, allow all experiments to be ranked equally. We might therefore add a fourth rationality condition that for some H, not all experiments have equal utility. This last condition rules out a large class of possible utility functions: $A_H(e, b)$ must be a non-degenerate function of e. Supposing otherwise, for all evidence e and e', $A_H(e, b) = A_H(e', b)$. Letting $e_1, \ldots, e_n$ be the (exclusive and exhaustive) possible outcomes of experiment X, we would then have

$$(2.4.) \quad U_H(X) = \sum P(e_i \mid b) \cdot A_H(e_i, b) = A_H(e, b)$$

which (for a fixed H) is a constant for all experiments. This provides grounds for rejecting any theory of hypothesis evaluation which is not some function of the evidence: one could never prefer to do any experimentation. It follows that if simplicity is to be a serious basis for hypothesis evaluation, it must be relativized to evidence and not be merely some fixed function determined by the hypotheses alone. No function of hypotheses alone can form an adequate theory of hypotheses evaluation.

These three rationality conditions can be proven for utilities in general, assuming three things: (1) the hypotheses (H) are the states of nature, (2) the probabilities of the states of nature are probabilistically independent of the acts, and (3) the utilities of the acts under the states of nature do not depend on (are at most a degenerate function of) the evidence (see Appendix, Theorems 2, 3, 4). However, epistemic utilities cannot in general appeal to these theorems, for (3) is not generally true. Indeed, it has been argued that (3) should not be true (see the end of Section 4). If the derived preferences over experiments do not satisfy the rationality conditions, the underlying theory of hypothesis evaluation must be wrong. Let us consider how epistemic utilities fare.

3. EPISTEMIC UTILITIES AND PREFERENCES OVER EVIDENCE AND EXPERIMENTS (I)

The theory of epistemic utility has been developed for the following problem. For fixed or given evidence, what subset of H should we accept as strongest on the basis of the evidence (where to accept a subset is to accept that at least one element of it is true)? Let $U_H(r \mid d)$ be the utility of accepting as strongest a subset r of H where d is the (given) evidence, and $R(H)$ is the set of subsets of H. An r is selected which maximizes $U_H(r \mid d)$. Then

$$(3.1) \qquad F_H(d) = \max_{r \in R(H)} U_H(r \mid d)$$

represents the utility of having d as far as the decision made – here evaluating hypotheses in terms of acceptance.

In the approach of epistemic utilities, the form of analysis is as follows[2] (for a more motivated treatment, see Levi (1967), Hilpinen (1968), or Goosens (1970)). The states of nature are the elements of H itself. There is a $1-1$ correspondence between the acts and the subsets of H. Where $s \in H$, let $U_{H,e}(r, s)$ stand for the utility of the act of accepting r as strongest in evaluating H given evidence e and assuming the state of nature is s. Let $\sim r$ stand for the elements of H not in r. Then

$$(3.2.) \qquad U_{H,e}(r, s) = \begin{cases} 1 - g_H(r \mid e) & \text{if } s \in r \\ -g_H(r \mid e) & \text{if } s \in \sim r \end{cases}$$

where g is a numerical function depending on r and (possibly) e and (possibly) H. (Where r is a unit set $\{h\}$, we sometimes will write $g_H(h \mid e)$ for $g_H(r \mid e)$ and $P(h \mid d)$ for $P(r \mid e)$.) In practically all applications, the act chosen will not influence the probability of the state of nature. Assuming this, then by the expected utility hypothesis

$$U_H(r \mid e) = \sum_{s \in H} P(s \mid e) \cdot U_{H,e}(r, s)$$

$$= \sum_{s \in r} P(s \mid e) \cdot [1 - g_H(r \mid e)] + \sum_{s \in \sim r} P(s \mid e) \cdot$$

$$\cdot (- g_H(r \mid e))$$

$$= P(r \mid e) [1 - g_H(r \mid e)] - P(\sim r \mid e) \cdot g_H(r \mid e)$$

$$(3.3.) \quad U_H(r \mid e) = P(r \mid e) - g_H(r \mid e).$$

Thus

$$F_H(e) = \max_{r \in R(H)} (P(r \mid e) - g_H(r \mid e)).$$

For Levi's original epistemic utility, $g_H(r \mid e) = q \cdot M_e(r)/H_e$, where $M_e(r)$ is the number of elements of r consistent with e and H_e is the number of elements of H consistent with e, and q is a free parameter such that $0 < q \leqslant 1$. In the only other published major alternative, $g_H(r \mid e) = q \cdot P(r)$ for the same q, where $P(r)$ is the prior (unconditional) probability of r. This latter function does not depend on e, and was first considered by Levi (1963), advocated by Hintikka (1966; 1968) and developed by Hilpinen (1968). We shall refer to this function as Hintikka's.

A key to understanding the approach of epistemic utilities is to combine (3.3.) with the following two properties of g.

(C-1.) $0 \leqslant g_H(r \mid d) \leqslant 1.$

(C-2.)   Where $r_1, \ldots, r_k$ are the elements in r, $g_H(r \mid d) = \sum g_H(r_i \mid d).$

Together (3.3.) and (C-2.) entail

$$(3.4.) \quad U_H(r \mid e) = \sum (P(r_i \mid e) - g_H(r_i \mid e)) = \sum U_H(r_i \mid e).$$

It is then easy to ascertain $\max_{r \in R(H)} U_H(r \mid e)$. Each element h of H will have negative utility if and only if $P(h \mid e) < g_H(h \mid e)$. Since by (3.4.) the utility of r derives from the component utilities of the elements of r, the set r_0 of exactly those elements in H with non-negative utility must be maximal. Conditions (C-1.) and (C-2.) give an important clue to the nature of g. Since g is non-negative and more elements in r add factors to $g_H(r \mid e)$, g is a non-decreasing function of the logical weakness of set r. Since g is subtracted from $P(r \mid e)$ to obtain $U_H(r \mid e)$, the probability that the subset r contains the truth is weighted against the weakness of r. For a fixed probability of truth, stronger assertions are favored. For a fixed strength, more probable assertions are favored. More constrains than we have discussed are placed on g, but we need not be concerned with them here. Our only concern with g is its behavior at the maximum of (3.3.), as this is what will contribute to preferences over evidence and experiments.

How could the analysis of epistemic utility be applied to forming preferences over possible evidence and evaluating experiments? In standard utility theory, the only value that coming to have evidence has (i.e. $I_H(d)$) is its influence on the decision made, i.e. the value of cost free evidence is numerically identical to the utility of the act chosen on that evidence. This is tantamount to the assumption that $I_H(d) = F_H(d)$. However, the approach of epistemic utility need not be committed to this, for a rather subtle reason. $F_H(d)$ represents the value of *having* evidence d. But when one is considering the possibility of doing experiments and obtaining information, his position is not that of having evidence, but possibly *coming to have* evidence. Assuming $I_H(d) = F_H(d)$ would be to assume that as far as the decision problem at hand (here, evaluating hypotheses), coming to have the evidence contributes nothing over and above actually having the evidence. In standard applications of decision theory, this distinction does not arise because it is assumed that the evidence influences the decision only through the probabilities and not the utilities. Then, preferences over (cost and error free) experiments are trivially determined: the value of coming to have an experimental outcome is the utility of the decision made after the outcome is obtained, and the expectation over its outcomes determines the value of the experiment. We have already noted that in epistemic utility this key assumption is false: the utility of an act depends on what the evidence is. However, the expectation of the utility of the act chosen would still be the desirability of the experiment if the only value of coming to have evidence for the decision is having the evidence.

Since epistemic utility has not been confronted with this issue as a make-or-break problem, proponents have not had to reveal their options. A cautious and yet fair response for the committed theorist would be to see if any of the options could mathematically handle the problem. A natural first choice would be the standard option. For argument's sake, let us assume

$$(3.5) \qquad I_H(d) = \max_{r \in R(H)} U_H(r \mid d).$$

After having detailed the difficulties for epistemic utility within 3.5, we shall return in the penultimate section to the other option.

4. Do epistemic utilities satisfy
the rationality conditions?

The following is a sufficient condition for the three rationality conditions to be satisfied (see Appendix, Theorem 6):

(C-3.)   For all d and d', $g_H(r \mid d) = g_H(r \mid d')$.

Hintikka's epistemic utility then satisfies the rationality conditions, simply because it satisfies (C-3.). Moreover, there are many possible alternative epistemic utilities that likewise take advantage of (C-3.). (C-3.) requires that g be a degenerate function of the evidence. A large class of possible epistemic utilities can be generated as follows. With each element h of H, we associate a number $M_H(h)$ such that

(a) $M_H(h) \geqslant 0$,
(b) $M_H(h \vee h') = M_H(h') + M_H(h')$ if $(h \,\&\, h')$ is inconsistent,
(c) $\sum M_H(h_i) = 1$ where $h_1, ..., h_n$ are the elements of H.

M_H is just what is called a normalized measure over H. Letting $g_H(r \mid e) = = q \cdot M_H(r)$ then satisfies (C-3.) (with q as before). M may or may not be a degenerate function of H. Letting $K(z)$ be the cardinality of z, a suitable measure that is a non-degenerate function of H is $g_H(r \mid e) = K(r)/K(H)$: here each h in H has equal measure.

Letting r_e be the elements in r consistent with evidence e, and similarly for H_e, Levi took $g_H(r \mid e) = q \cdot K(r_e)/K(H_e)$, which is a non-degenerate function of e, and hence (C-3.) fails. Within (C-3.) and Hintikka's approach, the only way that evidence influences how well off we are epistemically in evaluating H (i.e. $U_H(r \mid e)$) is through the probability that assertions are true (i.e. $P(r \mid e)$). Are there any compelling considerations against (C-3.)?

Levi (1967b, p. 374) has given an argument against Hintikka's measure, which easily generalizes against (C-3). The objection is that where d entails that h and k are equivalent, then for d as our evidence, h and k ought to be epistemically indistinguishable. Let us call a g which is a non-degenerate function of the evidence *conditionalized*, and otherwise *prior*. Obviously h

and k may be distinguished in their prior measures even if given d they are equivalent. By (3.3.), then, $U_H(h \mid d) \neq U_H(k \mid d)$. If this consideration carries weight (I find it convincing) we must explore conditionalized measures.

5. CONDITIONALIZED MEASURES

Levi's conditionalized measure g does not satisfy (C-3.), and hence our sufficient condition for the rationality conditions does not apply to his measure. And it is not too difficult to show that the rationality conditions fail. Very striking is the failure of the ideal experiment to be maximally preferable. What is provable (see Appendix, Theorem 8) is that for any q such that $0 < q \leqslant 1$, there exists an H and an experiment E such that $U_H(E) > U_H(B)$, where B is an ideal experiment (and E is not). Where Levi's epistemic utility goes wrong is perhaps clear. Conclusive outcomes are effectively penalized by the relativization of g to e. When only one hypothesis h is left, $M_e(h)/H_e$ is maximal, and hence a maximal amount is subtracted from the posterior probability in $U_H(h \mid e)$. An h entailed by e is treated no differently than a logical truth with maximal probability and maximal weakness.

How might g in general be relativized to e? Notice that the relativized correlate of Hintikka's approach, where $g_H(r \mid e) = q \cdot P(r)$, would be $g_H(r \mid e) = q \cdot P(r \mid e)$. But then $U_H(r \mid e) = (1-q) P(r \mid e)$, which maximizes where $P(r \mid e)$ maximizes. Since $P(H \mid e) = 1$, the strongest r with maximal utility would be one entailed by e. If this is to be the strongest acceptable on the evidence, relativizing probability to e leads to the undesirable consequence of accepting only the deductive consequences of e. The same argument applies to any linear transformation of posterior probability.

To overcome this difficulty, Levi's approach has been to define a measure function M which is not in general posterior probability but does vary with e. Then g is taken to be some linear transformation of M. The avoids the unpalatable consequence of relativizing g to $P(r \mid e)$. Without additional constraints, however, this approach has almost no hope of satisfying the rationality conditions for preferences over experiments.

The key consideration is satisfying the rationality conditions is how M is conditionalized under additional information. Levi has adopted what I call the eliminative re-normalization strategy for conditionalization.

First, with b as background information, additional information e revises M only when $(b \& e)$ entails that an element of H is false that b alone does not entail to be false. Additional information affects M beyond b only through elimination of elements of H. Second, one recomputes M by taking those elements of H consistent with $(b \& e)$ and re-normalizing, i.e. where $h_1, \ldots, h_k$ are exactly those elements of H compatible with $(b \& e)$ and $j \in \{1, \ldots, k\}$

$$(5.1) \qquad M_H(h_j \mid (b \& e)) = M_H(h_j \mid b) / \sum_i^k M_H(h_i \mid b).$$

And in general where S is the set of those elements of H compatible with $(b \& e)$ and r is a subset of H,

$$(5.2.) \qquad M_H(r \mid (b \& e)) = M_H(r \cap S \mid (b \& e))$$

The re-normalization strategy flatly violates the rationality conditions. It can be proved (see Appendix, Theorem 7) that with (3.3.), any choice of q such that $0 < q \leqslant 1$ combined with re-normalization, violates (R-3.).

In another paper (1967b, pp. 379, 383), Levi advocates an epistemic utility with a slightly different form[3]: $U_H(r \mid e) = \alpha \cdot P(r \mid e) - (1 - \alpha) \cdot M_H(r \mid e)$, where $1 > \alpha > \frac{1}{2}$. The same objectionable result holds for this approach. What can be shown (see Appendix, Theorem 10) is that for any choice of α multiplied by $P(r \mid e)$ and β multiplied by M_H where $\alpha > \beta > 0$, and $U_H(r \mid e)$ has the revised form and re-normalization is used, (R-3.) is violated.

Thus, all of Levi's published conditionalized epistemic utilities, combined with (3.5.) and re-normalization, violate rationality conditions for preferences over experiments. Notice, however, what happens if we keep the general form of conditionalization (4.4.) but do not re-normalize. As far as the act with maximal utility, this is equivalent to taking g to be some degenerate function of the evidence. For when an element of H is eliminated, it contributes no probability, but add a non-negative g to be subtracted; so the maximum occurs on some subset of H that includes only those compatible with the evidence. But for these, by the above strategy, g is unchanged. So there is a prior g which maximizes on the same elements of H with the same utility. The two therefore are equivalent as far as the act chosen and the resultant preferences over experiments. Let us call the prior $g_H(r \mid e) = q \cdot K(r) / K(H)$ the revised Levi utility – it satisfies (C-3.)

and hence the rationality conditions, and preserves the equal element treatment of H originally found in Levi (1967a).

Is there then no way to satisfy the rationality conditions with a conditionalized g? What we have shown does not merit such a strong conclusion. But before detailing the major remaining possibilities, let us go on to a second group of objections to existing epistemic utilities.

6. Efficiency (I)

In the previous sections we examined epistemic utilities from the viewpoint of general preferences over experiments. In the next two sections we will examine subclasses of particular experiments where the interest is in evaluating hypotheses, and we have sharp and defensible intuitions about preferences.

Let $H=(h_1, h_2, ..., h_n)$ stand for the (exclusive and exhaustive) hypotheses under investigation. We wish to know which one of the h's is true, and we choose experiments to further this goal is isolating the truth. Now consider experiments of the following form. We divide all the elements of H into k disjoint subsets $S_1, S_2, ..., S_k$ for $2 \leqslant k \leqslant n$. The outcome of each experiment is which subset the truth is in: there are k possible outcomes. For a fixed k, each way of dividing H among the k subsets defines an experiment. What should our preferences be among the various experiments?

The universally accepted analysis of this problem is as follows. What determines our preferences is our probability distribution over $H:P(h_i \mid b)$, where b is our evidence. Then

(6.1.) Any way of dividing the elements of H such that for each S_j there is exactly one element h in S_j where $P(h \mid b) > 0$, receives maximal preference.

(6.2.) If there is no experiment satisfying (1), the second best is any way of dividing H into k equally probable subsets, i.e. where for each S_j, $\sum_{h \in S_j} P(h \mid b) = 1/k$.

Of course, there may be no experiment satisfying (2), either. How then to choose an experiment we shall not here consider.

Based on this analysis, we can divide the defined experiments (nondefined experiments are possible which are combinations of these) into three exclusive and exhaustive categories: type I, satisfying condition (1); type II, satisfying condition (2) but not (1); and type III, satisfying neither (1) nor (2). Then the following are the general ranking intuitions.

(6.3.) All experiments of type I are ranked together (identically).
(6.4.) All experiments of type II are ranked together (identically).
(6.5.) All experiments of type II are ranked strictly below those of type I.
(6.6.) All experiments of type III are ranked strictly below those of type II.

The mathematical apparatus of epistemic utilities is applicable to this problem, which is a special case of a purely epistemic interest in evaluating hypotheses. If epistemic utilities do not preserve these ranking intuitions, then they are inadequate as an analysis of the epistemic evaluation of hypothesis. Again, how do epistemic utilities fare?

A type I experiment is an ideal experiment. Knowing that Levi's original function does not satisfy (R-3.), we expect that type I experiments will not be maximal, and it is easy to show that a type III experiment is ranked above a type I. What of the revised Levi, in which $g_H(r \mid e) = q \cdot M_H(r) = q \cdot K(r)/K(H)$? We know that all ideal experiments receive maximal ranking. What can be shown, though, is that some type III experiments are ranked with some type II experiments.

Suppose H has four elements and two subsets are picked, each subset with two of the elements. Where u and z are the probabilities of the elements paired in a subset, the posterior distribution over H induced by the outcome that the truth is in the subset is a permutation of $(u/(u+z), z/(u+z), 0, 0)$. In the revised Levi, $q \cdot (\frac{1}{4})$ is subtracted from each posterior over H, and the non-negative elements summed to obtain the utility of the best act. Assuming that both posteriors exceed or equal $q \cdot (\frac{1}{4})$, the utility of the outcome is identically $1 - q/2$. Assuming the same is true for each outcome, the utility of all such experiments is $1 - q/2$. Thus all experiments are ranked together where the induced positive posteriors lie between $q \cdot (\frac{1}{4})$ and $1 - q \cdot (\frac{1}{4})$. Hence where the initial distribution is $(\frac{3}{8}, \frac{3}{8}, \frac{1}{8}, \frac{1}{8})$, the pairing $(\frac{3}{8}, \frac{3}{8})$, $(\frac{1}{8}, \frac{1}{8})$ is ranked with the pairing $(\frac{3}{8}, \frac{1}{8})$, $(\frac{3}{8}, \frac{1}{8})$. Hence the latter, a type II experiment, is ranked with the former, a type III experiment.

The same example works against Hintikka's epistemic utility since in

$g_H(r \mid e) = qP(r)$, $P(r)$ is the logical or absolute prior. (It cannot be the relative prior, i.e. $P(r \mid b)$ where b is the background evidence, because then $I_H(b)$ would reduce to a linear transformation of posterior probability.) It is trivial to set up the priors so that they are all equal, and the background information induces $(\frac{3}{8}, \frac{3}{8}, \frac{1}{8}, \frac{1}{8})$. Then Hintikka's function would be the same as the revised Levi. Any preference possible on the revised Levi is possible on Hintikka's epistemic utility. A fortiori, any objection based on possible preferences against the revised Levi counts against Hintikka.

7. EFFICIENCY (II)

Thus far we have concentrated on the derived preferences over experiments. Now we wish to apply the objections directly to the utility of the fixed evidence for evaluation: the function $U_H(d)$. Suppose the prior distribution of probabilities over H is $(\frac{1}{4}, \frac{1}{4}, \frac{1}{4}, \frac{1}{4})$, and that outcome e induces the posterior distribution $(\frac{1}{2}, \frac{1}{2}, 0, 0)$ and that e' induces $(0, 0, \frac{3}{4}, \frac{1}{4})$. On all of the unconditionalized epistemic utilities the outcomes are equally good: $U_H(e) = U_H(e') = 1 - q/2$. Notice why this is so: an element of h contributes to $U_H(d)$ only when $P(h \mid d) > q \cdot g_H(h \mid d)$.

When several elements of H contribute, all that matters is the *magnitude* of the difference between $P(h \mid e)$ and $q \cdot g_H(h \mid e)$, and not how the probability in excess of $q \cdot g(h \mid e)$ is *distributed* among the contributing h's. One can add or subtract from the posterior probabilities on e of the contributing h's without affecting $U_H(e)$ so long as any amount subtracted from h is less than $P(h \mid e) - q \cdot g(h \mid e)$. *The additive nature of epistemic utilities makes it within the limits indifferent to the distribution of the posterior probabilities.*

This indifference to posteriors is counterintuitive, for intuitively how well off we are epistemically with evidence for the evaluation of hypotheses does vary with the distribution of the posteriors. We desire outcomes which concentrate probability in fewer outcomes, which allow us to *discriminate* more among the hypotheses. Indeed, a general principle is that where A and B are probability distributions over H, if B can be generated from A by taking two elements x and y in A, $x > y$, and subtracting a portion of $(x - y)$ from x and adding it to y, then A is preferable to B as a distribution induced by evidence.

This indifference to posteriors can easily be taken advantage of to generate counterintuitive preferences over experiments. Any distribution can be adjusted in two ways without altering the utility of the experiments. First, among elements of H contributing to the utility of an outcome, probabilities can be redistributed among the contributing elements as long as the same elements remain contributors. Among non-contributing elements of H, the probabilities can be likewise redistributed. One other constraint must be satisfied: since $P(h) = \sum P(h \mid e_i) P(e_i)$, the posteriors must be altered in such a way that the prior probabilities of the H's remain unaffected.

Let the probabilities of the outcomes be equal, H have four elements, the experiment two outcomes, and each row represent the induced posteriors. Then for $q = 1$ an experiment with

$$
\begin{array}{cccc}
0.375 & 0.375 & 0.24 & 0.01 \\
0.125 & 0.125 & 0.74 & 0.01
\end{array}
$$

would be ranked equally with an experiment with

$$
\begin{array}{cccc}
0.25 & 0.5 & 0.24 & 0.01 \\
0.25 & 0 & 0.74 & 0.01
\end{array}
$$

Yet the second is a more conclusive experiment, as every outcome has more discrimination. As q approaches 0, the amount by which non-contributing factors can be altered decreases, but the leeway for contributing factors increases. No value of q is then quantitatively less susceptible to these objections.

Existing epistemic utilities then do not discriminate correctly among experiments according to their desirability. The experiments we have in the last two sections been calling preferable are easily shown to be more efficient in terms of the expected number of experiments required to get at truth. Where H is finite, through iteration of the opportunity to divide H into k subsets we can eventually isolate the truth. Strategies for the sequential choice of experiments can be shown to vary in the expected number of experiments required to get to the truth: this I call their efficiency. The last rationality condition we require is this: using the derived preference ranking of experiments to choose among experiments ought in general to lead to efficient choices. The indifference of epistemic utilities to the posterior distribution appreciably dulls their efficiency.

8. EPISTEMIC UTILITIES AND PREFERENCES OVER EVIDENCE AND EXPERIMENTS (II)

In Section 3 we indicated that how to form preferences over experiments on the basis of epistemic utilities is open to analysis. We showed in ensuing sections the difficulties the program gets into if the utility of coming to have evidence is identified with the utility of the act of hypothesis evaluation given that evidence. The alternative is to deny this, of course. However, the alternative is difficult to criticize, simply because the class of functions possibly measuring the value of coming to have evidence is left indeterminate.

From the viewpoint of epistemic utilities, how could this utility problem be analyzed? Epistemic utilities would surely be committed to at least the position that the value of coming to have evidence is a partial function of the utility of the act chosen given the evidence. And it would seem reasonable to assume that the value of coming to have evidence for evaluation is consistent with evaluation on fixed evidence in the following sense: the ranking of hypotheses given fixed evidence is the same as the ranking of hypotheses using the value of coming to have that evidence for evaluation. Aside from the act of evaluation, the only things left to determine the utility of coming to have evidence are the evidence itself and the hypotheses under evaluation. These considerations could all be satisfied if

$$(8.1.) \qquad I_H(e) = S(e, H) + T(e, H) \cdot F_H(e)$$

where S and T are real-valued functions of the evidence (e) and the hypotheses (H) under investigation, and T is positive-valued. Whether or not the rationality conditions can thus be incorporated into epistemic utility remains to be investigated.

9. THE FUTURE OF EPISTEMIC UTILITY AND THE NATURE OF EVALUATION

We have presented serious challenges to epistemic utilities from the viewpoint of the value of evidence and preferences over experiments. The challenge to the approach of epistemic utilities is whether or not it forms an adequate theory of hypothesis evaluation. What has been shown is that we have general intuitions about preferences over experiments where the only interest is cost free experimentation used to epistemically evaluate hypotheses. These intuitions form adequacy conditions for any theory of hypothesis evaluation.

The possible failure of epistemic utilities to satisfy the adequacy conditions is particularly significant when it is realized that they are the best candidates for a decision-theoretic approach to hypothesis evaluation. What does evaluating hypotheses have to do with making decisions? The classical Carnapian program is to treat evaluation as change in a confirmation function. The Bayesian analogue would be to treat evaluation in terms of posterior probabilities. The formal outcome of evaluation is then not a decision, but a probability distribution. By then defining a function on probability distributions which measures their desirability (as the outcome of evaluation), its expectation over experiments defines preferences over experiments. This approach has been mathematically worked out within information theory, and can be developed in such a way that the rationality conditions we have discussed are provable consequences[4]. Thus a rival, non-decision-theoretic analysis of hypothesis evaluation exists within which intuitive preferences over experiments can be formed. The challenge clearly faces the approach of epistemic utilities to form adequate preferences over experiments.

Nothing we have shown demonstrates that the approach of epistemic utilities cannot satisfy the intuitions about preferences. The large class of epistemic utility functions with the form of (8.1.) remains to be investigated for their preferences over experiments: those which multiply existing utility functions by positive-valued functions of the evidence, and add any real-valued function of the evidence. This paper details both the intuitions that need to be satisfied and necessity for their satisfaction, if epistemic utility is to be anything approaching an adequate theory of hypothesis evaluation. If epistemic utilities can be so extended is an open question subject to proof.

From the viewpoint of the Bayesian program for the analysis of knowledge and science, the potential of the program of epistemic utility is especially interesting. Already the approach has shown the Bayesian how to extend his program so as to develop a classificatory concept, if the need arises (e.g. in knowledge, explanation, justification). The approach to hypothesis evaluation in terms of some function measuring degree of support is one major part of the Bayesian program. Whether or not decision theory can likewise figure significantly in the Bayesian foundations of hypothesis evaluation (and of the statistical testing of hypotheses) remains to be seen[5].

University of Virginia

APPENDIX

The first two theorems derive from Good (1967). Theorems 6–10 assume familiarity with the discussion of epistemic utility in the paper.

THEOREM 1. *Let A and B be finite sets where the elements of A are indexed by A_j, and the elements of B by B_k. Let f be any real-valued function of $A \times B$.*
Then

$$\sum_k \max_j f(A_j, B_k) \geqslant \max_j \sum_k f(A_j, B_k).$$

Proof. Let A^* maximize $\sum_k f(A_j, B_k)$. Clearly $\max_j f(A_j, B_k) \geqslant f(A^*, B_k)$ for each k.

THEOREM 2. *Let the states of nature be finite and indexed by s_i. Let r stand for an arbitrary act. Suppose that the states of nature are probabilistically independent of the acts. Let $P(s_i \mid d)$ stand for the probability of s_i given that d is the evidence. Suppose that the utility of choosing r depends only on the state of nature – for this utility we write $U(r, s_i)$. Then with $F_H(b)$ as in Section 3 and $U^*_H(X)$ as in Section 2, $U^*_H(X) \geqslant F_H(b)$.*
Proof. Distributing $P(e_j \mid b)$ in and applying Theorem 1 to $U^*_H(X)$, we obtain

$$U^*_H(X) \geqslant \max_r \sum_j P(e_j \mid b) [\sum_i P(s_i \mid (b \& e_j)) U(r, s_i)].$$

Multiplying $P(e_j \mid b)$ in and interchanging the summations, the above resolves to $F_H(b)$.

THEOREM 3. *Under the same antecedent as Theorem 2, let C be some sure experiment with outcome e where $P(e \mid b) = 1$. Then $U^*_H(X) \geqslant U^*_H(C)$.*
Proof. All outcomes of C except e have 0 conditional probability on b. Hence

$$U^*_H(C) = \max_r \sum P(s_i \mid (b \& e)) U(r, s_i)$$
$$= \max_r \sum P(s_i \mid b) U(r, s_i)$$
$$= F_H(b)$$

Theorem 3 then follows by Theorem 2.

THEOREM 4. *Define $U^*_H(X)$ as in Theorem 2 for exclusive and exhaustive e_i, and $U^*_H(Y)$ analogously for exclusive and exhaustive d_j. For each i and j, (e_i & d_j) likewise defines a group of exclusive and exhaustive conditions: let $U^*_H(XY)$ be as in Theorem 2 for these conditions. Then under the same antecedent as Theorem 2, $U^*_H(XY) \geqslant U^*_H(Y)$.*

Proof.

$$U^*_H(XY) = \sum_i \sum_j [P((e_i \& d_j) \mid b) \max_r \sum_k P(s_k \mid (b \& e_i \& d_j)) \cdot U(r, s_k)]$$

$$= \sum_j \sum_i \max_r P((e_i \& d_j) \mid b) [\sum_k P(s_k \mid (b \& e_i \& d_j)) \cdot U(r, s_k)]$$

Applying Theorem 1 to the above we obtain that

$$U^*_H(XY) \geqslant \sum_j \max_r \sum_i P((e_i \& d_j) \mid b) [\sum_k P(s_k \mid (b \& e_i \& d_j)) U(r, s_k)]$$

which resolves to $U^*_H(Y)$ by multiplying $P((e_i \& d_j) \mid b)$ in and interchanging summations.

THEOREM 5. *Let $U^*_H(B)$ be defined as in Theorem 2 for exclusive and exhaustive outcomes d_j, and $U^*_H(X)$ for outcomes e_i. Suppose that each d_j entails one and only one of the states of nature. Then, under the same conditions as for Theorem 2, $U^*_H(B) \geqslant U^*_H(X)$.*

Proof. Let t_j be that state of nature entailed by d_j. Then $U^*_H(B)$ is

$$\sum_j P(d_j \mid b) \max_r U(r, t_j),$$

$$= \sum_j [\sum_i P(d_j \mid (b \& e_i)) P(e_i \mid b)] \max_r U(r, t_j),$$

$$= \sum_i P(e_i \mid b) \sum_j \max_r P(d_j \mid (b \& e_i)) U(r, t_j).$$

Applying Theorem 1 to be above we obtain that $U^*_H(B) \geqslant$

$$\sum_i P(e_i \mid b) \max_r \sum_j P(d_j \mid (b \& e_i)) U(r, t_j).$$

Now, $U^*_H(X) = \sum_i P(e_i \mid b) \max_r \sum_j P(s_j \mid (b \& e_i)) U(r, s_j)$.
The theorem then follows if we can show that

$$\sum_j P(s_j \mid (b \& e_i)) U(r, s_j) = \sum_j P(d_j \mid (b \& e_i)) U(r, t_j).$$

Suppose $P(s_j \mid (b \& e_i)) = 0$. Then s_j contributes nothing to the left summation. On the right s_j contributes nothing if no d_k entails s_j. But if some d_k entails s_j, then $P(d_k \mid (b \& e_i)) = 0$, else $P(s_j \mid (b \& e_i)) \neq 0$. So s_j contributes nothing to the right either.

Suppose $P(s_j \mid (b \& e_i)) \neq 0$. Then some d_k must entail s_j. Then $U(r, s_j)$ appears in the right summation for exactly those d's that entail s_j. But then $P(s_j \mid (b \& e_i))$ equals the sum of the probabilities on $(b \& e_i)$ of those d's that entail s_j. So if s_j contributes to the left summation, the identical amount is contributed to the right summation.

But the cases cover all states of nature. Since the terms in each summation derive from the states of nature, the two summations are equal.

THEOREM 6. *If an epistemic utility satisfies (C-3.) and (3.5.), then it satisfies (R-1.), (R-2.), and (R-3.).*

Proof. The probabilities of the hypotheses (the states of nature) are assumed to be independent of the acts of acceptance. By (C-3.) and (3.2.), the utilities of acts depend only on the state of nature and the acts. Hence the conditions for Theorems 2, 3, and 4 apply. Then using (2.3.) and (3.5.), (R-1.), (R-2.), and (R-3.) follow respectively from Theorems 2, 3, and 4 as special cases.

THEOREM 7. *For an epistemic utility where the expected utility has the form of* (3.4.), $g_H(r \mid e) = q \cdot M_H(r \mid e)$, *M is a measure computed according to re-normalization* (5.1.), *preferences over evidence have the form of* (3.5.) *and B is an ideal experiment, then for every q such that $0 < q \leqslant 1$ there exists an H and an experiment E (not ideal) such that* $U_H(E) > U_H(B)$.

Proof. By (2.3,), $U_H(E) > U_H(B)$ if and only if $U*_H(E) > U*_H(B)$. Since B is an ideal experiment, each outcome entails exactly one element of H. The elements of H not entailed by an outcome have posterior probability ($P(h \mid (b \ \& \ e_i))$) of zero. Since g is non-negative, their utility is non-positive. The element entailed has a posterior probability of 1 and the normalized M is 1, hence its utility is $1 - q$. Thus the act with maximal utility for an outcome is to accept the element of H the outcome entails. Since for each outcome, the utility maximizes at $1 - q$, $U*_H(B) = 1 - q$.

The hypotheses H and experiment E can then be generated as follows. Suppose $x + 2$ objects, including w, are randomly distributed over four cells, where $x \geqslant 1$. The resultant distribution is given to be

	M	F
E	x	0
B	1	1

Let the hypothesis h under investigation be that $w \in FB$. H is then $\{h, \sim h\}$. Let both h and $\sim h$ have positive prior measure M. Let E be the experiment with the following outcomes: $w \in EM \cup FB$, $w \notin EM \cup FB$.

In the case of the latter outcome, $\sim h$ is entailed and thus the act of accepting $\sim h$ as strongest has maximal utility of $1-q$. In the case of the outcome e that $w \in EM \cup FB$, $P(e) = (x+1)/(x+2)$ and $P(h \mid e) = 1/(x+1)$. For any q such that $0 < q \leqslant 1$, it is possible to set x large enough so that $P(h \mid e) < q \cdot M_H(h \mid e)$. Letting $m = M_H(\sim h \mid e)$, simply take

$$(*) \qquad x > (1/(q \cdot (1-m))) - 1.$$

For such a choice of x, $P(\sim h \mid e) > q \cdot m$ and the act with maximal utility is to accept $\sim h$. Hence one factor in $U*_H(E)$, namely $P(e) [P(\sim h \mid e) - qm]$, is

$$(x/(x+2)) - ((x+1)/(x+2)) \cdot qm$$

If it is possible to choose x so that this factor alone is greater than $1-q$, we are done. Setting the above to be greater than $1-q$, the inequality resolves to

$$(**) \qquad x > ((2/q) - (2-m))/(1-m).$$

Obviously, it is possible to set x to satisfy both (*) and (**). For such x, $U*_H(E) > (1-q) = U*_H(B)$. Yet B is ideal, and E is not.

THEOREM 8. *Let K be the cardinality function, r_e be the set of elements of r consistent with e, and H_e be the number of elements of H consistent with e. For an epistemic utility where the expected utility has the form of* (3.4.), $g_H(r \mid e) = q \cdot K(r_e)/K(H_e)$, *and B is an ideal experiment, then for every q such that $0 < q \leqslant 1$ there exists an H and experiment E (not ideal) such that* $U_H(E) > U_H(B)$.

Proof. Clearly the defined M satisfies re-normalization, hence the theorem is only a special case of Theorem 7.

THEOREM 9. *Theorem 7 holds if the expected utility has the form, not of (3.4), but of* $\alpha \cdot P(r \mid e) - g_H(r \mid e)$, *where* $g_H(r \mid e) = \beta \cdot M_H(r \mid e)$, *and* $\alpha > \beta > 0$.

Proof. By the same argument in the proof of Theorem 7, $U^*_H(B) = \alpha - \beta$. The same, experiment works, only the inequalities to satisfy for x are

$$x > (\alpha/((1-m)\,\beta)) - 1,$$
$$x > ((2(\alpha - \beta)/\beta) + m)/(1 - m).$$

THEOREM 10. *Theorem 7 holds of the expected utility has the form, not of (3.4.), but of* $\alpha \cdot P(r \mid e) - g_H(r \mid e)$, *where* $1 > \alpha > \frac{1}{2}$, *and* $g_H(r \mid e) = (1 - \alpha) \cdot M_H(r \mid e)$.

Proof. This is just a special case of Theorem 9.

NOTES

[1] Each rationality condition is intuitively necessary, given that experiments are to be ranked only for their contribution to evaluation. An experiment with a sure outcome cannot result in any change of evidence, and therefore contributes nothing more to evaluation (R-1.). Likewise, an experiment where each outcome entails which hypothesis is true must of necessity definitively settle the problem of evaluation, and therefore has maximal preferability (R-3.). Given that we are interested in experiments only for their information for evaluation, doing more can never decrease the expected information (R-2.). Notice that from the viewpoint of gathering information, the sure experiment amounts to doing nothing: it is a dummy experiment for not really experimenting. (R-1.) then rules out ever preferring to do nothing at all to genuinely experimenting. (R-3.) rules out a flagrant form of inefficiency for getting at the true hypothesis.

[2] Epistemic utility is usually developed in a slightly different way, by assuming that the utility of accepting r depends only whether r is true or false, not how exactly how it is true or false. The act of accepting r is then treated as if the relevant states of nature were r and $\sim r$. I develop the notion in the standard decision-theoretic framework where there are states of nature common to all acts. The result is the same.

[3] Actually, the form is $\alpha \cdot P(r \mid e) + (1 - \alpha)\, C(r \mid e)$, where $C(r \mid e) = M_H(\sim r \mid e)$. But $M_H(\sim r \mid e) = 1 - M_H(r \mid e)$, so the expected utility is $\alpha P(r \mid e) - ((1 - \alpha) \cdot M_H(r \mid e)) + (1 - \alpha)$. The last factor is a constant, not affecting the utility measure.

[4] The most detailed treatment of this is found in Goosens (1970). Many of the essentials can be gleaned from Lindley (1956). However, one seldom publicized result is that ranking experiments according to the expected negentropy in the induced posterior probability distribution is not an order-preserving transformation of efficiency. For the maximally preferable experiments, the two agree. But for less the ideal experiments negentropy is only provable *approximately* efficient.

[5] This paper has greatly benefitted from earlier correspondence and conversation with Isaac Levi. Sections 3 and 8 are especially indebted to his comments.

BIBLIOGRAPHY

Good, I. J.: 1967, 'On the Principle of Total Evidence', *The British Journal for the Philosophy of Science* 17, 319–321.

Goosens, William K.: 1970, *The Logic of Experimentation* (Ph.D. Dissertation, Stanford University).

Hempel, C. G.: 1960, 'Inductive Inconsistencies', *Synthese* **12**, 439–469.

Hempel, C. G.: 1962, 'Deductive-Nomological versus Statistical Explanation', in *Minnesota Studies in the Philosophy of Science*, Vol. 3 (ed. by H. Feigl and G. Maxwell), University of Minnesota Press, pp. 98–169.

Hilpinen, R.: 1968, 'Rules of Acceptance and Inductive Logic', *Acta Philosophica Fennica* **22**, North Holland.

Hintikka, J.: 1968, 'The Varieties of Information and Scientific Explanation', in *Logic, Methodology, and Philosophy of Science* III (ed. by van Rootselaar and Staal), North Holland, pp. 311–331.

Hintikka, J. and Pietarinen, J.: 1966, 'Semantic Information and Inductive Logic', in *Aspects of Inductive Logic* (ed. by J. Hintikka and P. Suppes), North Holland, pp. 96–112.

Levi, I.: 1963, 'Corroboration and Rules of Acceptance', *British Journal for the Philosophy of Science* **13**, 307–313.

Levi, I.: 1967a, *Gambling with Truth*, Alfred A. Knopf.

Levi, I.: 1967b, 'Information and Inference', *Synthese* **17**, 369–391.

Lindley, D. V.: 1956, 'On a Measure of the Information Provided by an Experiment', *Annals of Mathematical Statistics* **27**, 968–1005.

KEITH LEHRER

INDUCTION, CONSENSUS AND CATASTROPHE

The conception of induction we shall articulate and defend is based on a concern for truth and the constraint of social consensus. Inductive inference, if totally successful, would yield maxiverific results. The result of accepting statements in a language is maxiverific if and only if the set of statements accepted contains all true statements of the language and only true statements of the language. In a language adequate for all scientific purposes, there is no realistic method for finding the maxiverific set. We require some local method of induction consistent with the general objective. We shall propose such a method based on consensual probability assignments which determine the cost and benefits of accepting statements in the quest for the maxiverific.

I. MAXIVERIFICITY

A set of statements is maxiverific in language L if and only if every true statement of L is in the set and no statement in the set is untrue. A maxiverific set is maximally consistent.[1] A set is maximally consistent in a language when the set is consistent and the addition of any new sentence of the language to the set would render the set inconsistent. More technically, a set S is maximally consistent in L if and only if S is logically consistent in L and S is not a proper subset of any logically consistent set in L. A maxiverific set S in L is maximally consistent in L because to add a new sentence of L to a maxiverific set is, necessarily, to add a false statement. A maxiverific set in L contains all true statements in L, and, therefore, it contains the denial of any false statement in L. Adding a false statement of L to a maxiverific set in L is to add a statement whose denial is a member of the original maxiverific set. More formally, if M is maxiverific in L, M is a proper subset of S in L, and P is in S but not in M, then, since all true statements in L are in M, P is false. The denial of P is true and in M, and, therefore, S contains both P and the denial of P.

R. J. Bogdan (ed.), Local Induction, 115–143. *All rights reserved.*
Copyright © 1976 by D. Reidel Publishing Company, Dordrecht-Holland

Thus, an inductive method aiming at maxiverificity can succeed only if it directs us to accept a maximally consistent set of L by induction from our evidence. Some method is needed for aiming at this general objective in local contexts of scientific investigation and rational inquiry generally.

II. COMPETITION

Suppose a scientist or some other epistemically virtuous individual considers the question of whether to accept some hypothesis. As formulated, the question is elliptical. An expanded question is one in which the competitors of H are considered, and the question is whether to accept H or some competitor of H. The set of competitors include statements considered relevant to H. One way of explicating relevance is in terms of a probability assignment so that K is relevant to H on evidence E if and only if $\mathbf{p}(H, K \& E) \neq \mathbf{p}(H, E)$. This account of relevance raises the question of what probability assignment is to be selected. We shall argue subsequently that the probability assignment should be a consensual assignment of a group of scientists linked by chains of positive respect. It should be noted, however, that the set of competitors generated by H should be restricted to those relevant statements members of the reference group consider worth investigation. For example, if K is relevant to H and J is not, the statement $(K \vee J)$ might be relevant to H simply because K is relevant. Members of the reference group, while acknowledging the relevance of the disjunction, may consider it unworthy of investigation. Any statement relevant to H on E in terms of the consensual probability assignment of the group which any member of the group considers worthy of investigation should, however, be counted as a competitor of H. The finite set S of statements including H and the competitors of H we call the *competition set* for H. From this set, which is part of a language L, we can generate an epistemic field F of statements by forming all truth functional combinations of members of S. The resultant set F will be logically inconsistent. For every member of S both that member and the denial of that member will be included in F. The field F thus becomes a sublanguage of L closed under truth functional operations.

Considering F, the field generated from the competition set for H, we obtain a notion of local maxiverificity, or maxiverificity in F. A set is maxiverific in F if and only if it contains all true statements in F and only true statements in F. A set C is maximally consistent in F if and only if C is a logically consistent set of members of F, and no set of statements of F of which C is a proper subset is logically consistent. Finding a maximally consistent set of the total language of science to achieve maxiverificity appears unrealistic, but a method for finding a maximally consistent set in F that it is reasonable to accept inductively to achieve maxiverificity in F is readily available.

III. A CONDITION OF ADEQUACY

If we could assign probabilities to maximally consistent sets of statements in an epistemic field, then we could obtain a condition of adequacy for the reasonable acceptance of sets of statements by induction to achieve maxiverificity. If, for example, one maximally consistent set in F was more probable on our evidence, then it would be reasonable to accept that set. Our reasoning is that if we accept no maximally consistent set in F, then we necessarily forego the opportunity to achieve maxiverificity in F. On the other hand, if one maximally consistent set in F is more probable on the evidence than any other such set, then there is a better chance that all the members of that set are true than that all the members of any other such set are true. Therefore, it would be more probable on the evidence that we would achieve maxiverificity in F by accepting that most probable maximally consistent set in F. The set of maximally consistent sets in F represent the alternative choices as we have for achieving maxiverificity. We choose one, or we fail to attain our objective. If it is more probable that we will achieve our objective by choosing one of these alternatives over the others, then it is reasonable to choose that alternative.

The foregoing argument entails the following condition of adequacy:

(CA)   If one maximally consistent set C in an epistemic field F is more probable on evidence E than any other maximally consistent set in F, then it is reasonable to accept C by induction from E.

There are a number of philosophers who have proposed rules of induction that violate this condition of adequacy. We shall consider their position below. Before doing so, it is essential to explain how we may assign probabilities to maximally consistent sets in a field of statements.

IV. THE PROBABILITY OF SETS

We shall assume that we have a method for assigning probabilities to single statements in the field and propose a method for extending such assignments to *sets* of statements. The most natural and least controversial method is to assign the same probability to a set of statements as we assign to a single statement logically equivalent to the set. A set of statements A is logically equivalent to another set B if and only if every logical consequence of A is a logical consequence of B and *vice versa*. A set of statements S is logically equivalent to a single statement P if and only if S is logically equivalent to $\{P\}$, the set of statements whose sole member is the single statement. Thus, if every maximally consistent set C in a field F is logically equivalent to some single statement in F, then we could extend a probability measure to assign probability values to maximally consistent sets in F. We assume the following principle for extending probability assignments to sets of statements.

(PE)   If S_H is a set of statements in L logically equivalent to the statement H, S_E is a set of statements in L logically equivalent to the statement E, and $\mathbf{p}$ is probability function on the statements of L, then $\mathbf{p}(H, E) = \mathbf{p}(S_H, E) = \mathbf{p}(H, S_E) = \mathbf{p}(S_H, S_E)$.

This principle is insufficient to determine the probabilities of every set of statements of L from the probabilities of single statements of L. It will, however, enable us to ascertain the probabilities of any maximally consistent set C in an epistemic field F, for, as we shall now show, every such set is logically equivalent to some single statement in F.

The members of every maximally consistent set C in F are logical consequences of a proper subset of C. Once again consider the set of statements composing the competition set S for touchstone hypothesis H from which F is generated. Let us call the members of S the *atoms* of F. Where A is an atom of F, the set $\{A, \sim A\}$ is an *atomic pair* of F. Let us call a set a *constitution* of F if and only if it contains exactly one member from each atomic pair of F and is a logically consistent set.[2] The constitutions of F thus constitute the set of truth conditions for members of F. For every sentence P of F there is a set of constitutions of F, C_P, such that P is true if and only if the statements of one constitution in C_P are all true. Every maximally consistent set M in F contains some constitution

C of F as a proper subset, and all the members of M are deductive consequences of C. Thus, every maximally consistent set in F is logically equivalent to some constitution in F. Finally, therefore, every maximally consistent set C in F is logically equivalent to a single statement in F, that statement being a conjunction of the members of the constitution included in C. Assuming that the original competition set was finite in membership, the conjunction will also be finite in length.

V. EXAMPLE: MAN ON EARTH

Let us consider hypotheses concerning how man came to exist on earth. One hypothesis, the most respectable one, is that man came to exist on earth as a result of evolving by genetic mutation from an ape. Let us call this hypothesis E. Another hypothesis that at least some intelligent men think worth consideration and relevant to E is that man came to exist on earth by divine intervention. I call this hypothesis G. Finally, there is the hypothesis that man came to the earth from some other world, that he is an invader on earth. Let us call this hypothesis I. To simplify matters, let us suppose that these are the only hypotheses that any intelligent person finds worthy of consideration. The hypotheses have been formulated in such a way that they are logically consistent with each other. It is logically consistent to suppose that evolution took place after apes were placed in a rocketship by men of another world to mutate and arrive on earth as men. It is logically consistent with this to imagine that divine intervention took the form of influencing the men of another world to dispatch the rocketship. Of course, the hypotheses of evolution, intervention, and invasion could be further elaborated so that they would be mutually exclusive in pairs.

The statements E, G, and I are the atoms of the epistemic field. The constitutions in the field F generated from these atoms are as follows:

$$(1) \ \{ \ E, \quad G, \quad I\}. \qquad (5) \ \{ \ E, \sim G, \sim I\}.$$
$$(2) \ \{ \ E, \sim G, \quad I\}. \qquad (6) \ \{\sim E, \quad G, \sim I\}.$$
$$(3) \ \{ \ E, \quad G, \sim I\}. \qquad (7) \ \{\sim E, \sim G, \quad I\}.$$
$$(4) \ \{\sim E, \quad G, \quad I\}. \qquad (8) \ \{\sim E, \sim G, \sim I\}.$$

Each of these is logically equivalent to a maximally consistent set in F, and those eight maximally consistent sets are all the maximally consistent sets in F. Each of the constitutions is logically equivalent to a conjunction of the members of the constitution. Let C_1, C_2 and so forth to C_8 be those conjunctions. The constitution and the maximally consistent set in F equivalent to it may be assigned a probability on our evidence equal to the probability of the corresponding conjunction by (PE). Surely E is more probable on our evidence than either of the other two hypotheses. Hypothesis E is logically equivalent to the disjunction of C_1, C_2, C_3, and C_5. However, a set of statements logically equivalent to E or to that disjunction is not maximally consistent in the field F. The maximally consistent sets in F correspond to the constitutions or to the conjunctions of members of such constitutions. According to (CA), if one maximally consistent set in F is more probable than any other, then it is reasonable to accept that set by induction from the evidence. Most people would consider the maximally consistent set equivalent constitution 5 to be the most probable. C_5 tells us that E is true and that both G and I are false. Assuming that conjunction to be more probable than the conjunction corresponding to any other constitution, it is reasonable by (CA) to accept the maximally consistent set corresponding to 5 by induction from our evidence.

Other philosophers have proposed rules of induction according to which it would not be reasonable to accept any maximally consistent set because of the probability of some other constitution.[3] Thus, for example, if we suppose that the probability of constitution 8, which tells us that none of the hypotheses is correct, were sufficiently high though less probable than constitution 5, then on such accounts it would not be reasonable to accept constitution 5. The intuitive justification for this position is that the evidence, though favoring 5, may fail to refute 8, and, therefore, we should suspend judgement between 5 and 8. Following this proposal, we would be reasonable to accept a set of statements whose members are the disjunction of C_5 and C_8 and the deductive consequences of that disjunction in F. Such a set is not maximally consistent in F, however. We could add statement C_5 to such a set without inconsistency.

Inductive rules that prohibit us from accepting a maximally consistent set in F more probable than other such sets do not aim at the induction of all and only true statements in F. Maxiverificity in F is not the objective of such rules. In almost any context of scientific inquiry, moreover, the constitution comparable to 8 should be expected to have a fairly high probability on our evidence. For there is a pretty high probability in most cases that none of the hypotheses we can articulate is exactly correct and that some other hypothesis, as yet unconceived, will turn out to be correct instead. Nevertheless, if some constitution is more probable than all the rest it would be unreasonable for us to refuse to accept that constitution simply because some other constitution is almost as probable. By such a refusal, we neglect our interest in truth, that is, in maxiverificity. For that objective can only be achieved if we conform to condition (CA) and accept some maximally consistent set that is more probable than any other on our evidence.

It should also be noted that by directing us to accept the most probable maximally consistent set in a field, condition (CA) leads us to the acceptance of one of the most definite sets of statements. A less definite set of statements might tell us that at least one of several hypotheses is true without telling us exactly which hypothesis is true and which are false. Such definiteness is also a form of informativeness. It does not, however, provide a complete explication of the concept of informativeness. Any C_i is more informative than the disjunction of that C_i with another, but some such conjunctions are *internally* more informative than others.[4] For example, C_8 is the least informative of all because it is simply negative, it tells us nothing positive about how man came to exist on earth. Purely negative statements are reasonably discounted because they do not enable us either to explain or to predict. One salient characteristic of men, in science and other areas of rational inquiry, is a readiness to accept some positive hypothesis rather than to withhold acceptance.

VI. MINIMAL INCONSISTENCY

We have formulated our condition of adequacy for induction aiming at maxiverificity in terms of the acceptance of maximally consistent sets of statements. It is customary and more natural to formulate rules of induction as directives to accept single statements by induction from our evidence. We shall now propose a rule of induction for single statements in a field F yielding results that satisfy condition (CA). We propose that a hypothesis in F may be inductively inferred from evidence when there is no set of statements in F that constitutes a cogent argument against F.

Thus in effect we assume that an inductive hypothesis is innocent unless there is a cogent argument to the contrary. An argument A against accepting H is cogent if each premiss of A is at least as probable as H and the denial of H is the valid conclusion of the argument. We restrict consideration to *essential* argument in which every premise is essential to the argument. No proper subset of the premises of such an argument yields the conclusion. An argument is valid if and only if the set of statements consisting of the premisses of the argument and the denial of the conclusion is logically inconsistent. A valid argument is essential in the required sense if and only if the set of statements consisting of the premises and the denial of the conclusion (or any truth functional transformation of it) is *minimally inconsistent*[5]. Finally, a set of statements is minimally inconsistent if and only if the set is logically inconsistent and every proper subset of the set is consistent.

There is no cogent argument against a statement H in F if and only if there is no essential deductive argument each premise of which is at least as probable as H on our evidence E and which has the denial of H as a conclusion. The latter is equivalent to saying that every minimally inconsistent set in F of which H is a member contains at least one statement that is less probable than H on E. From these considerations we obtain the following rule of induction:

(IR)   Accept H by induction from E in the field F if for any minimally inconsistent set M in F of which H is a member there is some statement K also a member of M such that $\mathbf{p}(H, E)$ exceeds $\mathbf{p}(K, E)$.

The result of (IR) can best be described by considering the conjunctions C_1, C_2, and so forth to C_n corresponding to the constitutions 1, 2, and so forth to n in the field F. We refer to them as *C-conjunctions*.[6]

If C_i is more probable than its denial, then we may accept C_i by induction from E by (IR). Every consistent statement in F is logically equivalent to a disjunction of C-conjunctions because, as we have noted, the constitutions, and hence, the C-conjunctions are truth conditions for the statements in F. In that probability is additive, the probability of a disjunction of C-conjunctions is equal to the sum of the probabilities of the individual C-conjunctions. The denial of a C_i is logically equivalent to the disjunction of all the remaining C-conjunctions. Any disjunction which contains C_i as a disjunct will be logically consistent with C_i. The most probable disjunction that is not consistent with C_i is, therefore, the one logically equivalent to the denial of C_i. Hence the consequence that a C_i more probable on the evidence than its denial may be accepted by induction from the evidence according to (IR).

Disjunctions of two or more C-conjunctions may be accepted by induction from the evidence according to (IR) if and only if it is more probable that the disjunction is true on the evidence than it is that at least one disjunct is false. Thus, if C_i is a disjunct of the disjunction D of C-conjunctions and D is more probable on E than the denial of C_i, then D may be accepted by induction according to (IR). On the other hand, if every C_i in D is such that its denial is at least as probable as D, then D may not be accepted. Every statement logically equivalent to a C-conjunction or disjunction of C-conjunctions accepted by (IR) may also be accepted by (IR).

The set of statements accepted according to (IR) is logically equivalent to a maximally consistent set in F if some maximally consistent set is more probable than any other such set in F. Some maximally consistent set in F is more probable than any other such set on our evidence E if and only if some C-conjunction in F is more probable than any other such conjunction on E. If some C-conjunction is more probable on E than any other C-conjunction, then the denial of every other C-conjunction in F will be accepted according to (IR). A C-conjunction in F is logically equivalent to the set of the denials of all other C-conjunctions in F. Therefore, if one C-conjunction in F is more probable on E than any other such conjunction in F, then the set of statements accepted by induction from E according to (IR) is logically equivalent to the most probable C-conjunction, and, therefore, to a constitution and maximally consistent set corresponding to that C-conjunction.

It is not the case, however, that (IR) directs us to accept all the members of the maximally consistent set that corresponds to the most probable C-conjunction. If, for example, the C-conjunction has a probability greater than any other C-conjunction in F but less than its denial, then though (IR) directs us to accept the denial of each C-conjunction other than the most probable, and, therefore, a set of statements logically equivalent to the most probable C-conjunction, it does not direct us to accept that C-conjunction. Rule (IR) is not deductively closed. By amending rule (IR) so that it tells us to accept the statements (IR) directs us to accept *and also* the deductive consequences of that set in F, we obtain a rule that satisfies (CA). The rule is as follows:

(AR)   Accept H by induction from E in the field F if and only if *either* (i) for any minimally consistent set M in F of which H is member, there is a member K of M such that $\mathbf{p}(H, E)$ exceeds $\mathbf{p}(K, E)$ *or* (ii) there is a set of statements S accepted by condition (i) and H is a logical consequence of S in F.

The set of statements consisting of a C-conjunction in F and the logical consequences in F of that C-conjunction is a maximally consistent set in F. There is a maximally consistent set in F more probable on our evidence than any other such set if and only if there is some C-conjunction in F more probable on our evidence than any other C-conjunction in F. Therefore, rule (AR) directs us to accept a maximally consistent set in F that is more probable on our evidence than any other such set, if there is such a set.

Suppose there is no C-conjunction in F that is more probable on our evidence than any other, and, therefore, no maximally consistent set is more probable on our evidence than every other such set. In that case, rule (AR) directs us to accept the disjunction of all those C-conjunctions that are most probable on the evidence, and, consequently, the intersection of the maximally consistent sets in F corresponding to those C-conjunctions. If probability gives us no basis for choosing between several maximally consistent sets in F, then (AR) directs us to withhold accepting any such set and to accept only those statements in F common to all the most probable maximally consistent sets in F. An alternative rule for dealing with those cases in which no maximally consistent set in F is more probable than all other such sets on our evidence might tell us to accept any one of those most probable sets. The objection to such a procedure, however, is that we are directed to accept such a set when we have no reason for accepting that maximally consistent set rather than another. If two maximally consistent sets are equally probable, then we have no reason to believe that one rather than the other of those sets is maxiverific. Therefore, a rule directing us to accept a maximally consistent set under such conditions is not a prescription of rationality. Such a rule would be one of arbitrary caprice.

VII. COSTS AND BENEFITS

To show that rule (AR) is rational we shall now demonstrate that the rule directs us to a policy yielding a maximum of expected utility when the utilities express our interest in the maxiverific. There are two possible outcomes of accepting a statement, accepting a true statement, or accepting a false one. When our objective is maxiverificity, these are the relevant outcomes. The expected utility of accepting a statement by induction may be computed as the sum of the utility of each outcome times the probability of that outcome. Letting '$uT(H, E)$' stand for the utility of accepting H by induction from E when H is true, '$uF(H, E)$' stand for the utility of similarly accepting H when H is false, and '$e(H, E)$' stand for the expected utility of accepting H by induction from E in the interests of maxiverificity, we obtain the following equation:

$$e(H, E) = p(H, E)uT(H, E) + p(\sim H, E)uF(H, E).[7]$$

We shall assume a probability assignment and subsequently consider how it is to be ascertained. Assuming a probability assignment, we shall propose a method for assigning values to the utilities.

The utilities in question may be explicated in terms of the benefits and costs of acceptance for maxiverificity. The benefit of accepting a hypothesis H from evidence E when H is false is obviously zero. One obtains no benefit in terms of accepting all and only true statements when the statement one accepts is false. When one accepts a true statement, on the other hand, the benefit is maximal. Letting '$bT(H, E)$' stand for the benefit of accepting H by induction from E when H is true, '$bF(H, E)$' stand for the corresponding benefit when H is false, and normalizing the maximum benefit as unity, we obtain the following equalities:

$$bT(H, E) = 1,$$
$$bF(H, E) = 0.$$

The utility of accepting a hypothesis may be explicated as benefit minus cost. Thus, the utility of accepting a hypothesis on our evidence when the hypothesis is true is equal to the benefit of accepting such a true hypothesis minus the cost of such acceptance, where the benefits and costs are measured in terms of our interest in the maxiverific.[8] Letting '$k(H, E)$' stand for the cost of accepting H by induction from E, we obtain the following equations for utilities:

$$\mathbf{uT}(H, E) = \mathbf{bT}(H, E) \sim \mathbf{k}(H, E),$$
$$\mathbf{uF}(H, E) = \mathbf{bF}(H, E) \sim \mathbf{k}(H, E).$$

Our remaining problem is to find a method for assigning values to our cost function.

In economics, according to Alchian, the cost of adopting an alternative is the highest valued opportunity necessarily forsaken.[9] Thus, for example, a person who invests in a stock should *not* think of the cost of his investment as being equal simply to the amount of money he invests. By investing his money in stock he necessarily forsakes the opportunity to invest it in government bonds with a fixed rate of interest. Thus, his cost is at least the amount he invested *plus* the interest he would have received from the bounds. This conception of cost, converted from economics to epistemology, is also appropriate for the solution of our problem.

The highest valued opportunity necessarily forsaken in accepting a hypothesis may be determined by considering what a person necessarily forsakes in accepting a hypothesis. If maxiverificity in F is our objective, then maximally consistent sets in F represent a set of alternatives. The acceptance of a hypothesis H in F by induction from E necessitates that we forsake the acceptance of all those maximally consistent sets in F of which H is not a member. If H is a C-conjunction in field F, then in accepting H we necessarily forsake the opportunity to accept any other C-conjunction in F because a C-conjunction corresponds to, and is a member of, exactly one maximally consistent set in F. Hence, the cost of accepting a C-conjunction by induction from E is equal to the highest value of accepting any other C-conjunction in F.

With respect to the acceptance of other statements in F, specifying the highest valued opportunity necessarily forsaken is somewhat more complicated. Every consistent statement in F is, we have noted, logically equivalent to a disjunction of C-conjunctions. If some C-conjunction, C_i, is accepted by induction from E, then a corresponding maximally consistent set in F containing C_i and the deductive consequences of C_i in F is accepted. In the event that no C-conjunction in F is accepted by induction because no such conjunction is more probable than every other such conjunction in F, then we must consider whether any disjunction of two or more different C-conjunctions in F should be accepted. Let us, to simplify matters, consider only disjunctions of C-conjunctions in which no C-conjunction occurs more than once and lower numbered C-conjunctions precede higher numbered ones in the disjunction. We shall call these *D-disjunctions*. Every consistent statement in F is logically equivalent to one and only one D-disjunction in F. We shall refer to the number of disjuncts in a D-disjunction as the *constitution number* of the disjunction because that number represents the number of constitutions in which the disjunction is true. Two-membered D-disjunctions have the constitution number 2, three membered ones the number 3 and so forth. The set of D-disjunctions having the same constitution number, for example, the number 2, is minimally inconsistent. Since maxiverificity is our objective, we cannot accept all the members of such a set. Therefore, we must necessarily forsake the opportunity to accept some member of such a set. In accepting a D-disjunction by induction, we not only forsake the opportunity to accept those C-conjunctions and corresponding maximally consistent sets that are not included in the disjunction, we also forsake the opportunity to accept at least one other D-disjunction of the same constitution number. For example, if F contains three C-conjunctions, C_1, C_2, and C_3, then it contains exactly three D-disjunctions, $(C_1 \vee C_2)$, $(C_2 \vee C_3)$, and $(C_1 \vee C_3)$ with the constitution number 2. The set of D-disjunctions is logically inconsistent, and, therefore, in accepting one of these D-disjunctions, for the sake of maxiverificity, we must forsake the opportunity to accept at least one of other D-disjunctions.

How should we measure the value of forsaken opportunities to accept hypotheses by induction from our evidence? Since what we value is the acceptance of all and only true statements, it is most natural to measure such value as the chance for accepting a true statement we have forfeited. The cost of accepting a C-conjunction by induction from E is the probability on E of the most highly probable other C-conjunction. The cost of accepting a D-disjunction D_i is again a more complicated matter. That cost is at least as great as the most highly probable C-conjunction not included in D_i, and, indeed, as the disjunction of all those C-conjunctions not included in D_i, because all those C-conjunctions are inconsistent with D_i. However, the cost of accepting D_i by induction from E may be even greater than the chance for truth forsaken in accepting the disjunction of all the remaining C-conjunctions in F. At least one D-disjunction of the same constitution number as D_i must be forsaken as well. Hence, D_i must be more probable on E than at least one other D-disjunction of the same constitution number. The probability of the most probable D-disjunction of the same constitution number represents an opportunity necessarily forsaken. If that probability is higher than the probability of the disjunction of all those C-conjunctions not in D_i, then that probability represents the highest valued opportunity necessarily forsaken in accepting H by induction from E.

VIII. MEASURING COST

From this analysis, we may specify a method for finding the cost of accepting H by induction from E. Let H^* represent the statement of the highest probability, or in case more than one is equally high, any of the highest, whose acceptance is necessarily forsaken by accepting of H on E in the field F. We then obtain the equality

$$k(H, E) = p(H^*, E).$$

We may calculate $p(H^*, E)$ as follows:

(i)   if H is a C-conjunction in F, then $p(H^*, E) = p(C_i, E)$, where C_i is a C-conjunction in F other than H, and for any C-conjunction, C_j, other than H, $p(C_i, E)$ is at least as high as $p(C_j, E)$.

(ii)   if H is a D-disjunction in F, then either $p(H^*, E) = p(C_i, E)$ or $p(H^*, E) = p(D_i, E)$. If $p(C_i, E) = p(D_i, E)$, then $p(H^*, E) = p(C_i, E)$. But if $p(C_i, E) \neq p(D_i, E)$, and $p(C_i, E)$ is higher than $p(D_i, E)$, then $p(H^*, E) = p(C_i, E)$. If $p(D_i, e)$ is higher then $p(C_i, E)$, then $p(H^*, E) = p(D_i, E)$. We let C_i be any C-conjunction in F and not in H such that for any C-conjunction, C_j, in F and not in H, $p(C_i, E)$ is as least as high as $p(C_j, E)$. We let D_i be any D-disjunction in F other than H with the same constitution number as H, such that for any D-disjunction, D_j, in F other than H with the same constitution number as H, $p(D_j, E)$ is not less than $p(D_i, E)$.

Conditions (i) and (ii) enable us to compute the cost of accepting any statement in F that is neither contradictory nor tautologous because every such statement is logically equivalent to either a C-conjunction or a D-disjunction in F. The tautologous disjunction of all C-conjunctions in F is also a D-disjunction. Moreover, there are no other D-disjunctions in F with the same constitution number. Therefore, no cost is assigned to that tautologous disjunction by (ii). We shall assign the minimum cost of zero to accepting a tautology in F since such a tautology belongs to every maximally consistent set in F, and we shall assign the maximum cost of one to accepting a contradiction in F since a contradiction belongs to no maximally consistent set in F.

We then obtain the following specification of cost:

(CS) $k(H, E) = p(H^*, E)$, where $p(H^*, E) = 1$ if H is a contradiction in F, $p(H^*, E) = 0$ if H is a tautology in F, and, if H is neither a contradiction nor a tautology in F, $p(H^*, E) = p(G^*, E)$ where G is a C-conjunction or a D-disjunction in F logically equivalent to H.

IX. EXPECTED UTILITY

From this specification together with the equality $p(\sim H, E) = 1 - p(H, E)$ and our earlier equations, we obtain the following equation for expected utility:

$$e(H, E) = p(H, E)(1 - p(H^*, E)) +$$
$$+ (1 - p(H, E))(0 - p(H^*, E))$$

which reduces to

$$e(H, E) = p(H, E) - p(H^*, E).$$

Thus expected utility of accepting a statement by induction from E is equal to the probability of that statement minus the probability of the statement representing the highest valued opportunity necessarily forsaken by accepting that statement. The expected utility of accepting a contradiction will always be -1, of accepting a tautology, 1, and of accepting other statements some number between these two values.

If restricted to accepting a single statement, one should accept that statement whose expected utility is maximal. However, we are concerned, not with accepting a single statement, but with accepting a set of statements to promote maxiverificity. We obtain a maximum of expected utility by accepting the set containing all those statements whose expected utility is positive. We do not obtain any increase in expected utility by accepting any statement whose expected utility is zero or negative, and, therefore, we eschew the acceptance of such statements. We therefore propose the following decision theoretic principle of acceptance:

(DR)   Accept H by induction from E in F if and only if $e(H, E)$ is greater than 0.

This rule is equivalent in the results it yields to the earlier acceptance rule (AR). Therefore, a person aiming at maxiverificity and seeking to accept a set of statements to attain that end, is reasonable just in case he conforms to (AR). If he does not conform to that principle, then either he is unreasonable or he is attempting to achieve some purpose other than that of accepting all and only true statements in an epistemic field.

X. OBJECTIONS AND REPLIES

Les us consider some objections that might naturally be raised against our theory. By meeting them we shall provide further clarification and justification for our methods. First it might be objected that our rule of acceptance is too lenient. Hempel, and others following him, have construed the acceptance of statements on antecedent evidence as a process whereby accepted statements are *added* to the evidence. Consequently, Hempel argues that no statement should be accepted on the basis of evidence unless it has at least a probability of $\frac{1}{2}$.[10] For, if a statement has a probability less than $\frac{1}{2}$, then the statement is more probably false than true and so should not be added to our evidence.

We agree with Hempel that no statement with a probability less than $\frac{1}{2}$ should be added counted as evidence. There is, however, an important distinction between accepting a statement as evidence and accepting a hypothesis by induction from evidence.[11] If what we accept as evidence is false, then inference based upon that evidence is founded on error. Hence it is critical that statements accepted as evidence be true, and, consequently, no statement should be accepted as evidence unless it is more probable that the statement is true than false. When we accept a statement as a hypothesis by induction from our evidence, however, it is accepted tentatively as a working hypothesis providing part of the best picture of world we can obtain from the evidence we have. It is a picture that might be altered by the very smallest addition of new evidence. When maxiverificity is our objective, it is reasonable to accept any statement we have some reason for accepting in order to obtain that complete description of the world contained in a maximally consistent set. We accept a set of statements, or a single statement, that is less probable than $\frac{1}{2}$ on our evidence because it is more probable on our evidence than any other maximally consistent set. Such acceptance is required by the condition of adequacy defended above.

Secondly, it may be objected that on our account a C-conjunction is accepted when it is only minutely more probable on our evidence than another C-conjunction. This is especially objectionable, it may be urged, when our evidence is positive evidence for the less probable C-conjunction. It might be concluded that such a minute difference does not justify accepting the slightly more probable C-conjunction at the expense of rejecting the other, and, therefore, that we should be more cautious and suspend judgment between the two C-conjunction. Theories of Levi, Hilpinen, and Hintikka imply that such suspension is the reasonable policy.[12]

Our reply is that the proposed policy abandons maxiverificity for some other objective. If we fail to accept a maximally consistent set, then we necessarily fail to achieve maxiverificity. Of course, when we have no reason to prefer one maximally consistent set to another, we may have to suspend judgement to avoid capriciousness. When, however, we have some reason, such as a difference in probability, to prefer one C-conjunction and corresponding maximally consistent set to another, then the objective of maxiverificity demands that we accept a maximally consistent set rather than suspend judgement.

Thirdly, it may be objected that our inductive rule is too adventuresome, and, therefore, that it fails to accord with scientific procedure. For example, Levi has noted that on the evidence that a fair coin is tossed one hundred times, our rule directs us to accept the statement that exactly half the tosses will fall heads. He concludes that our rule fails to conform to statistical practice because we would only be warranted in accepting the statement that approximately half the tosses fall heads.[13]

Our reply is based on a distinction between two forms of statistical estimation.[14] One form of statistical estimation is *point* estimation, and this requires an exact estimate. If we are asked to estimate exactly what percentage of the tosses will fall heads, then our rule yields the correct point estimate, namely, that 50 will fall heads and 50 will be tails. Another form of statistical estimation is *interval* estimation. If the field is properly chosen, we again obtain suitable results with our rule. For example, if we are concerned with interval estimates at a confidence level of 0.95, then a C-conjunction should tell us that an interval (40,60) is optimal at that confidence level. An optimal interval is the shortest interval at a given confidence level. If a statement affirming that an interval is optimal is more probable on our evidence than any other statement of optimality, then that statement should be accepted on standard statistical methodology.[15] That is precisely the result our rule yields. Thus, once the distinction between point estimation and interval estimation is clearly drawn, the objection is readily met. It should be noted, moreover, that some scientific practice may not have maxiverificity as an objective, and some statistical methods may be among such practices. We are not attempting to offer a monolithic account of scientific inquiry. The objectives of science are multiple and various. The pursuit of the whole truth, of maxiverificity, is, however, fundamental to all rational inquiry.

Finally, it may be objected that the cost of accepting a statement is not properly measured in terms of the probability of a statement whose acceptance is necessarily forsaken. Instead it should be measured in terms of the real cost to science, in experimentation for example, of accepting a statement. Such a theory as that propounded by Marschak is an attempt to capture this more practical conception of cost.[16] In reply, we only note once again that we have been concerned with one objective, maxiverificity, and we do not claim that our conception of cost is adequate to any other purpose. The practical costs to science, even those directly related to the conduct of scientific investigation, are different from the purely epistemic costs of accepting a statement by induction in the attempt to accept that maximally consistent set which is the most probable picture of the world we can construct on the basis of our present evidence.

These objections do not by any means exhaust the ones that could be raised against our theory, but we believe they are the most important with one exception. It may be objected that our methods and our rule can only be as sound as the probability assignment on which they are based. We now turn to the problem of finding an appropriate probability assignment.

XI. PROBABILITY

We have assumed a probability assignment for the statements of L from which the statements in F are drawn. It should be noted, however, that rule (AR) depends only on our being able to *compare* the probabilities and to decide whether one statement is more or less probable than another on our evidence. The numerical assignment of probabilities is, therefore, unneeded for the application of the rule. Moreover, our initial condition of adequacy (CA) was also formulated in terms of the comparison of the probabilities of maximally consistent sets. Thus, we need not assume that we have any method for selecting a single probability assignment as the basis for applying our rule. Any probability assignment yielding the same comparative ordering of probabilities on our evidence will yield the acceptance of the same set of statements. For the purposes of applying the rule, a numerical probability assignment may be construed as a convenient fiction.

A problem remains nonetheless. The comparative ordering of probabilities made by an individual may be unreasonable. For example, they may be internally inconsistent. A person who estimates that H_1 is more probable on his evidence than H_2, that H_2 is more probable than H_3, and that H_3 is more probable than H_1, is unreasonable because of the transitivity of the comparative probability relation. Moreover, even if his estimates are internally consistent, they may be unreasonable when they conflict with evidence that is in the public domain. A person whose views are internally consistent may be unreasonable because he ignores the opinions of scientific experts. In short, a person may be reasonable to accept a set of statements in terms of his *personal* probability estimates but unreasonable in terms of the *interpersonal* probability estimates of the experts.

If we could find some single probability assignment that would be objective in the sense that any reasonable person would employ that assignment to determine what statements to accept by induction from his evidence, that assignment would be interpersonal. The work of Carnap and Hintikka represent an attempt to find a probability assignment that gives proper weight to statistical and logical factors.[17] This important work does not, however, lead to the selection of a single probability assignment on the basis of such factors. Hintikka has shown that we may assign a general hypothesis corroborated by the observation of large number of favorable instances and no unfavorable ones any probability from zero to unity on the basis of such evidence.[18] We conclude, without claiming to have proven this, that such factors will not suffice for the selection of a single objective probability assignment.

As an alternative, there is the theory of subjective or personal probability articulated by Savage and Jeffrey based on a coherent preference function of an individual.[19] We may think of subjective probabilities as the estimates an individual makes of the chances a statement has of being true. Given the lack of any satisfactory method for selecting a single objective probability assignment, subjectivists argue, we are left with subjective or personal probabilities. We shall, however, propose a method for finding an interpersonal probability assignment for members of certain groups on the basis of their personal probability assignments.

XII. CONSENSUAL PROBABILITY

A group of scientists in the quest for the maxiverific may respect, either directly or indirectly, the personal probability assignments of other members of the group. When maxiverificity is our objective, the probability assignments of others may be considered as indicators of truth. We shall consider the probability assignments of some members of the group better indicators of truth than others. One person directly respects the probability assignment of another when he gives positive weight to his probability assignment. A person m_1 indirectly respects the probability assignment of another m_2 when m_1 gives positive weight to the probability assignment of some other person, that person in turn gives positive weight to the probability assignment of another person, and this chain of positive respect extends from m_1 to m_2 by the positive weight each person gives to the probability assignment of the next person in the chain.

Now let us suppose that every member of a group of scientists respects, directly or indirectly, every other member of the group. The weight each person gives to the probability assignment of another in the group may be interpreted by a number w_{ij} which is the weight person i gives to the probability assignment of person j. The weights in question may be conceived of as second level estimates or probabilities of first level probabilities, all of which we assume to be subjective. A person may estimate that another in the group has no chance of being correct, or he may simply have no estimate because of ignorance concerning the person in question. In such cases, we may interpret the weight given as equal to 0. It may seem unfortunate to assign zero weight to the probability assignment of a person simply because we are uninformed about him, but, as we shall see, this assumption is quite acceptable. For those members of the group, m_1, m_2 and so forth to m_n to whom person i gives some positive weight, there will be a second level probability, $\mathbf{p}_j$, that i assigns to the probability assignment of person j. Let $\mathbf{p}$ be the sum of the probabilities $\mathbf{p}_1$, $\mathbf{p}_2$, and so forth to $\mathbf{p}_n$ that i assigns to the members m_1, m_2, and so forth to m_n. To normalize the weights a person gives to probability assignments of others, we let $\mathbf{w}_{ij}$ equal $\mathbf{p}_i/\mathbf{p}$. We assume that one of these weights, $\mathbf{w}_{ii}$, is the weight person i gives to his own probability assignment.

Now assume that the group is in an initial state 0 in which each member of the group has a weight for every other member of the group. If a person takes others to be indicators of truth, and his weights are estimates of how reliable an indicator of truth others are, then, if he is reasonable, he should shift his probability assignment in the light of his estimates of the reliability of others. If we let '$\mathbf{p}_j^i(H, E)$' stand for the probability of H on E for a person j in state i, person i in state 0 should shift his probability assignment to state 1 by the following method of aggregation: $\mathbf{p}_i^1(H, E) = \mathbf{w}_{i1}\mathbf{p}_1^0(H, E) + \mathbf{w}_{i2}\mathbf{p}_2^0(H, E)\cdots + \mathbf{w}_{in}\mathbf{p}_n^0(H, E)$. If each person shifts from the initial state 0 to state 1, he will arrive at a new probability assignment, his state 1 probability assignment. By adding 1 to each superscript we obtain a formula for shifting from state 1 to state 2 using the same weights. If we increase the superscripts by one again, we obtain a formula for shifting from state 2 to state 3, and so on for each increase in the superscripts. When we calculate the results of shifting probability assignments by this method, we find that as the superscript increases to infinity, the probability assignments converge toward an interpersonal or consensual probability assignment.

This method for finding an interpersonal or consensual probability assignment rests on three conditions. First, that every member of the group respects, either directly or indirectly, every other member of the group; secondly, that the method of aggregation proposed is used to shift probability assignments; and, thirdly, that the weights remain constant through shifts. It would be beyond the scope of this paper to undertake a detailed defense of these methods.[20] Our contention is that if an impartial and rational observer with no probability assignment of his own were to choose a probability assignment solely on the basis of complete information about the initial state of members of the group every one of which respects, at least indirectly, every other, he should adopt the probability assignment toward which all probability assignments converge by the proposed method of aggregation. Moreover, our impartial observer need not observe any actual shifting from the initial state, he need only calculate the results of such repeated shifts to fully exploit his information concerning that state.

Finally it should be noted that the convergence results even if some members of the group have positive respect for only one other member, indeed, it could even result when each member of the group assigns positive weight to only one other member. Further, the probability assignments might converge toward exactly the same probability assignment as they would converge toward if each member of the group assigned a positive weight to every other member. Even if most members of the group are ignorant of the views of others, as might be the case in a large group, and thus assign zero weight to them, our method still applies. The positive weights that are assigned to some members of the group can convey positive weight through the process aggregation to other members of the group. Consequently, the same consensual probability assignment may result that would result if each person were aware of the probability assignments of all other members and attached some positive weight to them.

The probability assignment toward which subjective probability assignments converge by application of the method of aggregation is an intersubjective or interpersonal probability assignment. It has two uses. First, someone seeking to find a single objective probability from the infinity of possibilities might reasonably base his choice on the interpersonal probability assignment of a group of experts. Secondly, suppose a field F represents an area of expertise of some group, and there is a consensual probability assignment for that group. Then, for a person seeking maxiverificity, it is intersubjectively or interpersonally reasonable to accept a hypothesis H by induction from E in the field F if and only if rule (AR) directs such acceptance on the basis of the consensual probability assignment. We may, therefore, appeal to a consensual probability to explain why a person who accepts a hypothesis by induction according to (AR) on the basis of his *personal* probability assignment may be unreasonable nonetheless. He is unreasonable because the consensual probability of a relevant group of experts whose opinions are available to him would not countenance the acceptance of the hypothesis in question by induction according to (AR). A reasonable person in the quest for the maxiverific does not ignore consensus among the experts in forming his probability assignments and proceeding to accept hypotheses on the basis of them.

XIII. CATASTROPHE

The role of consensus in reasonable acceptance, especially in science, helps to account for what appear to be discontinuous and catastrophic shifts in what is accepted. Historians of science, Kuhn especially, have represented basic shifts as being essentially revolutionary and thus falling outside of any supervening scientific methodology.[21] Part of what is implied by Kuhn is that such changes alter our conception of what is to count as evidence, what was formerly accepted as evidence may be repudiated, and hence that science does not just accumulate more and more evidence in the storehouse of knowledge. Occasionally the storehouse gets a thorough cleaning. It is not our intention to rebut this contention. We do claim, however, that our model offers an explanation of why such catastrophic changes occur without appealing to the metaphor of revolution. Our rule provides a method which supervenes special theories and yields an account of rational acceptance. We contend that shifts may be both catastrophic and reasonable. Our analysis is adopted from catastrophy theory as articulated by Zeeman.[22]

To understand catastrophic shifts in science and other areas of inquiry, assume that rule (AR) gives us a rough approximation to actual practise in some domain. Moreover, assume that some approximation to a consensual probability assignment is used as a basis for such acceptance of hypotheses and theories. Though the personal probability assignments of members of a group of experts would not exactly coincide, we assume there is some awareness of where consensus lies. We suppose, moreover, that they consider it reasonable to accept hypotheses that would be accepted by (AR) on the consensual probability assignment for the group.

Now imagine that in some domain of science a hypothesis H_s is accepted by most of the experts. It is the standard theory. There is, however, another hypothesis H_a which is accepted by a smaller group of experts. It is the alternative theory. The first group is G_s and the second G_a. Finally, imagine that not every member of the combined group of G_s and G_a respect every other member of the group, even indirectly. Thus, we have no consensual probability assignment for the combined group. However, we suppose that there is a consensual probability assignment for each of the small groups, every member of G_s respects at least indirectly every other member of that group, and the same is true of the members of G_a. In this situation, we may imagine that the consensual probability assignment of group G_s assigns a high probability on the evidence to H_s and a very low, practically zero perhaps, to hypothesis H_a. By (AR) and their consensual probability assignment, they accept H_s by induction from the evidence. The opposite is true of the consensual probability assignment of group G_a, which assigns a high probability to H_a and, consequently, they accept H_a by induction from the evidence.

Finally, let us imagine that some members of each group, perhaps the younger and more explorative ones, come to respect the opinions of members of the other group as a result of experiment, theoretical ratiocination and improved communication. As group crossing respect increases, respect may be extended *indirectly* from every member of the combined group to every other member of the combined group. Moreover, this state can, in principle, be achieved by two people, one from each group, who have direct positive respect for each other. A consensual probability for the combined group supplants the original pair of consensual probability assignments of the groups G_s and G_a. The new consensual probability assignment for the combined group is influenced by the personal probability assignments of members of G_a and, consequently, may assign a much higher probability to H_a on the evidence than would have been assigned to H_a on the original consensual probability assignment of group G_s. Indeed, the probability distribution curve for the original consensual probability assignment of group G_s might have a single mode representing the high probability assigned to H_s sloping rapidly down to indicate the low probability assigned to competing hypotheses. By contrast, the subsequent consensual probability assignment of the combined group may be bimodal representing the higher probability assigned to each of the hypotheses, H_s and H_a, in contrast to the low probability assigned to other hypotheses.

The bimodal probability distribution representing the consensual probability assignment may, however, not be perceived by the majority of members of the original group H_s. They may, as a result, continue to base their acceptance of hypotheses on the original consensual probability assignment of the group G_s and, therefore, continue to accept H_s and reject H_a. Thus, there may be a lag between the time at which a new consensual probability assignment emerges from the new chains of respect that weld the two groups together and the time at which most of the members of the standard group become aware of this consensus. When this consensus is based on an exchange of information and resulting respect between a small number of members of both groups, it may be easily overlooked. Natural resistance to giving up a theory one accepts may also be influential.

As respect for members of group G_a among members of G_s increases, the consensual probability assignment for the combined group shifts from a bimodal distribution to one in which hypothesis H_s has a low probability and H_a a very high probability. When members of G_s who failed to perceive the earlier bimodal consensual probability distribution become aware of the later consensual probability distribution with a single mode assigning a high probability to H_a, the change may appear catastrophic. They may look upon the change as revolutionary and irrational. This is a total misperception. The shift is a reasonable adjustment of a probability assignment based on the acquisition of information.

Members of group G_s come to respect the probability assignments of some members of G_a, that is, they come to believe that there is some chance, however small, that some members of G_a may be right. These estimates both reflect an exchange of information and themselves constitute information. What a qualified expert believes in his area of expertise is relevant information. The first and second level probability assignments of the experts represent the information they possess. The method of aggregation enables us to compute the agreement or consensus contained in the personal probability assignments of members of the combined group. The consensual probability assignment represents, therefore, the fullest exploitation of the information available for accepting hypotheses inductively in the quest for the maxiverific. What appears to be a catastrophic revolution is, in fact, no more than a reasonable adjustment to new information. Those members of the standard group who suddenly confront a shift from a probability distribution assigning a high probability to the standard theory they accept and a low probability to the alternative, to a new consensual probability distribution assigning a low probability to the standard theory and a high probability to the alternative theory may regard the change as catastrophic. However, contrary to what Kuhn and others have suggested, the change is a reasonable response to new information contained in the probability estimates of experts. Our conclusion is that such apparently catastrophic shifts are simple applications of the maxim that no relevant information to which we have access should be ignored in determining probabilities and the acceptance of hypotheses. Consensual probability assignments summarize accessible information.

Though we shall not elaborate this contention in detail, shifts in what we accept as evidence as well as in what we accept as hypotheses by induction from evidence may result from the emergence of new consensual probability assignments. We have shown elsewhere that the acceptance of statements as evidence may be based on antecedent probability assignments in a manner analogous to that in which the acceptance of hypotheses by induction from evidence is based on conditional probabilities.[23] The rule for the acceptance of evidence must be more restrictive to yield a better safeguard against accepting false statements as evidence. Nevertheless, the acceptance of statements as evidence results when a statement has a higher antecedent probability than those with which it competes for the status of evidence. On this account, a shift in antecedent probabilities can change what we accept as evidence; we accept some new statement as evidence or repudiate some statement formerly accepted.

Assuming that the acceptance of statements as evidence is based on a consensual probability distribution arrived at by an analogous method of aggregating personal probability assignments, shifts in what it is reasonable to accept as evidence may occur as a result of new chains of respect welding groups together. Shifts in what we count as evidence may again appear as catastrophic and unreasoned to members of the standard group. In fact, the change results from nothing more than new information contained in the probability assignments of members of the group. Apparently catastrophic and revolutionary changes in what we count as evidence are, once again, simple adjustments made in response to new information embodied in the probability assignments of experts. Treating experts as indicators of truth, it is reasonable in seeking maxiverificity to accept hypotheses by induction from evidence in accord with our rule of acceptance based on a consensual probability assignment of those experts.

XIV. SUMMARY

We have argued that a person who seeks to accept all and only those statements of a language that are true, that is, who has maxiverificity as an objective, must accept some maximally consistent set to succeed. To pursue maxiverificity in a local context of investigation, we restrict consideration to statements in an epistemic field. For the purpose of achieving maxiverificity in a field it is reasonable to accept a set of statements that is both maximally consistent in the field and more probable than any other maximally consistent set in the field. A principle was proposed for computing the probability of such maximally consistent sets. We formulated a rule for the acceptance of a set of statements by induction from our evidence and showed that the rule is equivalent to one instructing us to accept a set of all those statements having positive expected utility calculated in terms of the costs and benefits of accepting such a set in the pursuit of maxiverificity. The costs and benefits were defined in terms of a probability assignment, and the rule of acceptance was defended against objections.

A distinction was introduced between what it is personally reasonable for a person to accept on the basis of his subjective or personal probability assignment and what it is intersubjectively or interpersonally reasonable for a person to accept on the basis of a consensual probability assignment of a group of experts. It was shown that a consensual probability assignment can be derived from the personal probability assignments of members of the group together with the weights each attach to the assignments of others in the group when every member of the group gives some positive respect, at least indirectly, to the probability assignment of every other member of the group. The latter may occur even if most members of the group give zero weight to most other members.

Finally, it was argued that apparently catastrophic and irrational shifts in what is accepted both as hypotheses and as evidence can be accounted for in terms of new positive respect among members of different groups welding new groups together with chains of indirect respect. Such chains produce consensual probability assignments for the combined group. Those who fail to perceive the initial stage of such a consensus may regard the eventual outcome as revolutionary and unreasoned. Our model reveals that such shifts are simply adjustments to the new information contained in the new consensual probability assignment. They conform to the maxim that we alter our acceptance of statements as evidence and hypotheses by induction from evidence when new information warrants that we do so. Our theory shows how, when we treat others as indicators of truth and respect them accordingly, the quest for maxiverificity can be made intersubjectively reasonable by being based on consensual probability assignments. The objective of acceptance is maxiverificity, acceptance is based on probability, probability is based on consensus among the experts, and the objective of the experts is maxiverificity. The wheel of inquiry spins in search for truth.

University of Arizona

NOTES

[1] For an elementary account of maximally consistent sets, see Mates [18].

[2] I borrowed the term *constitution* from John Vickers. He uses the term in his monograph (unpublished) on the probability of sets of statements.

[3] Cf., Levi [15], [16], Hilpinen [6], and Hintikka [8]. For a more detailed criticism, see Lehrer [14].

[4] Cf., Hintikka [8]. According to Levi [15], if we ask which C_i is true, then each C_i is equally informative. This is erroneous. Each C_i is an answer to the question, but they are *not* equally informative answers.

[5] In earlier articles, *essential* argument was called *relevant* argument [13]. The change in name is to avoid confusion with *relevance logic* articulated by Belnap and Anderson [2].

[6] The C-conjunctions are called *partitions* in earlier articles [11], [14].

[7] This approach to epistemic utility is due to Hempel [5]. It has been adopted and developed by Hintikka [8], Levi [15], and Hilpinen [6].

[8] The cost-benefit approach to utility is used by Marschak [17].

[9] Alchian [1].

[10] Hempel [5], also Hilpinen [6].

[11] For an elaboration of the distinction, see Lehrer [11], [14].

[12] Levi [15], Hilpinen [6]. Hintikka [8].

[13] Levi [16].

[14] It was suggested to the author by Ian Hacking.

[15] See Hilpinen [6].

[16] Marschak [17].
[17] Carnap [3], [4] and Hintikka [7].
[18] Hintikka [7].
[19] Jeffrey [9], and Savage [19].
[20] This method is defended in papers by the author forthcoming in *Nous* and *Synthese*.
[21] Kuhn [10].
[22] Zeeman [20].
[23] Lehrer [11], [12] and [14].

BIBLIOGRAPHY

[1] Alchian, A., 'Cost', in D. L. Sills (ed.), *International Encyclopedia of the Social Sciences*, New York, 1968, pp. 404–414.
[2] Anderson, A. R. and Belnap, N. D., 'Tautological Entailments', *Philosophical Studies* **2** (1974), 79–92.
[3] Carnap, R., *Logical Foundations of Probability*, second ed., Chicago, 1962.
[4] Carnap, R., Chaps. I and II, in R. Carnap and R. Jeffrey (eds.), *Studies in Inductive Logic and Probability*, Vol. I, Los Angeles, 1971.
[5] Hempel, C. G., 'Deductive-Nomological vs. Statistical Explanation', in *Minnesota Studies in the Philosophy of Science*, Vol. III, Minneapolis, 1962, pp. 98–169.
[6] Hilpinen, R., 'Rules of Acceptance and Inductive Logic', in *Acta Philosophica Fennica* **XXII** (1968)
[7] Hintikka, J., 'A Two-Dimensional Continuum of Inductive Methods', in J. Hintikka and P. Suppes (eds.), *Aspects of Inductive Logic*, Amsterdam, 1966.
[8] Hintikka, J., 'On Semantic Information', in J. Hintikka and P. Suppes (eds.), *Information and Inference*, Dordrecht, 1970, pp. 3–27.
[9] Jeffrey, R., *The Logic of Decision*, New York, 1965.
[10] Kuhn, T. S., *The Structure of Scientific Revolutions*, Chicago, 1962.
[11] Lehrer, K., 'Induction and Conceptual Change', *Synthese* **23** (1971), 206–225.
[12] Lehrer, K., 'Evidence, Meaning and Conceptual Change: A Subjective Approach', in G. Pearce and P. Maynard (eds.), *Conceptual Change*, Dordrecht, 1973, pp. 94–122.
[13] Lehrer, K., 'Relevant Deduction and Minimally Inconsistent Sets', *Philosophia* **1** (1973).
[14] Lehrer, K., 'Truth, Evidence, and Inference', *American Philosophical Quarterly* **11** (1974), 79–92.
[15] Levi, I., *Gambling with Truth*, New York, 1967.
[16] Levi, I., 'Truth, Content, and Ties', *Journal of Philosophy* **68** (1971), 865–876.
[17] Marschak, J., 'Information, Decision and the Scientist', in C. Cherry (ed.), *Pragmatic Aspects of Human Communication*, D. Reidel, Dordrecht, 1974.
[18] Mates, B., *Elementary Logic*, Oxford University Press, New York, 1965.
[19] Savage, L. J., *The Foundations of Statistics*, New York, 1954.
[20] Zeeman, E. C., 'Geometry of Catastrophe', *Times Literary Supplement* **10** (1971), 1556–7.

JAMES H. FETZER

ELEMENTS OF INDUCTION*

Perhaps no problem area within the philosophy of science has proven so resistant to successful explication as that of characterizing the principles and procedures of inductive (or non-demonstrative) inference. The purpose of this paper is to pursue attempts by Hempel, Hintikka, and Levi,[1] among others, where inductive procedures are examined by means of the concepts and methods that distinguish decision-theoretical analyses of principles of choice, given a finite set of mutually exclusive and jointly exhaustive courses of conduct.[2] The most important difference between these decision-theoretical analyses and those focusing upon the acceptance and rejection of hypotheses *per se* is that decisions determining behavior depend upon *practical* utilities, whereas those determining beliefs *per se* depend upon *theoretical* (or 'epistemic') gains and losses. Consequently, a distinctive characteristic of investigations of the kind contemplated here is that inductive procedures are explored within the framework of a decision-theoretical model employing 'epistemic' utilities.

Since a principal aim of this inquiry is to contribute to the theory of epistemic utility, it should be pointed out that this concept itself has not been without its able and articulate critics. Ian Hacking, in particular, has emphasized what he takes to be the significant differences between *actions* and *beliefs*, on the basis of which he is led to portray such 'utilities' as little more than the off-spring of a faulty analogy: having a belief, he claims, in general has no consequences, while performing an action in general does.[3] As a result,

It is reasonable for a man to perform a deliberate action only if he takes account of the consequences of performing that action, but it is in general reasonable for a man to have an explicit belief without his taking account of the 'consequences' of having that belief.[4]

What is most remarkable about this argument, of course, is that Hacking should have advanced it; for clearly the 'consequences' attending the acceptance of a belief are those of a *logical*, rather than a *causal*, kind. And it appears too obvious to argue that a man who holds an explicit

R. J. Bogdan (ed.), Local Induction, 145–170. All rights reserved.
Copyright © 1976 by D. Reidel Publishing Company, Dordrecht–Holland

belief without taking account of its logical consequences is in general *not* 'reasonable' at all. Indeed, as we shall discover but as Hacking overlooks, the theory of epistemic utility is syntactical and semantical to its core.[5]

The general conception to be developed in this proposal may be set forth along the following lines: On the basis of a preliminary explication of the concept of the knowledge context K for an individual x at a specified time t, the *estimated epistemic utility* of an hypothesis H_i for x relative to K at t, i.e., $EEU(H_i \mid Kxt)$, is explicated as the product of the epistemic probability of H_i on the data d for x relative to K at t, i.e., $EP(H_i \mid d \& Kxt)$, multiplied by the epistemic utility of H_i in language $\mathfrak{L}$ for x relative to K at t, i.e., $EU(H_i \mid \mathfrak{L} \& Kxt)$. The *epistemic probability* of H_i, moreover, is explicated as the product of the degree of confidence in the data d itself possessed by x relative to K at t, i.e., $C(d \mid Kxt)$, multiplied by the likelihood of H_i on d for x relative to K at t, i.e., $L(H_i \mid d \& Kxt)$; while the *epistemic utility* of H_i is explicated as the product of the logical scope of H_i for x relative to K at t, i.e., $S(H_i \mid Kxt)$, multiplied by the systematic power of H_i for x at t relative to $\mathfrak{L}$ and K, i.e., $E(H_i \mid \mathfrak{L} \& Kxt)$.

Since the estimated epistemic utility of an hypothesis is the product of a degree of confidence in d multiplied by its likelihood on d multiplied by its logical scope multiplied by its systematic power (all for x relative to K at t), the elaboration of a theory of this kind requires (i) the development of four sub-theories, i.e., for measuring confidence in data, for measuring likelihoods on data, and for measuring the logical scope and systematic power of an hypothesis, as well as (ii) the elaboration of principles of rational choice with respect to the acceptance and rejection of alternative hypotheses, with respect to the selection of suitable hypotheses for inductive investigation, and with respect to the objective of maximizing epistemic utility within a knowledge context. All of these considerations, however, presuppose the successful introduction of the underlying conception of the knowledge context K for an individual x at a specified time t, to which our attention now turns.

1. THE KNOWLEDGE CONTEXT Kxt

The concept of the knowledge context K for an individual x at a specified time t may be explicated as the set of beliefs $b_1, b_2, ..., b_j$, i.e., $\{B_j\}$, which

(a) individual x accepts at time t (where x accepts b_i at t if and only if x believes b_i at t, i.e., x believes that some sentence S_i in a language $\mathfrak{L}$ is true, where S_i asserts that b_i is the case); and which (b) satisfies as well the following three conditions of rational belief (where $\{S_j\}$ is a set of sentences in $\mathfrak{L}$ reflecting the set of beliefs $\{B_j\}$ of an individual x at time t):

(CR-1)   The set of sentences $\{S_j\}$ is logically consistent, i.e., it is not the case that for any sentence S_i belonging to $\{S_j\}$, its negation $\sim S_i$ belongs to $\{S_j\}$ as well;

(CR-2)   The set of sentences $\{S_j\}$ is deductively closed, i.e., any logical consequence S_k of any members of $\{S_j\}$ is also a member of $\{S_j\}$, so long as $\{S_{j+k}\}$ satisfies (CR-1); otherwise, it is not; and,

(CR-3)   The inferential acceptance of any sentence H_i into the set $\{S_j\}$ is determined with reference to $\{S_j\}$ itself or to that subset of $\{S_j\}$ whose members are relevant to determining the acceptability of H_i.

A set of belief-sentences $\{S_j\}$ of an individual x at a time t which satisfies (CR-1), (CR-2), and (CR-3) may be referred to as a (minimally) *rational set of beliefs. An individual x is in a knowledge context K_j at a time t, therefore, if and only if the set of belief-sentences representing the set of beliefs $\{B_j\}$ accepted by x at t forms a (minimally) rational set of beliefs.* Unless otherwise indicated, the knowledge context K_j for an individual x at a specified time t will be symbolized as 'Kxt' rather than as 'K_jxt'.

The three conditions of rationality that have been adopted here, of course, are precisely the same as those advanced by Hempel with one minor and one major modification: first, the requirement of logical consistency, rather than that of deductive closure, is taken as (CR-1); second, the requirement of deductive closure is qualified to prevent the acceptance of mutually incompatible beliefs.[6] For as Kyburg has demonstrated, the employment of Hempel's extremely intuitive and seemingly obvious conditions of closure and consistency, when combined with a probabilistic rule for inductive acceptance – for example,[7]

> *Tentative Rule for Inductive Acceptance*: accept or reject h, given K, according as [the degree of confirmation] $c\,(h, k) > 0.5$ or $c\,(h, k) < 0.5$; when $c\,(h, k) = 0.5$, h may be accepted, rejected, or left in suspense. –

logically entails one or another version of Kyburg's lottery paradox; that is, consistent application of such an acceptance rule leads to violation of either the condition of consistency or the condition of closure.[8] An illustration of this paradox would be a 1000 ticket lottery with only one winning ticket, such that for each of the 1000 individual tickets T_1, $T_2, ..., T_{999}, T_{1000}$, the probability that that ticket will win the lottery is only 0.001. Consequently, the probability that that ticket will *not* win the lottery is 0.999, leading to the acceptance of the conclusion that ticket n will not win for every one of the 1000 tickets, thereby contradicting the assumption one of them will win.

In his analysis of this difficulty, Kyburg has pointed out what he takes to be an ambiguity in Hempel's formulations, namely: the option of taking the closure condition as applying to each member of K *individually* rather than to their collective *conjunction*. As one possible solution, therefore, he offers the following proposal:

> Let us take CR1 * to mean *not* that every logical consequence of the *conjunction* of the statements of K belong to K, but only that every logical consequence of each *single* element of K belongs to K. Thus if P and Q belong to K, their conjunc- may not belong to K, unless it is included in K on independent grounds. In conjunction with a probabilistic rule of detachment this interpretation of CR1 * is very natural – it is clear that P can be overwhelmingly probable and Q can be overwhelmingly probable without their conjunction being overwhelmingly probable.[9]

Although this maneuver is entirely reasonable under the circumstances, it leads, in effect, to the imposition of restrictions upon the *logical apparatus itself* of the language $\mathfrak{L}$ rather than to a limitation upon the *conditions of acceptance* for specified individual hypotheses. Yet, however arbitrary it might seem to be, no logical paradox is generated by the acceptance of 999 sentences asserting of each of 999 tickets that ticket T_1 will not win, ticket T_2 will not win, and so forth; nor does any paradox arise from accepting the conjunction of any number of these 999 sentences. Once 999 such belief-sentences have been accepted into the knowledge context K, however, inconsistency would arise *either* from accepting the sentence, 'Ticket T_{1000} will not win', *or* from assigning a probability other than *one* to the sentence asserting ticket T_{1000} will win or a probability other than *zero*, e.g., 0.999, to the sentence asserting ticket T_{1000} will lose.

Once this situation has been clarified, therefore, the semblance of paradox tends to disappear; for surely the acceptance of an hypothesis H_i into a knowledge context K requires consideration for the logical consequences attending its acceptance. Rather than restricting the principles of deduction, therefore, it seems to be far preferable to delimit the conditions of acceptance, such that an hypothesis H_i may be accepted into a knowledge context K *only if* its acceptance is logically compatible with the truth of the other members of $\{K_j\}$. The implicit conception, therefore, is one according to which a knowledge context K for an individual x at a time t is *a rational set of beliefs possessing maximal consistency and closure*. And it is significant to note that, although these deliberations are reflected by the formulation of (CR-2), no grounds have been discovered for rejecting, in principle, the conception of a probabilistic rule of detachment.[10]

2. THE OBJECT LANGUAGE: $\mathfrak{L}$ OR $\mathfrak{L}^*$?

Our characterization of the knowledge context Kxt is incomplete, however, without a specification of the object language in which it is embedded, since each belief-sentence S_i is a well-formed formula of that language.[11] Choosing one language framework rather than another, moreover, is a matter of considerable theoretical significance, as the following alternatives will demonstrate: Let $\mathfrak{L}$ be the standard first-order predicate calculus with identity (employing the negation, disjunction, and identity signs as syntactical primitives); let $\mathfrak{L}^*$ be $\mathfrak{L}$ enriched by the addition of the subjunctive and causal conditionals as syntactical primitives as well. Then the differences between $\mathfrak{L}$ and $\mathfrak{L}^*$ may be described as follows:

(i) the expressive-completeness properties of $\mathfrak{L}$ and $\mathfrak{L}^*$ differ vastly with respect to natural and scientific languages, since $\mathfrak{L}^*$, but not $\mathfrak{L}$, provides for the formalization of causal and subjunctive conditionals (as well as material conditionals); moreover,

(ii) this difference facilitates the (partial) explicit definition of dispositional predicates within the framework of $\mathfrak{L}^*$, while these same predicates receive only (partial) implicit definition by means of reduction sentence meaning postulates within the framework of $\mathfrak{L}$; and, finally,

(iii) $\mathfrak{L}^*$, but not $\mathfrak{L}$, provides a sufficient foundation for distinguishing lawlike from accidental generalizations on the basis of syntactical properties, while the employment of $\mathfrak{L}$ for differentiating between sentences of these kinds is essentially dependent upon extra-linguistic pragmatical considerations.[12]

These distinctive features are manifestations of the fundamental difference between $\mathfrak{L}$ and $\mathfrak{L}^*$, namely: $\mathscr{L}$ is an *extensional* language, while $\mathfrak{L}^*$ is not (where a language is extensional if and only if (a) its molecular sentences are truth-functional and (b) the truth-values of its atomic sentences are determined, in principle, by the history of the world).[13] Consequently, any conditional that is expressible in $\mathfrak{L}$ will be true or false insofar as it describes or fails to describe (some segment of) the history of the world; while the truth-values of conditionals expressible in $\mathfrak{L}^*$ may or may not be likewise determinable.

In order for $\mathfrak{L}$ or $\mathfrak{L}^*$ to be adequate for the purposes of scientific inquiry, of course, they require supplementation by certain mathematical principles, such as those of the theory of sets and the calculus of probability. Although these principles, in general, possess the same syntactical significance in $\mathfrak{L}$ and in $\mathfrak{L}^*$, the differences between them may be illustrated by the alternative explications they support for the concept of statistical probability itself which, as should be expected, receives an extensional interpretation within $\mathfrak{L}$ and an intensional interpretation within $\mathfrak{L}^*$. Let us assume that an *experimental arrangement E* is an object or collection of objects upon which trials of a certain kind T may be conducted, where each trial results in one of at least two possible outcomes O. A (simple) *probability statement* may then be defined as a sentence attributing a certain probability P to the occurrence of an outcome of kind O on a trial of kind T with an experimental arrangement of kind E, where the magnitude of P is represented by the assignment of a numerical value r. A *probability hypothesis*, therefore, is a probability statement under one or the other of the following interpretations:

(I) $P[O \mid E \cap T] = r =_{df}$ the limiting frequency with which outcomes of kind O occur on trials of kind T with experimental arrangements of kind E during the history of the world is r; or,

(II) $P^*[O \mid E \cap T] = r =_{df}$ the strength of the dispositional tendency for a single trial of kind T with an experimental arrangement of kind E to produce an outcome of kind O is r.

These formulations may be referred to as the (long-run) *frequency* and (single-case) *propensity* interpretations of probability hypotheses, respectively.[14]

Let us assume that a (simple) *frequency distribution* statement provides a description of the results of n trials of kind T with arrangements of kind E in the form of the relative frequency f for outcomes of kind O within that n trial sequence as follows:

(III) $F_n[O \mid E \cap T] = f =_{df}$ over a sequence of n trials of kind T with experimental arrangements of kind E, outcomes of kind O occurred with a relative frequency equal to f.

Then one difference between the (extensional) frequency and the (intensional) propensity formulations is exhibited by the divergent logical relations that obtain between probability hypotheses and *long run* frequency distributions; for as the number of trials n increases without bound, the frequency f for outcomes of kind O will converge toward the value of its probability r as a matter of *logical necessity* on the frequency conception; while this same convergence will occur as a matter of *overwhelming probability* on the propensity conception. Thus, limiting frequency probabilities and infinite frequency distributions coincide necessarily under interpretation (I); while infinite frequency distributions and single case propensities only probably coincide under interpretation (II).[15] It should be evident, therefore, that the choice between $\mathfrak{L}$ and $\mathfrak{L}^*$ is an important one, indeed, since it will determine the sense and the significance of hypotheses themselves. Intriguingly, this decision apparently depends upon (what may be referred to as) *ultimate epistemic utilities*; i.e., upon a preference for epistemic security as opposed to systematic significance, or conversely (as we shall subsequently ascertain).

3. RANDOMNESS AND RELEVANCE

It should come as no surprise, therefore, that these alternative concepts of probability are accompanied by similarly differing concepts of randomness; for the concept appropriate to the frequency analysis is based upon the limit properties of certain subsequences of sequences of trials, while the concept appropriate to the propensity analysis concerns the strength of the tendency for certain outcomes from trial to trial, that is:

(A)   Under the frequency construction, a sequence of trials of kind T with experimental arrangements of kind E is *normal* if and only if (i) its outcomes are free from aftereffect, i.e., the frequencies for outcomes of various trials do not depend upon the outcomes of their preceding trials; and, (ii) its outcomes are insensitive to ordinal selection, i.e., the frequency for outcomes within the entire sequence is equal to the frequencies for outcomes of those kinds within any subsequence selected by taking every kth trial.[16]

(B)   Under the propensity construction, a sequence of trials of kind T with experimental arrangements of kind E is *random* if and only if (i) the strength of the tendency for outcomes of kind O is equal from trial to trial; and, (ii) the strength of the tendency for an outcome of kind O_1 on trial T_m and an outcome of kind O_2 on trial T_n is equal to the product of the strength of the tendency for O_1 on trial T_m times the strength of the tendency for O_2 on trial T_n.[17]

These definitions are significant insofar as they specify sufficient conditions for the applicability of Bernoulli's theorem; for it is not logically necessary on either interpretation for *finite* frequencies to vary if and only if certain probabilities likewise vary. Indeed, the logical implications of both concepts with respect to the results of finite trial sequences are identical, namely:

>   *The Short Run Principle*: Let n be a sequence of normal (or random) trials of kind T with experimental arrangements of kind E; then if the probability for an outcome of kind O is r, then if n is a relatively long sequence, it is practically certain

that the relative frequency F_n of O-outcomes will be approximately equal to r; and,

The Single Case Principle: Let n be a single trial belonging to a sequence of normal (or random) trials of kind T with experimental arrangements of kind E; then if the probability for an outcome of kind O is very close to one (or to zero), it is practically certain that outcome O will (or will not) occur.[18]

The Short Run Principle itself (as well as its corollary), of course, has been advanced by no less a figure than Cramér as his formulation of the frequency interpretation of probability hypotheses; however, these principles should not be confused with either interpretation, since they are implied by them both.[19]

It is possible to demonstrate, quite generally, that a trial sequence is normal (or random) if and only if the experimental arrangements generating it are essentially homogeneous, i.e., their statistically relevant properties are the same from trial to trial. Let us assume that *a property F is statistically relevant to the occurrence of an outcome O* if and only if the probability for that outcome within the reference class $R \cap F$ differs from the probability for that outcome within the reference class R itself.[20] Since these probabilities may be construed either as frequencies or as propensities, this criterion may be formulated either as *the long-run relevance rule*, i.e.,

(1) $$P[O \mid E \cap F \cap T] \neq P[O \mid E \cap T],$$

for employment under the frequency interpretation of probability hypotheses; or as *the single-case relevance rule*, i.e.,

(2) $$P^*[O \mid E \cap F \cap T] \neq P^*[O \mid E \cap T],$$

for employment under the propensity interpretation. Then the demonstration in terms of propensities is perfectly straightforward, for if any such trial were conducted with an arrangement whose propensity to produce an outcome of kind O on a trial of that type differed from that of any other member of the sequence, the probabilities for each of the outcomes would not be equal and independent; and conversely. The demonstration in terms of frequencies is similarly plain, on the convenient assumption that trials with arrangements of kind $E \cap F$ may be introduced at regular intervals such that they form a subsequence selected by taking every kth trial. If this assumption is regarded as *too* convenient, however, the demonstration that a sequence is normal if and only if it is also homogeneous may be viewed with suspicion; although, in any case, the propensity proof that *a sequence of trials is random if and only if the statistically relevent properties effecting each of the possible outcomes do not vary from trial to trial* remains entirely unimpaired.[21]

It is important to recognize, however, that – theoretical differences notwithstanding – a property F will be assumed to be statistically relevant to the occurrence of outcome O on either interpretation, in general, if and only if:

$$(3) \qquad F_n[O \mid E \cap F \cap T] \neq F_n[O \mid E \cap T];^{22}$$

that is, when and only when the frequency distribution for outcomes of kind O over n trials of kind T with arrangements of kind E differs from the frequency distribution for outcomes of that kind over trials of that kind with arrangements of kind $E \cap F$. Statistical differences of this kind, of course, provide no logical guarantee of statistical relevance; nevertheless, as Cramér's *Short Run Principle* declares, there is a *probabilistic* connection of this very sort. In order to appraise the significance of any frequency distribution, therefore, it is necessary to ascertain whether or not it records a random sample, i.e., whether or not that data actually describes the results of n trials of kind T with arrangements of kind E. Surprising as it may seem, this problem is one of the most difficult confronting the theory of non-demonstrative inference.

4. EPISTEMIC PROBABILITIES

Let us assume that the epistemic probability EP of an hypothesis H_i with respect to a frequency distribution d and a knowledge context Kxt is equal to the product of the degree of confidence C in the data d itself possessed by x relative to K at t multiplied by the likelihood L of H_i on d for x relative to K at t, i.e., $EP(H_i \mid d \& Kxt) = C(d \mid Kxt) L(H_i \mid d \& Kxt)$. This conception reflects the plausible intuition that the strength of the evidence in support of an hypothesis should depend not only upon the likelihood of that hypothesis on relevant evidence but upon confidence in the reliability of that evidence as well. The theory of epistemic probability, therefore, may be explicated by developing a theory of confidence and a theory of likelihood, which we shall now consider.

4.1. *The Theory of Confidence*

In the exploration of principles of induction, it is extremely important to differentiate clearly between the world itself and our beliefs about the world; for although our beliefs will be true just in case the world possesses the properties our beliefs attribute to it, the only reasonable measure of the truth of our beliefs is the extent to which the evidence at our disposal lends them support. Thus, although the concept of a random sample may be provided an ontic definition *with respect to the world itself* as follows,

(D-1)   *A frequency distribution F_n is a random sample (of size n)* if and only if it records the results of a sequence of n random trials;[23]

the elaboration of principles of induction requires employment of the derivative concept of a sample that is random *with respect to our beliefs about the world.* Following Hempel's general procedure, let us assume the definition of a maximally specific reference predicate with respect to a specified outcome predicate within the knowledge context Kxt, i.e.,

(D-2)   *A predicate 'M' is a maximally specific reference predicate relative to an outcome predicate 'O_i' within the knowledge context Kxt* if and only if (i) 'M' is logically equivalent to a conjunction of predicates that are statistically relevant to the outcome O on trial i within the knowledge context Kxt; (ii) 'M' does not entail either 'O' or '$-O$'; and, (iii) no predicate expression stronger than 'M' satisfies (i) and (ii), i.e., if 'M' is conjoined with a predicate that is statistically relevant to 'O_i' within the knowledge context Kxt, the resulting expression entails 'O' or '$\sim O$', or else it is logically equivalent to 'M'.[24]

With respect to a frequency distribution, i.e., '$F_n[O \mid E \cap T] = f$', therefore, 'M' should be logically equivalent to '$E \cap T$' with respect to each predicate 'O_i' describing a possible outcome of trial i conducted with arrangements of that kind. Then the concept of a random sample may be provided an epistemic definition relative to the knowledge context *Kxt* as follows,

(D-3)   *A frequency distribution statement,* '$F_n[O \mid E \cap T] = f$', *records a random sample (of size n) within the knowledge context Kxt if and only if any predicate 'M' that is a maximally specific reference predicate relative to an outcome predicate 'O_i' where i is a member of the sample of n trials is logically equivalent to any other predicate 'M*' that is a maximally specific reference predicate relative to outcome predicate 'O_j' where j is any other member of that sequence; and 'M*' as well as 'M' are logically equivalent to 'E ∩ T' in Kxt.*

A frequency distribution statement belonging to *Kxt* records a random sample of size *n*, therefore, when and only when each of those *n* trials is conducted under precisely the same relevant conditions to the best of our knowledge. Thus, let us assume that our belief that this is the case may be measured by an *a priori* degree of confidence in data *d* relative to *Kxt* that is equal to one; and that, in the absence of such a belief, our *a priori* degree of confidence is unknown.

Although we have focused primarily upon (simple) frequency distributions, which reflect outcomes of one particular kind, we are obviously concerned with (complex) frequency distributions, which reflect all possible outcomes, as well. Let us assume that a *random variable X* is any outcome of any singular trial with any experimental arrangement such that, if the possible outcomes of such a trial are $O_1, O_2, ..., O_j$, i.e., members of the set of outcomes $\{O_j\}$, *X* may assume any of the corresponding values $X = O_1, X = O_2, ..., X = O_j$.[25] Consequently, a (complex) *frequency distribution* for a sequence of *n* trials of such a random variable will record the relative frequency f_i with which each possible outcome O_i actually occurred within that trial sequence, where the frequency statements '$F_n(X = O_1) = f_1$', '$F_n(X = O_2) = f_2$', ..., '$F_n(X = O_j) = f_j$', satisfy three conditions, namely:

(a)   each relative frequency f_i is a non-negative real number;
(b)   the numerical values of $f_1, f_2, ..., f_j$ sum to one; and,
(c)   for any two incompatible outcomes O_i and O_j, $f_{(i \vee j)} = = f_i + f_j$.[26]

Analogously, a (complex) *probability distribution* for a random variable X will report the probability r_i for each possible outcome O_i of that random variable, i.e., '$P(X=O_1)=r_1$', '$P(X=O_2)=r_2$',..., '$P(X=O_j)=r_j$', where, as before,

(a)   each probability r_i is a non-negative real number;

(b)   the numerical values of $r_1, r_2,..., r_j$ sum to one; and,

(c)   for any two incompatible outcomes O_i and O_j, $r_{(i \lor j)}=$
         $=r_i+r_j$.[27]

Furthermore, the corresponding *empirical distribution function* $\mathscr{F}_n$ for the random variable X (whose values may be represented by non-negative real numbers x) and the corresponding *probability distribution function* $\mathscr{P}$ may be specified as follows, namely:

$$\mathscr{F}_n(x) = \sum_{i \leqslant [x]} f_i; \text{ and, } \mathscr{P}(x) = \sum_{i \leqslant [x]} r_i.\text{[28]}$$

Since our *a priori* degree of confidence in the reliability of an empirical distribution function for the purpose of determining the likelihoods of various probability distribution functions is based upon our beliefs both with respect to (i) the set of all properties that are statistically relevant to outcomes of these O_j kinds and (ii) the sub-set of those properties that were concurrently present on each of these n trials, it should be obvious that, even though our *a priori* degree of confidence happens to have the value of one, that degree of confidence itself reflects a *conjecture*, namely: that each and every member of that n trial sequence was actually conducted under the same relevant conditions. It would be extremely useful, therefore, to have a means for measuring posterior as well as prior degrees of confidence on the basis of the characteristics of the empirical distribution function itself; and, indeed, there appears to be suitable theoretical support for such a conception in the form of *the central limit theorem*. For let $X_1, X_2,..., X_n$ be a sequence of n trials with a random variable X with probabilities for outcomes that are equal and independent, i.e., $\{X_i\}$ is a sequence of n independent identically distributed random variables. Let θ be the theoretical mean and σ^2 the theoretical variance of the experimental population. Then if $S_n = X_1 + X_2 + \cdots + X_n$ and $\bar{X} = S_n/n$, as n increases without bound the distribution function $\sqrt{n}(\bar{X}-\theta)/\sigma$ converges to a normal distribution with mean 0 and variance 1; that is, $\sqrt{n}(\bar{X}-\theta)/\sigma \to \mathscr{N}(0, 1)$ in distribution.[29]

Consequently, a suitable measure of our posterior degree of confidence would appear to be the extent to which the empirical distribution function $\mathscr{F}_n$ diverges from our *a priori* expectation; for surely our confidence in the reliability of our data should be shaken if the distribution function $\mathscr{F}_n$ does not converge toward a normal distribution $\mathscr{N}$. As it happens, Paul Lévy has provided a suitable measure for our purpose; for a necessary and sufficient condition for the (weak) convergence of a distribution function $\mathscr{F}_n$ toward a normal distribution function $\mathscr{N}$ is as follows: $D(\mathscr{F}_n, \mathscr{N}) \to 0$, where the distance $D(\mathscr{G}, \mathscr{H})$ between two distribution functions $\mathscr{G}$ and $\mathscr{H}$ is defined as the infimum of all ε such that for all x,

$$\mathscr{H}(x - \varepsilon) - \varepsilon \leqslant \mathscr{G}(x) \leqslant \mathscr{H}(x + \varepsilon) + \varepsilon.^{30}$$

The distance $D(\mathscr{G}, \mathscr{H})$, moreover, satisfies the conditions for a metric; that is,

(a) $D(\mathscr{G}, \mathscr{H}) = 0$ if and only if $\mathscr{G} = \mathscr{H}$;
(b) $D(\mathscr{G}, \mathscr{H}) = D(\mathscr{H}, \mathscr{G})$; and,
(c) $D(\mathscr{G}, \mathscr{I}) \leqslant D(\mathscr{G}, \mathscr{H}) + D(\mathscr{H}, \mathscr{I})$.

Let us take as our measure of divergence from the normal distribution that ε which is the smallest ε satisfying the specified condition, such that the Lévy distance $D(\mathscr{F}_n, \mathscr{N}) = \varepsilon$. Then as our measure of confidence in the reliability of any frequency distribution function $\mathscr{F}_n$ as relevant evidence for ascertaining the likelihoods of various probability distribution functions we may adopt

> *The Principle of Confidence*: When the prior degree of confidence in the frequency distribution function $\mathscr{F}_n$ is equal to one, the posterior degree of confidence in that same distribution function is equal to one minus ε, where ε is the smallest distance satisfying the Lévy measure of degree of divergence of the function $\mathscr{F}_n$ and the normal distribution function $\mathscr{N}$; thus, the degree of confidence in data $d = \mathscr{F}_n$ within the knowledge context Kxt is equal to one minus ε, i.e., $C(\mathscr{F}_n \mid Kxt) = 1 - \varepsilon$; otherwise, $C(d \mid Kxt)$ is unknown.[31]

The point of the final provision, therefore, is to explicitly acknowledge that, when these specified conditions are not satisfied, the degree of reliability of data d for the purpose of determining the likelihoods of probability hypotheses within the knowledge context Kxt cannot be measured (and epistemic probabilities may be calculated only with an unknown risk of error); in which case, of course, epistemic caution is clearly warranted.

4.2. *The Theory of Likelihood*

The elaboration of principles of likelihood is a considerably less complex affair. Let us assume that the likelihood of ψ, given χ, is equal to the probability of χ, given ψ; i.e., $L(\psi \mid \chi)=P(\chi \mid \psi)$. For each (simple) probability statement, '$P[O_i \mid E \cap T]$', therefore, there is a corresponding (simple) likelihood statement, '$L[E \cap T \mid O_i]$'; and for every (complex) probability distribution, '$P(X=O_1)=r_1 \cdot P(X=O_2)=r_2 \cdot \ldots \cdot P(X=O_j) =r_j$', there is a corresponding (complex) likelihood distribution, '$L(X \mid O_1)=r_1 \cdot L(X \mid O_2)=r_2 \cdot \ldots \cdot L(X \mid O_j)=r_j$'. In other words, if the probability for an outcome of kind O_i on a single trial of kind $E \cap T$ is equal to r_i, then the likelihood that a single trial was a trial of kind $E \cap T$, given an outcome of kind O_i, is equal to r_i as well; and if the probabilities for outcomes of kind O_1, $O_2, \ldots$, O_j with the random variable X are r_1, $r_2, \ldots, r_j$, respectively, then the likelihoods that outcomes of kinds O_1, $O_2, \ldots$, O_j which have, in fact, occurred, were outcomes of X are equal to $r_1, r_2, \ldots, r_j$, respectively.

Let us further assume a set of j mutually exclusive probability hypotheses $\{H_j\}$ that is jointly exhaustive in the following sense: with respect to random variable X, the set $\{H_j\}$ includes at most eleven hypotheses per possible outcome O_i such that these hypotheses attribute probabilities to the first decimal place for outcomes of that kind, i.e., 0.0, 0.1, ..., 0.9, 1.0; or, assuming the first decimal place is known, attribute probabilities to the second decimal place for outcomes of that kind, i.e., 0.10, 0.11, ..., 0.19, 0.20; or, ...; and so on. The appropriate level of specificity of the members of such a set, therefore, is determined on the basis of empirical and theoretical considerations within Kxt; for example, the number of possible values of that variable and the frequencies with which those outcomes have occurred in the past to the best of our knowledge. Provided the data already at our disposal has enabled us to determine the first decimal place with a suitable degree of confidence, we may proceed to determine the second decimal place with a suitable degree of confidence, and so forth, by what is essentially a method of successive approximation, eventually advancing to any degree of specificity desired. The principal benefit of this procedure, of course, is that it allows for the systematic investigation of an arbitrarily large number of hypotheses of arbitrary specificity a finite number at a time.

Following Hacking, let us define a *joint proposition* as an ordered couple consisting of a probability hypothesis and a frequency distribution, such that both statements concern outcomes of trials of kind $E \cap T$; then, if both concern only the outcome O_i, they form a *simple* joint proposition; otherwise, they form a *complex* joint proposition.[32] Thus, for each hypothesis H_i, the corresponding distribution statement will record all of the data in Kxt that is relevant to determining the likelihood of H_i [in conformity with the condition of rationality (CR-3)]. The likelihood of an hypothesis H_i (attributing a probability r_i to outcomes of kind O_i on trials of kind $E \cap T$), therefore, is measured by the probability of the frequency distribution of its joint propositional component (recording the empirical frequency for outcomes of kind O_i during n trials of kind $E \cap T$), given that hypothesis, i.e., $L(H_i \mid d \ \& \ Kxt) = P(d \mid H_i \ \& \ Kxt)$, that is: $L[P(X=O_i)=r_i \mid F_n(X=O_i)=f_i] = P[F_n(X=O_i)=f_i \mid P(X=O_i)=r_i]$. Let us label this principle *The Principle of Likelihood*. We shall also assume, *à la* Hacking, (a) that data d supports hypothesis H_i better than hypothesis H_j if the likelihood of H_i exceeds that of H_j, given d; and, (b) that data d supports hypothesis H_i better than H_j if the likelihood ratio of H_i to H_j exceeds one.[33]

The Principle of Confidence and *The Principle of Likelihood*, significantly, provide necessary and sufficient conditions for calculating the epistemic probability EP of an hypothesis H_i, where $EP(H_i \mid d \ \& \ Kxt) = C(d \mid Kxt) \times L(H_i \mid d \ \& \ Kxt)$. Taken together, they provide a theoretical foundation for the following probabilistic rule of detachment:

> *Rule for Tentative Inductive Acceptance*: if H_i is the best supported member of a set of mutually exclusive and jointly exhaustive hypotheses $\{H_j\}$, accept or reject H_i as the epistemic probability $EP(H_i \mid d \& Kxt) > 0.5$ or $EP(H_i \mid d \& Kxt) < 0.5$; when $EP(H_i \mid d \& Kxt) = 0.5$, H_i may be accepted, rejected, or left in suspense.[34]

In appraising the adequacy of this rule of inductive inference, at least three of its important features deserve explicit recognition, namely:

(i) it provides for the expansion, contraction, and replacement of members of the set of belief-sentences $\{S_j\}$ accepted by an individual x at time t; i.e., the knowledge context Kxt reflects those beliefs best supported by the evidence in our possession at time t, in conformity with condition of rationality (CR-3);

(ii) it promotes the acceptance of that hypothesis H_i belonging to any set of mutually exclusive and jointly exhaustive hypotheses $\{H_j\}$ meeting a minimal condition of rational acceptability whenever possible, while still conforming with conditions of rationality (CR-2) and (CR-1); and, finally,

(iii) it exhibits a reasonable standard for the acceptance and rejection of hypotheses when no measure of the practical or theoretical consequences of their acceptance is taken into consideration, i.e., it supplies a criterion of *pure rational credibility*; yet it may obviously be supplemented by appropriate utility functions of either kind to yield a measure of estimated epistemic (or practical) utility, as the following account is intended to display.

5. Epistemic utilities

Let us assume that the epistemic utility EU of an hypothesis H_i with respect to a knowledge context Kxt is equal to the product of its logical scope S relative to K at t for x multiplied by its systematic power E relative both to K at t for x and to the language $\mathfrak{L}$ in which it is expressed (which may be either $\mathfrak{L}$ or $\mathfrak{L}^*$); i.e., $EU(H_i \mid \mathfrak{L} \ \& \ Kxt) = S(H_i \mid Kxt) \cdot E(H_i \mid \mathfrak{L} \ \& \ Kxt)$. This conception too reflects a plausible intuition, namely: that hypotheses of broad logical scope possess greater epistemic value than do hypotheses of narrow logical scope; and that hypotheses possessing great systematic significance are epistemically preferable to those possessing slight systematic significance. Thus, the theory of epistemic utility may be explicated by developing a measure of logical scope and a measure of systematic power, which we shall now consider.

5.1. *The Concept of Logical Scope*

The theory of logical scope qualifies as the least complicated aspect of our inquiry; for the logical scope of an hypothesis H_i within the knowledge context Kxt may be measured by a function based upon the proportion of those members of the world W with respect to which that hypothesis provides specific information.[35] Let us assume that there are w individuals in the world W within the knowledge context Kxt such that (i) if H_i is logically equivalent to a conjunction of m atomic conjuncts, each of which is itself an atomic sentence attributing a property F to an individual a or the negation of such a sentence, then the logical scope of H_i is equal to m/w; and that (ii) if H_i is logically equivalent to a disjunction of m atomic disjuncts, each itself an atomic sentence attributing a property F to an individual a or the negation of such a sentence, then the logical scope of H_i is equal to $1/mw$; providing,

(a) $S(H_i \mid Kxt)$ is a non-negative real number;

(b) $S(H_i \mid Kxt)=0$ if and only if H_i is logically true; and,

(c) $S(H_i \cdot H_j \mid Kxt)=S(H_i \mid Kxt)+S(H_j \mid Kxt)$ if $H_i \vee H_j$ is logically true.[36]

Consequently, if H_i is a universal generalization, $S(H_i \mid Kxt)=w/w=1$; while, by contrast, if H_i is an existential generalization, $S(H_i \mid Kxt)=$ $=1/ww=1/w^2$; so that as $w \to \infty$, $1/w^2 \to 0$, while, of course, as $w \to \infty$, w/w remains the same.

The principles of this theory, of course, are conspicuously applicable to sets of alternative hypotheses $\{H_j\}$, where each hypothesis H_i belonging to $\{H_j\}$ attributes identical properties to different classes of individuals in W. Note especially that, although a (simple) frequency distribution statement will be logically equivalent to a disjunction consisting of $nCk=$ $$=\frac{n!}{(n-k)!k!}$$ distinct disjuncts (where n is the size of the sample, $k=n$ times the frequency f, and each disjunct is an n-membered conjunction) and possesses a logical scope of $n/(nCk)\,w$, the logical scope of the corresponding probability hypothesis will be equal to w/w regardless of the interpretation it is given. For although frequency hypotheses attribute probabilities as properties of sequences of trials *collectively*, while propensity hypotheses attribute probabilities as properties of trial sequences *distributively*, nevertheless, under either interpretation, these hypotheses may be construed as possessing universal scope; for, in effect, both assert of every member m of W that either m is not a member of reference class $E \cap T$ or else the probability that m is a member of outcome class O is equal to r. Consequently, the logical scope of a probability hypothesis

H_i will be logically equivalent to a conjunction of w conjuncts, where $S(H_i \mid Kxt) = w/w = 1$, on either construction. Although universal generalizations, frequency distributions, finite conjunctions, singular statements, finite disjunctions, and existential generalizations display successively lesser and lesser degrees of logical scope respectively, in general, therefore, the logical scope of probability hypotheses will be equal to that of universal generalizations under either interpretation.

At least three features of the theory of logical scope adumbrated here may deserve explicit mention, namely:

(1) the logical scope of an hypothesis is not fixed once and for all, but rather varies with the precise content of K for x at t (except as noted below);

(2) the logical scope of an hypothesis is a purely formal property of that hypothesis within the knowledge context Kxt, regardless of its content; and,

(3) as the size of the world increases, the logical scope of any universal generalization is constantly maximal, while the logical scope of any existential generalization is constantly minimal.[37]

The epistemic utility of an hypothesis H_i clearly depends upon its content as well as its form, however, which leads to consideration of the systematic power of alternative hypotheses possessing the same logical form.

5.2. *The Concept of Systematic Power*

Since the systematic power of an hypothesis H_i may be construed to be its explanatory and predictive significance with respect to the occurrence of singular events, one appropriate measure would appear to be a function based upon the proportion of events of specified kinds that might occur during the course of the world's history for which that hypothesis possesses explanatory and predictive significance. Let us assume that any such explanation consists of a set of sentences S_j, known as the *explanans*, and another sentence S_k describing the event to be explained, known as the *explanandum*, if and only if:

(a) the explanandum is either a deductive or a probabilistic consequence of its explanans;

(b) the explanans contains at least one general law of either universal or statistical character actually required for the derivation of the explanandum;

(c) the general law(s) invoked in the explanans satisfy the requirement of maximal specificity;[38] and,

(d) the sentences constituting the explanation – both the explanans and the explanandum – belong to the knowledge context Kxt.

From this point of view, a set of sentences satisfying conditions (a), (b), and (c) may be regarded as a 'potential' explanation, while those satisfying all four are 'adequate' explanations, within the knowledge context Kxt. Thus, if the laws invoked in the explanans are essentially universal, the logical properties of the relationship between the sentences constituting the explanans and its explanandum will be those of complete entailment; if they are essentially statistical, they will be those of only partial entailment. Explanations of these two kinds may be called 'universal-deductive' and 'statistical-probabilistic', respectively.[39]

Precisely what statements qualify as 'general laws', however, depends upon the language $\mathfrak{L}$ or $\mathfrak{L}*$ within which the knowledge context Kxt is embedded. For if Kxt is embedded within the intensional language $\mathfrak{L}*$, the only statements qualifying as general laws will be logically general disposition sentences attributing dispositions of universal or statistical strength to experimental arrangements; while, if Kxt is embedded within the extensional language $\mathfrak{L}$, those statements qualifying as general laws will be historical descriptions attributing limiting frequencies of universal or statistical strength to sequences of trials. Thus, sentences that would qualify as general laws within $\mathfrak{L}$ would qualify, not as general laws, but as frequency distributions within $\mathfrak{L}*$; while sentences that would qualify as general laws within $\mathfrak{L}*$ are not even expressible within $\mathfrak{L}$. For one of the distinguishing features of an intensional, as opposed to an extensional, language is that a disposition-predicate χ is explicitly definable within $\mathfrak{L}*$ by employing the 'causal' conditional '$\ni_n$' as follows,

$$(\mathfrak{L}*) \quad x \text{ is } \chi \text{ at } t =_{df} (T^1 xt \ni_n O^1 xt) \cdot (T^2 xt \ni_n O^2 xt) \cdot \ldots;[40]$$

which, in effect, asserts, x is χ at t if and only if the occurrence of events of kind T^1, T^2, ... at time t would (invariably, if $n=u$; or probably, when $n=r$) bring about the occurrence of events of kind O^1, O^2, ... at time t as well (so long as x is χ at t); while that same disposition-predicate χ is only implicitly definable within $\mathfrak{L}$ by means of reduction sentence meaning postulates employing the material conditional as follows,

$$(\mathfrak{L}) \quad [T^1 xt \supset (\chi xt \supset O^1 xt)] \cdot [T^2 xt \supset (\chi xt \supset O^2 xt)] \cdot \ldots;$$

which, in effect, asserts, either events of kind T^1, T^2,... do not occur at time t or else either x is not χ at t or events of kind O^1, O^2,... occur at t during the history of x (so long as x is χ at t).

By contrast, therefore, any sentence obtained by quantifying or instantiating the sentential function 'χxt' within $\mathfrak{L}$ will describe only what *actually does occur* during the history of x, while within $\mathfrak{L}^*$ that same sentence instead will describe what *would occur* if events of specified kinds *were to occur*; where the difference between them may be perceived most clearly by observing that $(\mathfrak{L}^*)$ embraces a primitive *brings-about* relation such that $(\mathfrak{L})$ is surely a necessary but not a sufficient condition for $(\mathfrak{L}^*)$. Indeed, since the strongest sentences expressible in $\mathfrak{L}$ are historical descriptions, disposition-predicates within $\mathfrak{L}$ designate mere pseudo-dispositions providing descriptive summaries of (segments of) an object's history, from the point of view of $\mathfrak{L}^*$.

One suitable measure of the systematic power of alternative hypotheses of the same logical form, therefore, would measure the systematic power of the predicate expression constituting the sentential function from which that sentence was derived; such that (i) if (the sentential function implicit in) H_i attributes an *event-property* (or a sequence of m event-properties) to x, where each event-property χ describes an event of duration t during the history of the world t^*, then the systematic power of H_i is equal to mt/t^*; and that (ii) if (the function implicit in) H_i attributes a *disposition-property* (or a complex of m disposition-properties) to x, where each disposition-property χ is a tendency (of strength n) to display an outcome of kind O_i when subjected to a trial of kind T_i at any time that object possesses that disposition-property, then the systematic power of H_i is equal to $mnt^*/mt^* = n$; providing (for the appropriate language),

(a) $E(H_i \mid \mathfrak{L} \,\&\, Kxt)$ is a non-negative real number;

(b) $E(H_i \mid \mathfrak{L} \,\&\, Kxt) = 0$ if and only if H_i is logically true; and,

(c) $E(H_i \cdot H_j \mid \mathfrak{L} \,\&\, Kxt) = E(H_i \mid \mathfrak{L} \,\&\, Kxt) + E(H_j \mid \mathfrak{L} \,\&\, Kxt)$ if $H_i \vee H_j$ is logically true.

Consequently, if H_i attributes a disposition of universal strength u to display an outcome of kind O_i on each and every trial of kind T_i, then $E(H_i \mid \mathfrak{L} \,\&\, Kxt) = 1$; and if H_i attributes a disposition of statistical strength r to display an outcome of kind O_i on each every trial of kind T_i, then $E(H_i \mid \mathfrak{L} \,\&\, Kxt) = r$. Nevertheless,

(1) the systematic power of an hypothesis H_i attributing a 'disposition' χ to an object x at t in $\mathfrak{L}$ is properly understood as attributing an *event-property* of a complex kind such that, at each moment t, either a trial event of kind $T^1, T^2, \ldots$ does not occur or an outcome event of kind $O^1, O^2, \ldots$ actually does occur (so long as x is χ at t); and,

(2) since probability hypotheses under the frequency construction provide statistical summaries of event-properties of the world's history, while the same statements attribute statistical disposition-properties to classes of objects on the propensity construction, the systematic power of probability hypotheses and *long-run* frequency distributions will differ in $\mathfrak{L}^*$ but not in $\mathfrak{L}$; and,

(3) the systematic power of an hypothesis H_i is therefore relative to the language in which that hypothesis is expressed, where the choice between $\mathfrak{L}$ or $\mathfrak{L}^*$ apparently depends upon one's *ultimate* epistemic utilities, specifically: upon a preference for systematic power as opposed to epistemic security or conversely, a manifestation of which will be the adoption of an essentially *realistic* language $\mathfrak{L}^*$ or of an essentially *instrumentalistic* language $\mathfrak{L}$, respectively.[41]

6. EPISTEMIC STRATEGIES

Regardless of one's ultimate epistemic utilities, however, the principles of the theory of logical scope and of the theory of systematic power provide a sufficient condition for determining the epistemic utility of an hypothesis H_i where, of course, $EU(H_i \mid \mathfrak{L} \ \& \ Kxt) = S(H_i \mid Kxt) \ E(H_i \mid \mathfrak{L} \ \& \ Kxt)$. Clearly, those sentences possessing maximal epistemic utility will possess broad logical scope and high explanatory and predictive significance, namely: universal generalizations and probability hypotheses; while those sentences possessing minimal epistemic utility will possess narrow logical scope and low explanatory and predictive power, namely: existential generalizations. Although it would be mistaken to assume that hypotheses should not be accepted into the knowledge context Kxt when they possess scant epistemic utility, the epistemic utility of an hypothesis should receive primary consideration in at least two circumstances, namely:

(a) when choosing from among a set of exclusive and exhaustive alternative hypotheses $\{H_j\}$ for the purpose of inductive investigation, the suitable choice would appear to be that hypothesis possessing maximal epistemic utility; that is:

> Rule for Inductive Hypothesis Selection: given a set of hypotheses $\{H_j\}$ as specified, the rational choice for inductive investigation is that hypothesis H_i possessing maximal epistemic utility, i.e., whose epistemic utility is not exceeded by that of any alternative member of that set.

Although the Rule for Inductive Hypothesis Selection is not an effective decision procedure for the invention of hypotheses of broad scope and systematic power, it does fulfill, at least in part, the objectives of a logic of scientific discovery; and,

(b) when confronted by the choice between two alternative hypotheses H_i or H_j possessing the same epistemic probability on the data d within the knowledge context Kxt, the suitable choice would appear to be that hypothesis affording the maximal epistemic ulitility; that is,

> Rule for Maximizing Epistemic Utility: given two alternative hypotheses H_i and H_j possessing the same epistemic probability on data d within knowledge context Kxt, the rational choice is to prefer that hypothesis H_i affording maximal epistemic utility, i.e., whose epistemic utility is not exceeded by that of any alternative hypothesis possessing equal epistemic probability.

In effect, therefore, the principles of the theory of epistemic utility, employed together with the principles of the theory of epistemic probability, provide what may well be envisioned as a standard of applied rational credibility. While the Rule for Tentative Inductive Acceptance provides a standard for ascertaining the rational acceptability of hypotheses generally, regardless of considerations of epistemic utility, therefore, we have discovered that that rule may clearly be supplemented by principles of epistemic utility to adopt, in effect, the general strategy of maximizing estimated epistemic utility.[42] Thus, the conception of knowledge which emerges from such an account at last is one of a set of beliefs possessing maximal content, consistency, and closure.

University of Kentucky

NOTES

* The author is indebted to Peter Purdue and Zakkula Govindarajulu of the Department of Statistics for useful suggestions concerning Section 4.1., the theory of confidence.

[1] Carl G. Hempel, 'Deductive-Nomological vs Statistical Explanation', in *Minnesota Studies in the Philosophy of Science*, Vol. III (ed. by H. Feigl and G. Maxwell), University of Minnesota Press, Minneapolis, 1962; Jaakko Hintikka and Juhani Pietarinen, 'Semantic Information and Inductive Logic', in *Aspects of Inductive Logic* (ed. by J. Hintikka and P. Suppes), North-Holland Publishing Company, Amsterdam, 1966; and Isaac Levi, 'Induction and the Aims of Inquiry', in *Philosophy, Science, and Method* (ed. by S. Morgenbesser, *et al.*), St. Martin's Press, New York, 1969.

[2] H. Chernoff and L. Moses, *Elementary Decision Theory*, John Wiley & Sons, Inc., 1959, is a useful introduction to statistical decision-theoretical principles.

[3] Ian Hacking, *Logic of Statistical Inference*, Cambridge University Press, Cambridge, 1965, p. 164; cf. also pp. 29–32, p. 75, and esp. pp. 164–66.

[4] Hacking, *op. cit.*, p. 164.

[5] Levi, however, holds a different point of view; *op. cit.*, p. 99.

[6] Cf. Hempel, *op. cit.*, pp. 150–151.

[7] Hempel, *op. cit.*, p. 155. Note that to reject h is to accept $\sim h$.

[8] Henry Kyburg, Jr., 'Probability, Rationality, and a Rule of Detachment', in *Logic, Methodology, and Philosophy of Science* (ed. by Y. Bar-Hillel), North-Holland Publishing Company, Amsterdam, 1964, pp. 304–308.

[9] Kyburg, *op. cit.*, p. 307.

[10] Since the reformulation of the closure condition may be envisioned as merely reinforcing the consistency condition, it might reasonably be argued that no lottery paradox arises even on the basis of Hempel's original conditions when properly applied, i.e., when applied as has been outlined here.

[11] Levi seems to hold a contrary view; cf. Isaac Levi, *Gambling with Truth*, Alfred A. Knopf, New York, 1967, p. 25.

[12] James H. Fetzer, 'The Likeness of Lawlikeness', *Boston Studies in the Philosophy of Science*, Vol. XXXII, D. Reidel Publishing Company, Dordrecht-Holland, forthcoming.

[13] The sentences resulting from the instantiation of extensional sentential schemata, in other words, are *eternal* rather than *occasion* sentences, in the sense of W. V. O. Quine, *Word and Object*, The MIT Press, Cambridge, 1960, pp. 191–195.

[14] James H. Fetzer, 'Statistical Probabilities: Single Case Propensities vs Long Run Frequencies', in *Developments in the Methodology of Social Science* (ed. by W. Leinfellner and E. Kohler), D. Reidel Publishing Company, Dordrecht-Holland, 1974.

[15] Cf. R. N. Giere, 'Objective Single-Case Probabilities and the Foundations of Statistics', in *Logic, Methodology, and Philosophy of Science* (ed. by P. Suppes *et al.*), North-Holland Publishing Company, Amsterdam, 1973, pp. 477–478; and James H. Fetzer, 'Dispositional Probabilities', in *Boston Studies in the Philosophy of Science*, Vol. VIII (ed. by R. Buck and R. Cohen), D. Reidel Publishing Company, Dordrecht, Holland, 1971, pp. 475–476.

[16] The rationale for this formulation of the concept of normality is discussed in Fetzer, 'Statistical Probabilities', p. 389 and esp. p. 396, fn. 5.

[17] Cf. Harold Freeman, *Introduction to Statistical Inference*, Addison-Wesley Publishing Company, Reading, Mass., 1963, pp. 167–168.

[18] These formulations are discussed, for example, by Carl G. Hempel, 'Aspects of Scientific Explanation', in *Aspects of Scientific Explanation*, The Free Press, New York, 1965, pp. 386–387; and Hempel, 'D-N vs SE', pp. 129–130.

[19] This may account, in part, for the practice of statistical inference taking place without the necessity of resolving theoretical foundational problems.

[20] Cf. Fetzer, 'Statistical Probabilities', pp. 392–393. Note that here $R = R \cap \sim F$.

[21] Possibly this difficulty underlies those reservations concerning the use of the concept of randomness with respect to physical sequences expressed, for example, by J. Alberto Coffa, 'Randomness and Knowledge', and Henry Kyburg, Jr., 'Randomness', in *Boston Studies in the Philosophy of Science*, Vol. XX (ed. by K. Schaffner and R. Cohen), D. Reidel Publishing Company, Dordrecht-Holland, 1974; which, so far as I am able to discern, are problems only for the frequency conception, being logically irrelevant to the propensity view.

[22] For suitable n, of course. As Salmon has observed, these principles may be supplemented with an appropriate 'screening-off' rule to eliminate mistaking symptoms for causes; e.g., Wesley C. Salmon, *Statistical Explanation and Statistical Relevance*, University of Pittsburg Press, Pittsburgh, 1971, p. 55.

[23] Cf. Hacking, *op. cit.*, pp. 120–121.

[24] Cf. Carl G. Hempel, 'Maximal Specificity and Lawlikeness in Probabilistic Explanation', *Philosophy of Science* 35 (1968), 131.

[25] Cf. William Feller, *An Introduction to Probability Theory and Its Applications*, Vol. I, John Wiley & Sons, Inc., New York, 1950, pp. 212–213.

[26] Cf. J. L. Hodges, Jr. and E. L. Lehmann, *Basic Concepts of Probability and Statistics*, Holden-Day, Inc., San Francisco, 1964, pp. 16–17.

[27] Hodges and Lehmann, *loc. cit.*

[28] Cf. Feller, *op. cit.*, p. 213.

[29] Cf. Feller, *op. cit.*, p. 244.

[30] B. V. Gnedenko and A. N. Kolmogorov, *Limit Distributions for Sums of Independent Random Variables*, Addison-Wesley, Reading, Mass., 1954, p. 33.

[31] The Lévy distance is employed here as an illustration of the kind of measure that could appropriately be relied upon for this purpose. Alternatives are discussed, for example, by B. V. Gnedenko, *The Theory of Probability*, Chelsea Publishing Company, New York, 1962, pp. 444–458; and by Jaroslav Hájek, *Non-Parametric Statistics*, Holden-Day, Inc., San Francisco, 1969, Chs. 1 and 2.

[32] Hacking, *op. cit.*, pp. 57–58.

[33] Hacking, *op. cit.*, p. 62 and p. 71.

[34] Cf. Hempel, 'D-N vs SE', pp. 154–155.

[35] Alternative measures of the amount of information provided by an hypothesis are discussed, for example, by Hintikka and Pietarinen, *op. cit.*, pp. 99–100.

[36] Cf. Hempel, 'D-N vs SE', p. 154. As Hempel observes, if (a), (b), and (c) are satisfied, then (d) $S(H_i \mid Kxt) = 1 - S(\sim H_i \mid Kxt)$; (e) if H_i implies H_j, then $S(H_i \mid Kxt) \geq \geq S(H_j \mid Kxt)$; and, (f) if H_i and H_j are logically equivalent, then $S(H_i \mid Kxt) = = S(H_j \mid Kxt)$. But (d) holds strictly only in an infinite universe.

[37] Since the logical scope of an hypothesis varies with an individual's beliefs, the views set forth here may conflict with Levi's assumption that epistemic utilities do not vary with changing evidence; cf. Isaac Levi, 'Decision Theory and Confirmation', *Journal of Philosophy* 58 (1961), 623.

[38] James H. Fetzer, 'A Single Case Propensity Theory of Explanation', *Synthese* 28 (1974), 191–192. Note that the appropriate requirement within a knowledge context

Kxt is indeed Hempel's requirement of maximal specificity; cf. Hempel, 'Maximal Specificity', p. 131, less the requirement of high probability. See, for example, James H. Fetzer, 'Statistical Explanations', *Boston Studies in the Philosophy of Science*, Vol. XX (ed. by K. Schaffner and R. Cohen), D. Reidel Publishing Company, Dordrecht-Holland, 1974.

[39] Fetzer, 'A Single Case Propensity Theory of Explanation', pp. 192–193.

[40] Fetzer, 'The Likeness of Lawlikeness', *forthcoming*.

[41] The distinction may be drawn on the basis of three criteria, namely: (i) the thesis that theoretical terms denote; (ii) the thesis that theoretical statements are either true or false; and, (iii) the thesis that science attempts to explain as well as to predict the phenomena of experience. In each case, the realistic position affirms, while the instrumentalistic position denies, the thesis in question. A useful discussion of the realist-instrumentalist issue is provided by Sidney Morgenbesser, 'The Realist-Instrumentalist Controversy', in *Philosophy, Science, and Method,* (ed. by S. Morgenbesser *et al.*), St. Martin's Press, New York, 1969.

[42] An individual's knowledge context *K* at any time *t*, therefore, should maximize the epistemic utility of his set of accepted beliefs by adherence to the *Rule for Maximizing Epistemic Utility*. Since the *Rule for Tentative Inductive Acceptance*, in effect, imposes a minimal condition of rational acceptability, while the *Rule of Maximizing Epistemic Utility* supports maximal content whenever possible, in conjunction they support the conception of a set of beliefs possessing maximal content, maximal consistency, and maximal closure.

KLEMENS SZANIAWSKI

ON SEQUENTIAL INFERENCE

1. Introduction

The idea of sequential inference is probably as old as that of induction
itself. All traditional accounts of induction by simple enumeration say
roughly this: elements of a class A are examined, one by one, for a property B; the process is continued until either (1) an element of A turns out
to be not-B (in which case the hypothesis 'all A are B' is rejected), or (2)
the inference maker is satisfied that he has accumulated enough evidence
to accept the hypothesis 'all A are B'. In spite of its crude form (no criterion of sufficient evidence is provided), the above procedure has certain
properties characteristic of sequential inference. The accumulation of
evidence on which the conclusion is to be based proceeds in consecutive
steps; their number is not determined in advance.

Like all inductive procedures, sequential inference can exactly be
described in probabilistic language. It is, therefore, not surprising that
both the concept and the term to denote it were first introduced within
the theory of mathematical statistics. The definition, due to Wald (1947),
is as follows[1].

> Let $Z = \{h_1, h_2, \ldots, h_N\}$ be a complete set of pairwise exclusive
> hypotheses. Let $X_1, X_2, \ldots$ be a series of random variables,
> representing the outcomes of consecutive experiments. Let
> $e_n = \langle x_1, x_2, \ldots, x_n \rangle$ denote a specific series of outcomes of the
> first n experiments; the class of all e_n will be symbolized by
> E_n. Now, a sequential rule of inference consists in defining
> for all n ($n = 1, 2, \ldots$) a partition of E_n into $N + 1$ disjoint subsets
> $S_n^1, S_n^2, \ldots S_n^{N+1}$ (some of them may be empty). They have the
> following interpretation. If the outcome e_n, actually obtained,
> belongs to S_n^k ($1 \leqslant k \leqslant N$) the hypothesis h_k is accepted; if
> $e_n \in S_n^{N+1}$ no conclusion is drawn and $(n+1)$th experiment is
> performed.

R. J. Bogdan (ed.), Local Induction, 171–182. *All rights reserved.*
Copyright © 1976 *by D. Reidel Publishing Company, Dordrecht-Holland*

As Wald himself pointed out, the above is a generalization of the 'ordinary' rule of inductive inference, i.e. one with the number n of experiments fixed in advance. Such a rule is obtained if we assume $S_m^k = \phi$ for all $m = 1, 2, ..., n-1, k = 1, 2, ..., N$, and $S_n^{N+1} = \phi$.

Having defined the concept of sequential inference, Wald gave an example of such a procedure, viz. the sequential probability ratio test. A generalization of that rule will be discussed below.

Wald's research was motivated mainly by practical needs: sequential techniques make it possible to reduce the average size of experimental data, without impairing the conclusion's degree of validity. It seems, however, that sequential inference is also interesting from the point of view of general methodology, for it represents a natural model of the scientist's decision to stop accumulating (homogeneous) evidence and choose one of the available hypotheses. The nature of this decision and its possible justification can be conveniently studied by means of this model.

It will be assumed throughout that the variables X_i are independent and identically distributed under each h_k. We adopt the notation

$$(1) \qquad p(x \mid h_k) = Pr\{X_i = x \mid h_k\},$$

so that for $e_n = \langle x_1, ..., x_n \rangle$ we have

$$(2) \qquad p(e_n \mid h_k) = \prod_{i=1}^{n} p(x_i \mid h_k).$$

The last expression is the probability, under the assumption that h_k is true, of the sequence of outcomes $x_1, ..., x_n$. We will also call it the likelihood of h_k, given the evidence e_n. Prior probabilities of hypotheses h_k, if assumed to be known, will be denoted by $p(h_k)$.

Our discussion of sequential inference will begin by a rule defined in terms of likelihoods only (generalization of sequential probability ratio test). Then, assuming prior distribution, I will define a sequential rule based on average epistemic utility. The two approaches will be compared, under the assumption of prior equidistribution of hypotheses. Finally, I will interpret sequential inference as a search for information: the accumulation of evidence stops when a given level of information has been attained. The meaning of 'information' in this context may be either classical (i.e. based on entropy), or pragmatic (i.e. defined in terms of epistemic decision problem); the two approaches turn out to be closely related.

2. Strengthened Maximum Likelihood

Wald's work was based on the assumption, universally accepted at the time, that inductive inference should not presuppose the knowledge of prior probabilities of hypotheses. In accordance with this belief, his sequential rule of inference was defined in terms of likelihoods only. I will present it here in a generalized form (Szaniawski, 1961).

Let $[A_{rs}]$ be a $N \times N$ matrix, its entries satisfying $A_{rs} \geq 1$, $A_{rr} = 1$.

Rule R: Hypothesis h_r is accepted on evidence e_n if

$$(3) \qquad p(e_n \mid h_r) \geq A_{rs} p(e_n \mid h_s) \quad \text{for all } s.$$

If there is no h_r satisfying (3) then no conclusion is drawn and the $(n+1)$ th experiment is performed.

The rule R is an obvious generalization of Wald's sequential probability ratio test: Wald's test is obtained by postulating $N = 2$. And it has analogous properties which justify its use.

Thus, if we denote by $\beta_{rs}(r \neq s)$ the probability of accepting h_r when in fact h_s is true, we have

$$(4) \qquad \beta_{rs} \leq (1/A_{rs}) \quad \text{for all } s \neq r.$$

The proof of (4) is a straightforward generalization of Wald's argument. We will say that e_n is of the type r if the relation

$$(5) \qquad p(e_m \mid h_t) \geq A_{ts} p(e_m \mid h_s) \quad \text{for all } s \neq t$$

is not satisfied for $m = 1, \ldots, n-1$, $1 \leq t \leq N$, and is satisfied for $m = n$, $t = r$. According to the rule R, hypothesis h_r is accepted on the evidence e_n iff e_n is of the type r. By definition, e_n of the type r satisfies condition (3), which means that the probability of obtaining e_n of the type r is at least A_{rs} times greater under the assumption that h_r than under the assumption that h_s. Therefore, the probability measure of all e_n $(n = 1, 2, \ldots)$ of the type r under the assumption that h_r is at least A_{rs} times their probability measure under the assumption that h_s. Formally,

$$(6) \qquad \text{for all } s \neq r : \omega_r \geq A_{rs} \beta_{rs},$$

where ω_r is the probability of accepting h_r when h_r is true. Since (6) implies (4), this completes the proof.

The inequality (4) defines an upper bound for the probability of error consisting in the acceptation of hypothesis h_r when some other hypothesis h_s is true. Errors of that type may, in general, be of unequal importance, which will find expression in unequal values of the constants A_{rs}. However, the essential property of rule R, expressed by (4), is that the probability of *any* error can be made as small as desired by increasing the value of respective parameter of the rule. Of course, the average number of experiments needed to arrive at some conclusion would be thereby increased.

In analogy with any algorithm, there ought to be a guarantee that the procedure will terminate after a finite number of steps. Since the rule R is based on random events, the postulated property of R must also be probabilistic. It can be described as follows. The probability of accepting some hypothesis after at most n experiments tends to 1 as n increases. Whether this is the case or not, depends, of course, on the probability distributions $p(x \mid h_k)$. For at least some applications of R, the last mentioned property can be proved. Thus, for instance, let h_k be interpreted as stating that $\mu = \mu_k$, where μ is the mean of normal distribution with known variance. The proof (cf. Szaniawski, 1961) that R leads to the acceptance of some h_k with the probability equal to 1 in the limit, rests essentially on the law of large numbers.

The rule R has $N(N-1)$ parameters, viz. the constants $A_{rs}(r \neq s)$. Such, at least, is the case when the inference is intended to serve a practical purpose, well defined in advance. In terms of this purpose, there is, in general, a preference for some errors over certain others. If, however, the inductive procedure is governed by a purely theoretical goal there is no reason for treating the hypotheses in Z (hence, the errors) asymmetrically. The number of parameters can be then reduced to 1, by postulating

(7) $A_{rs} = A > 1$ for $r \neq s$.

The condition (3) takes on the form

(SML) $p(e_n \mid h_r) \geqslant A p(e_n \mid h_s)$ for all $s \neq r$.

It will be called 'strengthened maximum likelihood' condition, for rather obvious reasons. The maximum likelihood condition demands that this hypothesis be accepted which maximizes likelihood with respect to e_n. Allowing A in SML to be equal to 1, we obtain the statement that the likelihood of h_r is maximum.

There is one important consequence of strengthening the maximum likelihood principle by means of $A > 1$: it may well happen that no h_r satisfies the SML condition. Hence, the rule of acceptance based on SML is necessarily sequential, whereas the maximum likelihood principle leads to a conclusion for any evidence and is, therefore, connected with fixed-size experimentation.

The rule R simplified by means of (7) will be called R_A. Thus, the rule R_A recommends the acceptance of h_r on the evidence e_n if h_r satisfies the SML condition.

3. MINIMUM OF EPISTEMIC UTILITY

If we allow the use of prior probabilities of hypotheses a different approach is possible. We now can compute posterior probability distribution of hypotheses in Z, so that it makes sense to speak of the (average) utility of accepting h_r on the evidence e_n. An intuitively appealing sequential procedure would then consist in experimenting until one of the hypotheses, say h_r, satisfies the following condition: the utility of accepting h_r exceeds a preassigned threshold value.

The choice of the utility function is determined by the nature of the present analysis. Since we are treating induction as a purely cognitive process, its goal can be defined as that of accepting the true conclusion. Let $u(h_r, h_s)$ be the epistemic utility of accepting h_r when h_s is true. We have

$$(8) \qquad u(h_r, h_s) = \begin{cases} 1 & \text{for} \quad r = s \\ 0 & \text{for} \quad r \neq s \end{cases}$$

The prior probabilities $p(h_k)$, together with likelihoods, determine posterior distribution $p(h_k \mid e_n)$:

$$(9) \qquad p(h_k \mid e_n) = p(e_n \mid h_k) p(h_k)/C; \quad C = \sum_k p(e_n \mid h_k) p(h_k).$$

We average the utility function over all h_s, by means of $p(h_s \mid e_n)$. As the result we obtain (cf. Marschak, 1974; Szaniawski, 1974) $p(h_r \mid e_n)$. In other words, the epistemic utility of accepting h_r on the evidence e_n is the posterior probability of that hypothesis.

Let B, satisfying $0 \leqslant B \leqslant 1$, be the threshold value of $p(h_r \mid e_n)$, if h_r is to be accepted. This determines the following sequential rule of inference.

Rule S_B: Hypothesis h_r is accepted on the evidence e_n if

$$(10) \qquad p(h_r \mid e_n) = \mathrm{Max}_s\, p(h_s \mid e_n) \geqslant B.$$

If there is no h_r satisfying (10) no conclusion is drawn and the $(n+1)$th experiment is performed.

Condition (10) is, of course, the conjunction of the following two statements.

$$(11) \qquad p(e_n \mid h_r)\, p(h_r) \geqslant p(e_n \mid h_s)\, p(h_s) \quad \text{for all } s\,;$$
$$(12) \qquad p(e_n \mid h_r)\, p(h_r) \geqslant B {\textstyle\sum_s} p(e_n \mid h_s)\, p(h_s).$$

The first of the above conditions postulates maximization of posterior probability, hence also epistemic utility as defined by (8). Clearly, for any evidence e_n there exists a hypothesis h_r that satisfies (11). This condition is, therefore, not characteristic of sequential inference, for which (12) plays an essential role. One might say that (12) requires the best alternative to be *good enough*; if no hypothesis satisfies this requirement one is forced to increase the evidence until there is one that does.

It would be interesting to study the relation between the two approaches, i.e. between the rules R_A and S_B. The comparison cannot be carried out unless prior probabilities in S_B are fixed, since R_A was defined in terms of likelihoods only. The simplest way to do this is by assuming prior equidistribution:

$$(13) \qquad p(h_k) = 1/N \quad \text{for all } k.$$

Conditions (11) and (12) become now

$$(\text{ML}) \qquad p(e_n \mid h_r) \geqslant p(e_n \mid h_s) \quad \text{for all } s\,;$$
$$(\text{MU}) \qquad p(e_n \mid h_r) \geqslant B {\textstyle\sum_s} p(e_n \mid h_s),$$

respectively. The first condition is that of maximum likelihood, the second will be called the minimum utility condition.

4. The Relation Between the Rules R_A and S_B

We are now in the position to compare the two rules of inference, by studying the relation between SML on the one hand, and the conjunction of ML and MU on the other. Obviously, SML implies ML. What is the sufficient condition for SML to imply MU?

The answer is not difficult to find if we add up all the inequalities in SML. We then have

(14) $(N - 1) p(e_n \mid h_r) \geqslant A \sum_{s \neq r} p(e_n \mid h_s).$

On the other hand, MU is equivalent to

(15) $(1 - B) p(e_n \mid h_r) \geqslant B \sum_{s \neq r} p(e_n \mid h_s).$

It follows that

(16)   SML $\Rightarrow$ MU   if $B \leqslant A/(A + N - 1).$

The result can be put in words as follows. If a hypothesis h_r is accepted on evidence e_n according to the rule R_A then it is also accepted on evidence e_n according to the rule S_B, provided that B is at most equal to $A/(A+N-1)$. The last expression is an increasing function of A and a decreasing one of N. This accords with intuition: when the number of equiprobable hypotheses increases it becomes more difficult for the posterior probability of any hypothesis to reach the threshold value, so that the upper bound for B must be correspondingly lowered if MU is to be satisfied by some h_r.

In order to see what ensures the reverse relation between R_A and S_B, let us notice that in the light of (15), $p(e_n \mid h_r)$ is at least $B/(1 - B)$ times the sum of all the remaining likelihoods. Hence,

(17)   MU $\Rightarrow$ SML   if $\cdot$ $A \leqslant B/(1 - B).$

Thus, if A satisfies the inequality in (17) the rule S_B is stronger than R_A, in the sense that the acceptance of some h_r according to S_B implies its acceptance according to R_A.

As a corollary, we get that if $B > \frac{1}{2}$ then MU implies ML, so that the two conditions defining S_B merge into one. Incidentally, since $A > 1$, $B > \frac{1}{2}$ is a necessary condition for the inequality in (17) to hold.

Assuming $N = 2$ we can combine (16) and (17) into

(18)   if $N = 2$ and $A = B/(- B)$ then   SML $\Leftrightarrow$ MU.

It turns out that in the case of two hypotheses, equivalence of the rules R_A and S_B is possible; and only in such a case.

It will be recalled that both R_A and S_B had a more general form to start

with. The comparison was made possible by the simplifying assumptions
(7) and (13) which make all the parameters in R equal to A, and all the
prior probabilities in S_B equal to $1/N$.

5. SEQUENTIAL INFERENCE AS THE SEARCH FOR INFORMATION

In defining a sequential procedure, the crucial problem is, what shall be
the criterion according to which the evidence is judged conclusive. Two
possible answers to this question were discussed above. The first criterion
was defined in terms of likelihoods and justified as putting upper bounds
on the probabilities of errors. The second one interpreted the conclusive-
ness of evidence as making it possible to attain a given level of epistemic
utility.

Still another approach, perhaps even more intuitive, would consist in
adopting information as the required criterion. On this view, evidence is
considered conclusive if it provides sufficient amount of information
with respect to the initial problem (in our case, which of the hypotheses
in the set Z is true?). A measure of information once defined, the arbi-
trariness, present in any form of induction, reduces to the choice of the
minimum value of information.

Let us first consider the Shannon measure. If we adopt the usual symbol
for entropy, information [2] provided by e_n on Z is defined as

(19) $I(e_n, Z) = H(Z) - H(Z \mid e_n).$

The absolute entropy of Z being constant with respect to e_n, we may
dispense with it and make our rule of inference depend on the second
term only. Thus, we postulate that the process of accumulating evidence
continue until the conditional entropy of Z, given the evidence, falls
below a certain level D; the most probable hypothesis is then chosen.

Rule T_D: Hypothesis h_r is accepted on the evidence e_n if the following
conjunction holds:

(20) $p(h_r \mid e_n) = \text{Max}_s \, p(h_s \mid e_n)$
(21) $H(Z \mid e_n) = - \sum_s p(h_s \mid e_n) \log p(h_s \mid e_n) \leqslant D,$

where D is a constant, satisfying $0 < D < \log N$. Experimentation continues
if (21) does not hold.

One simple relation between T_D and S_B is given by the following assertion.

(22)   If $D < -\log B$, (21) implies (10).

In order to see this, let us assume that (21) is satisfied for $D < -\log B$. Let h_r be the most probable hypothesis, in accordance with (20). We then have

$$(23) \qquad -\log B \geqslant D \geqslant H(Z \mid e_n) \geqslant -\sum_s p(h_s \mid e_n) \log p(h_r \mid e_n) =$$
$$= -\log p(h_r \mid e_n).$$

Hence, $B \leqslant p(h_r \mid e_n)$, which is precisely what (10) asserts. We thus see that, for $D \leqslant -\log B$, T_D is stronger than S_B, in the sense that a hypothesis accepted according to T_D is also accepted according to S_B.

If prior equiprobability of hypotheses is assumed, condition (20) becomes ML, i.e. the postulate of maximum likelihood for h_r, while (21) assumes the following form.

(ME) $-(1/W) \sum_s p(e_n \mid h_s) \log p(e_n \mid h_s) + \log W \leqslant D,$

where $W = \sum_s p(e_n \mid h_s)$. ME here stands for: maximal entropy.

In view of (23), ME implies MU for $D \leqslant -\log B$, hence by (17) it also implies SML if $A \leqslant B/(1-B)$. All the three rules are, therefore, closely related when the hypotheses are initially equiprobable. This last assumption has, in terms of (19), a natural interpretation: other things being equal, prior equidistribution maximizes information on Z, provided by e_n. If $H(Z \mid e_n)$ were replaced by $I(e_n, Z)$ in (21) then out of all possible prior distributions, equidistribution would be the one for which (21) expresses the strongest requirement concerning the evidence.

A sufficient condition for the rule S_B to be stronger than T_D is slightly more complex. To fix the ideas, let us interpret log as $\log_e$. The function $-x \log x$ reaches its maximum for $x = 1/e < \frac{1}{2}$. Therefore, $B > 1 - 1/e$ implies

(24) $-x \log x \leqslant B \log B$ for all $x \geqslant B,$
$\qquad -x \log x \leqslant (1-B) \log(1-B)$ for all $x \leqslant 1 - B.$

On the other hand, (10) implies

(25) $p(h_s \mid e_n) \leqslant 1 - B$ for all $s \neq r.$

It follows from the above that if $B > 1/e$ then

(26)   (10) implies (21), for any $D \geqslant -B \log B - (N-1)(1-B)$
$\log(1-B)$.

Of course, the Shannon expression (19) is not the only existing explication of the concept of information concerning an exhaustive set Z of mutually exclusive hypotheses. An example of alternative approach is the concept of pragmatic information, relative to epistemic utility (cf. Szaniawski, 1974).

The idea is to compute the difference between the maximum average utility when e_n is present and the corresponding maximum when it is not, i.e. when average utility is determined by prior distribution on Z. In the case we are interested in, the decision consists in choosing a hypothesis out of Z, and the epistemic utility of any such choice is given by (8). Pragmatic information of e_n concerning Z, say $C(e_n, Z)$, is then defined simply as

(27) $C(e_n, Z) = \text{Max}_s \, p(h_s \mid e_n) - \text{Max}_s \, p(h_s).$

Since only the first term depends on e_n, the second one may be dropped in the definition of a sequential rule based on C (or it may be subtracted from the threshold value). We are then led to the condition (10), defining the rule S_B.

Thus, the analogies between the rules T_D and S_B are due to the similarity between the entropy-based concept of information and that of pragmatic information, when the last one is relativized to the epistemic problem, as defined by (8). The similarity was examined in some detail in Szaniawski (1974), for the case when both (19) and (27) are averaged over all $e_n \in E_n$. When this operation is performed its result defines, of course, a property of fixed-size evidence, viz. its average information concerning Z. The last mentioned quantity is non-negative, whereas both (19) and (27) can be negative for some e_n; the only exception is the case of prior equidistribution, relative to which any outcome e_n has some positive information.

6. CONCLUDING REMARKS

All the rules of inference discussed above had a common structure, characteristic of sequential inference in general. It can be described as follows.

If a hypothesis, say h_r, is to be accepted on evidence e_n a specified relation must hold between the two. Let it be symbolized by $W(e_n, h_r)$. As soon as e_n is such that W holds for some h_r, the procedure is terminated by accepting this particular h_r.

The relation W might consist in the fact that h_r is at least A times as probable on h_r than on any other hypothesis (SML), or in the probability of h_r given e_n exceeding a threshold value (MU), or it may represent some other requirement. The point is that it imposes a necessary condition on the evidence e_n. It follows that the rule of inference must provide for the case when this condition is not satisfied by e_n. Going on to e_{n+1} is the answer.

A non-sequential rule associates a conclusion with *all* possible premisses of the type e_n. Or if it allows, for some e_n, the 'no conclusion' verdict, no further steps are dictated by the rule. It follows that the number n of experiments can be specified in advance, since it depends exclusively on the subject's decision. The decision may be based on essentially the same considerations as those that govern the choice of the exact form of W. Let us, for instance, assume that the purpose of inference is adequately described by the function which associates with each e_n $\mathrm{Max}_s\, p(h_s \mid e_n)$, coupled with the probability distribution of e_n. How the choice of n is influenced by the shape of the two functions, is something I am not going to discuss here. Certainly, there is no unique solution to this problem.

In the case of sequential inference, the subject must decide on the actual value of at least one parameter in his rule. If, for example, his purpose is given by (8) he has to choose the threshold value B. Now the number of experiments needed to reach a conclusion becomes a random variable and the consequences of the subject's choice are represented by the probability distribution of that variable. The situation is, in a sense, reverse to the one described above when n was fixed and the degree of the goal's achievement varied randomly.

Although the cost of experimenting has, so far, been left out of account, there is the implicit assumption that it exists and puts restrictions on the parameters of the inductive procedure. Otherwise, there would be no limit to the process of gathering evidence. If the cost of experimentation is explicitly introduced (and the purpose of local induction defined in the same units) a kind of balance can, in principle, be established, leading to a rule which maximizes the average net gain. I shall not pursue the topic further, since it seems hardly possible to inquire into the nature of this balance without making specific assumptions about the probabilistic relation between the hypotheses and the data.

To conclude the present discussion, a few words on the so-called 'truncated' rules. There is always a limit to the gain in utility, associated with a correct choice of hypothesis. Now, a sequential rule of inference provides, in general, no guarantee that the procedure will terminate before this limit is attained by the increasing cost of experimentation. Such a possibility cannot be excluded, even if it is highly improbable. Clearly, to continue experiments beyond the above mentioned limit would be in-consistent with the meaning of utility. Hence it is necessary to modify the sequential rule by adding a proviso which at a certain point puts a stop to experimentation, even if the evidence does not satisfy the condition W that defines the rule. The use of 'truncated' rules can, therefore, be justified on purely theoretical grounds if both the purpose of inference and the limited character of data are explicitly stated.

University of Warsaw

NOTES

[1] Actually, Wald's work dates from 1943. For the duration of the war the results were classified, which delayed their publication.
[2] Hintikka (1968) calls it 'transmitted information', averaged over the elements of Z.

BIBLIOGRAPHY

Hintikka, J.: 1968, 'The Varieties of Information and Scientific Explanation', in B. van Rootselaar and J. F. Staal (eds.), *Logic, Methodology and Philosophy of Science* III, North-Holland Publishing Company, Amsterdam, 1968, pp. 311–332.
Marschak, J.: 1974, 'Information, Decision and the Scientist', in C. Cherry (ed.), *Pragmatic Aspects of Human Communication*, Theory and Decision Library, Vol. 4, D. Reidel Publishing Company, Dordrecht and Boston, 1974, pp. 145–178.
Szaniawski, K.: 1961, 'A Method of Deciding Between N Statistical Hypotheses', *Studia Logica* 12, 135–143.
Szaniawski, K.: 1974, 'Two Concepts of Information', *Theory and Decision* 5, 9–21.
Wald, A.: 1947, *Sequential Analysis*, J. Wiley, New York, 1947.

GÜNTER MENGES AND E. KOFLER

COGNITIVE DECISIONS UNDER PARTIAL INFORMATION

1. INTRODUCTION

Since Shannon's original work about information theory the notion of information and later on the notions of information content, semantic information, information value etc. have run through many definitions. Carnap (1966), Hintikka (1970), and others defined semantic information in a way (tracing back to Laplace (1812)) which virtually states that the more probable (or less surprising) an information is, the less valuable it is. This view is closely connected with Popper's (1959) work, in which less probable hypotheses are considered more valuable, more informative, and more easily accessible to falsification. The 'Surprise Thesis' of information theory was questioned mainly by Levi (1969), Menges (1972), and Marschak (1974). The latter, in particular, emphasized the importance of the distinction between prior and posterior probabilities.

Our own view may be summarized in a few simple statements. The 'Surprise Thesis' holds for Laplace's a priori probabilities based on the principle of insufficient reason. It is, however, untenable with respect to Cournot's a priori probabilities (based on the principle of sufficient reason) and equally untenable with respect to a posteriori probabilities of any kind. In a cognitive framework there are two basic modalities of information, viz. that of the *degree* of information and of the *value* of information. The latter is easily to be defined as the property of an information to possess an epistemic utility for the receiver of the information (he is considered as a 'cognitive decision maker' or an 'inquirer'). With respect to the degree of information, we distinguish three cases:

(a) Perfect information about the probabilities of the hypotheses to be true;

(b) null-information, i.e. no information at all about the probabilities of the hypotheses to be true;

(c) partial information, i.e. more than null-information but less than perfect information about the probabilities of the hypotheses to be true.

R. J. Bogdan (ed.), Local Induction, 183–189. All rights reserved.
Copyright © 1976 by D. Reidel Publishing Company, Dordrecht-Holland

Case (c) is certainly the most important one from a pragmatic aspect. In this and other regards we share the view of Levi (1969). It is, however, not the task of this paper to compare or link our view with that of Levi or the local induction set-up altogether. Instead we try to formulate the problem of partial information and inference in a decision-theoretical framework and examine in particular the role of probability in this context, and we hope that our considerations, although they are only of a preliminary character, may contribute to the local induction problem. (For the wider framework, see Kofler (1974) and Menges (1974)).

2. THE FRAMEWORK

Let $\bar{A}$ be a set of cognitive actions a_i $(i=1,\ldots,m)$ on which a Borel algebra Σ_1 is given. Σ_1 makes $\bar{A}$ a measurable space A. The adoption of an action $a \in A$ leads to an epistemic utility whose arguments are the a's and the hypotheses b_j $(j=1,\ldots,n)$. The b's fill a set $\bar{B}$ on which a Borel algebra Σ_2 is given which makes $\bar{B}$ a measurable space B. The epistemic utility function is a real-valued bounded function

$$u:A \times B \to \mathbb{R}.$$

On B, there is given a probability vector $p=(p_1,\ldots,p_n)$. It may be considered a point $\rho(p)$ in the n-dimensional euclidian space: $\rho(p) \in \mathbb{R}^n$. Such a point may be called a probability point. If the true probability point $\rho^*(p) \in \mathbb{R}^n$ is known to the inquirer case (a) of perfect information is given, and he will – according to the Bernoulli acceptance rule – choose as optimal the cognitive action $a^* \in A$ whose expected epistemic utility $U(a)$ is maximal (Menges, 1974):

$$(1) \qquad U(a^*) = \max_{a \in A} U(a)$$

where

$$U(a) = \sum_{j=1}^{n} u(a,b_j)\, p_j.$$

In case (b) of null-information the inquirer will – according to the maximin acceptance rule – choose as optimal the cognitive action $a^0 \in A$ for which

$$(2) \qquad \min_{b \in B} u(a^0, b) = \max_{a \in A} \min_{b \in B} u(a,b).$$

Since the knowledge of $\rho^*(p)$ is informative for the inquirer, the information value V of $\rho^*(p)$ can reasonably be measured by the difference between (1) and (2):

$$(3) \qquad V(\rho^*(p)) = U(a^*) - \min_{b \in B} u(a^0, b).$$

We shall now consider case (c) of the degree of information which lies between the two extremes indicated by (1) and (2) and whose information value therefore is smaller than in (3).

3. PARTIAL INFORMATION

In cases between the two extremes the possible probability points $\rho(p)$ form a $(n-1)$-dimensional simplex $S^{(n)}$ in $\mathbb{R}^n$ which we call distribution simplex:

$$(4) \qquad S^n = \left\{ (p_1, ..., p_n) \mid p_j \geq 0, \; \sum_{j=1}^{n} p_j = 1 \right\}.$$

In the case of complete information where the information value is $U(a^*)$, $\rho(p)$ identifies uniquely a point in $S^{(n)}$. In the case of null-information where the decision value is $\min_{a \in A} u(a^0, b)$, all points of $S^{(n)}$ have to be considered. In the case of partial information we can define accordingly:

DEFINITION 1. The information I on p is a partial or local information (PI), if it identifies a proper area or locality $T(PI)$ of the distribution simplex which contains more than one point:

$$(5) \qquad I \in \{PI\} \leftrightarrow T(PI) \subset S^{(n)} \wedge |T(PI)| \notin \{0, 1\}.$$

4. LINEAR PARTIAL INFORMATION

Since it is often sufficient to restrict the considerations on the class of convex polyhedrons, we may introduce the notion of linear partial information (LPI):

DEFINITION 2. The partial information PI on p is called linear (LPI) if the corresponding locality $T(LPI)$ of the distribution simplex $S^{(n)}$ is a convex polyhedron.

5. DECOMPOSITION AND INFERENCE

We first consider the case in which by Cournot's a priori or any a posteriori knowledge the hypotheses space B can be decomposed into r disjoint subspaces B_v

$$\left(v = 1, ..., r, \quad B = \bigcup_{v=1}^{r} B_v; B_v \cap B_\mu = \emptyset \quad \text{for} \quad v \neq \mu\right)$$

such that the probability of a B_v to contain the true hypotheses is p_v ($p_v \in [0, 1]$, $\sum_{v=1}^{r} p_v = 1$).

It is then rational (Menges, 1966) to accept the cognitive action $\tilde{a} \in A$ for which

$$(6) \qquad \sum_{v=1}^{r} p_v \min_{b \in B_v} u(\tilde{a}, b) = \max_{a \in A} \sum_{v=1}^{r} p_v \min_{b \in B_v} u(a, b),$$

and the information value of this decomposition $B^{(v)}$ is then, in analogy to the basic Definition (3),

$$(7) \qquad V(B^{(v)}) = \sum_{v=1}^{r} p_v \min_{b \in B_v} u(\tilde{a}, b) - \min_{b \in B} u(a^0, b).$$

If by means of inference a given decomposition $B^{(v)}$ can be made finer so as to become $B^{(\mu)}$ with

$$\mu = 1, ..., s \quad \text{and} \quad s > r,$$

(6) and (7) can be reformulated accordingly. It would then also be reasonable to ask for the conditional information value of $B^{(\mu)}$ given $B^{(v)}$ which may be denoted by $V(B^{(\mu)} \mid B^{(v)})$:

$$(8) \qquad V(B^{(\mu)} \mid B^{(v)}) = \sum_{\mu=1}^{s} p_\mu \min_{b \in B_\mu} u(\tilde{\tilde{a}}, b) - \sum_{v=1}^{r} p_v \min_{b \in B_v} u(\tilde{a}, b),$$

where the first expression of the right-hand side of (8) is equal to

$$\max_{a \in A} \sum_{\mu=1}^{s} p_\mu \min_{b \in B_\mu} u(a, b).$$

6. EMPIRICAL EVIDENCE AND BAYES' A POSTERIORI DISTRIBUTION

Let e be the empirical evidence, then a given (Cournot) a priori or former a posteriori probability $P(B_v)$ can by means of Bayes' theorem be transformed into the a posteriori probability $P(B_v | e)$:

$$(9) \qquad P(B_v | e) = \frac{P(B_v, e) \cdot P(B_v)}{\sum\limits_{v=1}^{r} P(B_v, e) \cdot P(B_v)}$$

$$v = 1, ..., r$$

$$B_v \subset B \forall v$$

$$\bigcup_{v=1}^{r} B_v = B$$

$$B_v \bigcap_{v \neq \mu} B_\mu = \emptyset$$

where $P(B_v, e)$ is the likelihood of B_v after the evidence e has been gained. The information value of the empirical evidence given the a priori distribution $p^{(v)} = (P(B_1), ..., P(B_r))$ denoted by $V(e | p^{(v)})$ is then, analogous to (8),

$$(10) \quad V(e | p^{(v)}) = \sum\limits_{v=1}^{r} P(B_v | e) \min_{b \in B_v} u(\tilde{a}, b) - \sum\limits_{v=1}^{r} P(B_v) \min_{b \in B_v} u(\tilde{a}, b)$$

where the first expression of the right-hand side of (10) is equal to:

$$\max_{a \in A} \sum\limits_{v=1}^{r} P(B_v | e) \min_{b \in B_v} u(a, b)$$

and the second expression of the right-hand side of (10) is equal to

$$\max_{a \in A} \sum\limits_{v=1}^{r} P(B_v) \min_{b \in B_v} u(a, b).$$

The connection of Bayesian inference with *LPI* leads to the following property: Let the *LPI* of the a priori distribution $p^{(v)}$ be denoted by *LPI* $(p^{(v)})$, the *LPI* of the likelihood function $L^{(v)}(e) = (P(B_1, e), ..., P(B_r, e))$ by *LPI* $(L^{(v)}(e))$, and the *LPI* of the a posteriori distribution $p^{(v)}(e) = (P(B_1 | e), ..., P(B_r | e))$ be denoted by *LPI* $(p^{(v)}(e))$, then the following theorem holds.

THEOREM 1. *From the existence of LPI* $(p^{(v)})$ *and of LPI*$(L^{(v)}(e))$ *there follows the existence of LPI* $(p^{(v)}(e))$.

Proof: We consider the left-hand side of (9) as function of $p^{(v)}$ and of $L^{(v)}(e)$:

$$P(B_v \mid e) = \varphi_v(p^{(v)}, L^{(v)}(e)).$$

From (9) and the assumed existence of $p^{(v)}$ and of $L^{(v)}(e)$ it follows that $\varphi_v(p^{(v)}, L^{(v)}(e))$ is a continuous function in a compact, convex area. The compactness and convexity is due to the fact that the partial localities of the distribution simplex which correspond to $LPI(p^{(v)})$ and $LPI(L^{(v)}(e))$ are compact and convex. Due to the well-known Weierstraß theorem the function φ_v assumes its maximal and minimal value within the compact area. From the existence of $\max \varphi_v = M_v$ and $\min \varphi_v = N_v$ there follows the inequality

$$N_v \leqslant P(B_v \mid e) \leqslant M_v$$

and therefore $LPI(p^{(v)}(e))$ exists.

Very often in practical problems, the exact likelihood function is known to the inquirer.

THEOREM 2. *For known likelihoods there follows from the existence of* $LPI(p^{(v)})$ *the existence of* $LPI(p^{(v)}(e))$.

Proof. The proof is analogous to that of Theorem 1.

In the process of scientific inquiry it may often be the case that a given a priori distribution is improved by empirical evidence; the resulting a posteriori distribution is again improved by new empirical evidence; and so on. If such a sequence is called a *Bayesian chain*, one can also prove:

THEOREM 3. *For known likelihoods there follows from the existence of* $LPI(p^{(v)})$ *the existence of a LPI at the end of the chain.*

Universität Heidelberg
Universität Zürich

BIBLIOGRAPHY

Carnap, R.: 1966, 'Probability and Content Measure', in *Mind, Matter, and Method* (ed. by P. K. Feyerabend and G. Maxwell), University of Minnesota Press, Minneapolis, pp. 248–260.

Hintikka, J.: 1970, 'On Semantic Information', in *Information and Inference* (ed. by J. Hintikka and P. Suppes), Reidel, Dordrecht, pp. 3–27.

Kofler, E.: 1974, 'Entscheidungen bei teilweise bekannter Verteilung der Zustände', *Zeitschrift für Operations Research* (Serie A: Theorie) **18**, 141–157.

Laplace, P. S. de: 1812, *Théorie Analytique des Probabilités*, Paris.

Levi, I.: 1969, 'Information and Inference', *Synthese* **17**, 369–391.

Marschak, J.: 1974, 'Prior and Posterior Probabilities and Semantic Information', in *Information, Inference and Decision* (ed. by G. Menges), Reidel, Dordrecht, pp. 167–180.

Menges, G.: 1966, 'On the Bayesification of the Minimax Principle', *Unternehmensforschung* **10**, 81–91.

Menges, G.: 1972, 'Semantische Information und statistische Inferenz', *Biometrische Zeitschrift* **14**, 409–418.

Menges, G.: 1974, 'Elements of an Objective Theory of Inductive Behaviour', in *Information, Inference and Decision* (ed. by G. Menges), Reidel, Dordrecht, pp. 3–49.

Popper, K. R.: 1959, *The Logic of Scientific Discovery*, Hutchinson, London.

HENRY E. KYBURG, JR.

LOCAL AND GLOBAL INDUCTION

I

In 1967 Isaac Levi introduced a distinction which has had a serious impact on the direction of research into inductive problems. The distinction itself was not new in principle; people for years had been aware that the problem of justifying a particular inductive conclusion in a practical scientific context was quite different from the general problem of justifying inductive conclusions wholesale. But Levi made the distinction sharper, and, more important, turned what had been taken to be the philosophical point of the distinction on its head. Traditionally, the point of the distinction was this: as philosophers we are not concerned with the particular problems of how a sociologist, for example, should justify his argument from an observed sample to a population; the sociologist, after all, can rely on a whole body of knowledge whose justification is not (for him) at issue. As philosophers (so the traditional view went) we are interested in the general and abstract problem of justification; to justify an inductive conclusion I by reference to principles which themselves are just as much in need of ultimate justification as I itself, is to accomplish nothing of philosophical interest. Levi argues, on the contrary, that there are profound and interesting philosophical problems in local induction: it is not at all a simple matter to say what it is that justifies the sociologist in accepting his inductive conclusion, even *given* his background knowledge. Although Levi neither claims nor argues that the concern with global induction is philosophically uninteresting, his language suggests that that concern is less than vital: "No attempt will be made to justify the criteria offered by deriving them in some globalistically impeccable manner from the incontrovertibly evident".[1]

The local problem of induction is the problem of providing an analysis of justification appropriate to a particular scientific problematic situation. Here the *evidence* "will consist of those of the investigator's findings and beliefs that are relevant to the problem at hand and are not likely to be

R. J. Bogdan (ed.), Local Induction, 191–215. All rights reserved.
Copyright © 1976 by D. Reidel Publishing Company, Dordrecht-Holland

questioned by any participant in the inquiry..."[2] As contrasted with this, the global problem is to show that a whole body of beliefs are justified; in dealing with this problem we wish to admit as evidence only those statements that are not subject to doubt: the direct evidence of our senses, for example, or intuitively apparent necessary truths. It is well known that there are problems associated with the global approach: while people tend to agree that necessary truth is limited to logical truth, the extent of logical truth is not altogether clear; the effort to find data that are suitably incorrigible and will serve as a basis for global justification of what we all know has been perpetually frustrated.

Under these circumstances, it is not surprising that Levi's suggestion that inductive logicians concentrate on the problems of local induction has been followed enthusiastically. Lehrer has written extensively from this point of view;[3] Hintikka and his students may also be construed as having adopted the same point of view. Most recently, Niiniluoto and Tuomela have both taken it for granted that theories may be counted among the evidential statements relative to which we assess probabilities, and for them, even the choice of a probability function is an empirical matter, guided by considerations outside of and prior to the problem of local justification. It is even a 'condition of adequacy' for the explication of hypothetico-inductive inference that the required probability function *not* be determined by logical considerations alone.[4]

The local-global distinction is an important one, and it is increasingly clear that there are interesting and important local problems. This is clear particularly in statistical inference, where foundational questions are a lively source of debate and disagreement even among practicing statisticians. Often what is at issue in questions concerning the acceptance of statistical hypotheses is precisely the question of what, in a given context, with a given common store of background knowledge, constitutes good inductive evidence. Since in every form of scientific inference we must take account of error – even in the most 'deductive' parts of celestial mechanics – these statistical considerations are not special or peripheral, but central to the whole inductive enterprise. I shall argue, in sections two and three, that most of the work that has been done on the problems of local induction has suffered from just the disease that the local-global distinction was supposed to cure us of: the disease of excessive generality. It thus provides relatively little help in cases in which there is genuine disagreement or puzzlement on a local level. If we are going to adopt a Deweyan stance, and claim that it is not in the abstract that inductive problems should be resolved, but in down-to-earth practical situations in which there is a felt problem, then the theory of local induction should make it possible (often, if not always) for two people who are holding conflicting views concerning the bearing of a specific body of evidence on a specific set of hypotheses, to resolve that conflict. This is not to say that the context may not have to be broadened in order to achieve a resolution, but that the theory should indicate the direction in which a resolution should be sought.

On the other hand, in sections four and five, I shall argue that if 'broadening the context' is an acceptable move, then we must take the global problems seriously. The traditional problems of induction – some of them, at any rate – are just the problems at which you arrive when you 'broaden the context' as much as you can. In order to be *sure* that a local conflict will be resolvable, you must be confident that no matter how the context must be broadened, it will be possible to apply the general inductive theory. But this in turn means that we must, even if only hypothetically and artificially, be capable in principle of justifying a whole body of beliefs.

There is another class of problems that pushes us toward global justification: the problems of epistemology. Here we have traditional philosophical problems, which, in view of their traditional philosophical generality, require an understanding of more than local induction for their resolution. It may be that they are false problems, in the sense that no resolution is possible. But as yet there has been no persuasive argument to that effect, and indeed the work that has been done on global approaches to induction has shed some light on these philosophical problems.

II. LOCAL NON-PROBABILISTIC INDUCTION

There are two reasons why local non-probabilistic induction has received relatively little attention.[5] The first is that, being demonstrative, it employs no novel or interesting logical relationships. It just employs the same old logic we know all about. Furthermore, even when a local inductive conclusion is entailed by the evidence for it combined with background knowledge, that reconstruction seems simplistic because it ignores the pervasive presence of error. I shall present and comment on the same old problem by way of illustration.

An organic chemist creates a new organic compound. It is crystalline at ordinary temperature and pressure. He performs an experiment to test the melting point. After one experiment, he concludes, with perfect generality, that all samples of this compound, past present and future, tested and untested, will melt under standard pressure at T degrees centigrade. No one will come up to him and say, "Aha! you have no evidence for that; your sample of one surely can't lead to such a strong conclusion; and even if you had done the experiment thousands of times, that wouldn't indicate that in the *future* the compound would continue to melt at that temperature; and of course you have no evidence at all that untested samples of the compound *would have* melted at that temperature, had they been tested". In that context, the argument (idealized though it may be) is deductive in character. We can reconstruct it as follows. (I use second order logic as being more intuitive here; but of course a first order reconstruction in set theory is just as easy.)

Before the experiment the chemist (and you and I, insofar as we believe chemical doctrine – and that's not what is at issue here) accepts the general statement:

(1) (F) $(F$ is a crystalline chemical compound $\supset (t)$ $[(\exists x)$ $(x$ is a specimen of $F \wedge x$ melts at $t) \supset (x)$ $(x$ is a specimen of $F \supset x$ melts at $t))]$.

Loosely translated: If something is a crystalline chemical compound and any sample of it melts at a certain temperature under standard conditions, then every sample of it melts at that temperature.

Before the experiment the chemist (and you and I, insofar as we are not taking issue with his laboratory techniques) accept that his new compound, c, is indeed a crystalline chemical compound.

(2) c is a crystalline chemical compound.

Finally, when the chemist performs his test, he concludes that a sample of c melts at standard conditions at T degrees. Again, assuming we are not taking issue with his techniques or his state of mind, we also accept that conclusion.

(3) a is a sample of c and a melts at T degrees.

From (1) and (2) and (3) it follows deductively that all samples of c melt at T degrees.

(4) (x) (x is sample of $c \supset x$ melts at standard pressure at T degrees).

There is nothing particularly exciting about the logic of this inductive inference, or about the fact that it entails its conclusions. One might well nevertheless be able to find, in journals devoted to chemistry or physics or even biology, arguments complex enough to have an interesting structure which could still be construed in this deductive fashion.

There is still something that seems to be left out: it isn't all this easy to establish a new physical constant. Surely the chemist would not be satisfied to conduct one melting point experiment, but would perform the experiment several times. The reason for this is not far to seek: it lies in the insecurity of statement (3). If we could be quite certain that one sample of c melted at T degrees, we would be quite certain of the general conclusion. But any process of physical measurement has associated with it a pattern of error, and thus all we can be sure of when we conduct the melting point experiment is that the reading for that run is T. We replicate the experiment in order to ensure that the melting point is (close to) T. We do not, in fact, accept statements of the form (3), but only statements of the form:

(3′) a is a sample of c, and a melts at T plus or minus ε degrees,

where ε is characteristic of experimental procedure. Statement (3′) in fact is a result of another inference, which is statistical and problematic, and which therefore calls for another analysis. Nevertheless, the deductive pattern does represent something that is going on in the scientific inference, which we may explicate by reference to a notion of practical certainty.

Statements (1), (2), and (3′) may all be challenged. The first statement may be defended as (partially) analytic of what we mean by a crystalline chemical compound, or it may be defended as part of a physical theory. In either event, it can (in some way or other, I am supposing) be defended as at least practically certain. It is not up for question in the same way that the melting point of this brand new chemical substance is open to question. The second statement is somewhat more open to question: it isn't hard to be mistaken in thinking that you have a chemical compound when in fact you have a mixture of chemical compounds. But there are various laboratory tests (one of which is sharpness and consistency of melting point – but for the purposes of this example we leave that to one side) for determining whether c is a crystalline chemical compound. In the local context of determining the melting point of c, we are accepting (2) as practically certain. It is tempting to construe 'practical certainty' in terms of probability, and in fact I think that approach, though complicated and requiring a good deal of clarification, is the right one. If we do that, and if the evidence (consisting of one or several tests of melting point) is such as to render (3') just as probable, or just as 'practically certain' as (2), then (4') will follow deductively from (1), (2), and (3'), and will do so in a relatively realistic way.

(4') (x) (x is sample of $c \supset x$ melts under standard pressure at T plus or minus ε degrees C.).

What has now become problematic is the statistical inference that renders (3') so probable as to be practically certain. We therefore turn to probabilistic inference in general.

III. LOCAL PROBABILISTIC INDUCTION

This is the area in which most of the work on local induction of recent years has been done. This is the area that Levi singled out, that has been taken as paradigmatic by Lehrer, that has been discussed by Niiniluoto and others of Hintikka's school. Our treatment of local probabilistic induction will reflect, of course, our views regarding probability itself. We shall therefore look at the local problems as they appear under each of three views regarding probability: the logical and subjectivistic, the objective, and the epistemological. Furthermore, there are two degrees of inference that we may have in mind: the degree of inference according to which one comes to *accept* a statistical hypothesis (or to choose one against an alternative, or to choose one set of hypotheses as against another set); and the degree of inference according to which one merely *assigns a probability* to a statement (general or particular), or, more generally, according to which one assigns a probability distribution to some quantity, which may be a parameter in a statistical hypothesis. In the first case, we might come to accept that the measure of tosses yielding heads is a half, or is in some interval around a half, or that a certain quantity is distributed normally (m, s) or is distributed approximately normally with mean between m_1 and m_2, and variance between s_1 and s_2. In the second case we might assign a distribution to the parameter p characterizing the coin and tossing apparatus.

The subjectivistic interpretation of probability seems tailor made for local probabilistic induction in the second degree. The situation is typically described as follows: We begin with a set of statements b representing our background knowledge. The prior probability of the hypothesis in question, h, is p. We accumulate evidence e. The conditional probability of h on e (and the background knowledge) is q. The conditional probability q may be very much greater than the prior probability p. More generally, we begin with a prior distribution D for a random quantity; we accumulate evidence e, and, by Bayes theorem, obtain a posterior distribution D_e for that random quantity.

Two familiar problems must be dealt with. First, since probabilities on this view represent personal degrees of belief, there is no "standard" prior probability or prior distribution. Two people may with equal right have widely differing prior probabilities. Their posterior probabilities will then also differ. Second, this approach throws no light on the question of acceptance. To *accept* a statement on this view, is to assign it probability one, and conditionalization cannot lead to unit probabilities except in trivial and uninteresting cases.

There are two sets of answers to these questions, one provided by the practicing statisticians who have adopted a Bayesian approach in their professional work, and the other provided by philosophers concerned with the problem of learning from experience.

(1) The statisticians' answers: There is no 'standard' prior distribution, but there is nevertheless rough agreement in many practical instances. There is a certain amount of conventional wisdom on which the statistician may draw. Furthermore, it is generally the case (subject to some very weak conditions), that as the amount of evidence increases, two people will come to agree more and more closely regarding their posterior distributions, even if they begin with quite divergent prior distributions. You may have a probability of one tenth for a coin to land heads; my probability may be nine-tenths; but after we have tossed it a thousand

times your conditional probability and my conditional probability for heads on the thousand and first toss will be very close together, unless one of us is adamantly pig headed. Finally, there are many circumstances under which the prior distributions may vary extremely widely, without having more than a very tiny effect on the posterior distribution. As for the other problem – the problem of acceptance – that is no problem at all: whatever the parties to an investigation agree to accept is acceptable as background knowledge, evidence, or whatever. Since they are concerned with statistical inference, the question of coming to accept hypotheses simply does not arise; they are concerned only with obtaining a posterior distribution on which to base a practical decision.

(2) The philosophers' answers: Philosophers also refer to the convergence of posterior distributions, but they also sometimes suggest constraints that prior distributions should satisfy. These constraints have been formulated only for first order (and, usually, monadic) languages. (Carnap is the classical figure in this approach, but it has also been followed by Hintikka and his students.) These constraints may be construed as purely logical (Carnap) or as reflecting empirical commitments (Niiniluoto and Tuomela). There are, in the construction of probability measures for these languages, one or two free parameters (λ of Carnap's λ-system; λ and η in his New System; α and λ in Hintikka's system). Although the parametrization of degrees of belief imposes certain restrictions on the prior distribution of belief, posterior probabilities will depend on the choice of the parameters. People choosing different values for the parameters cannot be expected to agree on their posterior distributions.

There are two philosophical approaches to the problem of acceptance. Richard Jeffrey shuns acceptance: there are certain experiences that may *cause* us to assign probability one to certain statements, but in general not even evidence statements can be assigned probability one. An evidence statement, on his view, is a statement in which a shift in probabilities *originates*; its probability may leap from 0.1 to 0.9 as a result of an observation. Jeffrey then gives us an algorithm for computing the effect of this shift on our other probabilities. William Harper has broadened this approach so that, given a certain shift in a set of basic probabilities, we move to a new probability function for all the statements of our language which is, in a sense yet to be made precise, the *closest possible* probability function to our original one. The other approach is that derived from

Hintikka's work. As Hilpinen and Hintikka show,[6] it is possible to define a rule of acceptance, in terms of both probability and a minimum sample size, which is demonstrably consistent. Levi formulates a rule of acceptance (rule A) which is also demonstrably consistent;[7] Lehrer has developed rules similar to Levi's.[8]

What is to be said about the bearing of all this on the problems of local probabilistic induction? The philosophical approach that devises an acceptance rule based on logical or constrained subjectivistic probability is (so far) applicable only to very simple languages. It cannot be applied directly to practical instances of scientific inference. In any such situation, we have a large body of information, relevant in varying ways, which cannot even be formulated in a first order monadic language. This holds of all the philosophical approaches to local probabilistic induction based on constrained subjectivistic or logical probability, whether they involve a rule of acceptance or not. Furthermore, these approaches do not provide the one thing that we might reasonably demand: a framework within which it is possible to discuss and come to agreement concerning whether a given posterior distribution (or given acceptance) is justified by the evidence, unless we begin with not only the same body of knowledge, but the same belief structure. To be sure, we should perhaps not demand a simple algorithm which will crank out a result automatically, but we should at least be able to focus on the kinds of items we have in our real bodies of knowledge that are relevant to the argument.

The out and out subjective Bayesians are in a slightly better position. They can take the prior probability measure to be defined over our native language. There may not be a natural way to do this, but in principle we can determine the degree of belief a person has in any arbitrary statement of his own language. (How large a set of statements one can test this way before running into incoherence is open to question. And how one handles the incoherencies that one does run into is also open to dispute – dispute, note, that the theory is powerless to adjudicate. Savage, the most thoroughgoing subjectivist, says that it is a matter of the person distinguishing between degrees of belief of which he feels 'more sure' and degrees of belief of which he feels 'less sure'.) Furthermore, subjectivistic Bayesian statisticians have in fact made important contributions to the solution of the very problem we are concerned with: much of what they have done is directly relevant to the problem of local probabilistic induction. There is just one constraint on the applicability of their techniques: it is required that the prior distributions of parties to an inductive episode be similar, in the sense that the differences among the prior distributions do not give rise to significant differences in the posterior distributions. (We have learned that the differences in prior distributions can be rather large in many cases without being dissimilar in the sense at hand.) But this doesn't amount to much: the thesis appears to be that if our probability distributions aren't too different to start with, they won't be too different afterwards. What we are looking for, as philosophers at least, is some way of resolving disagreements about what is how much evidence for what, in this local situation. This is just what the subjectivistic approach fails to give us. Talk about how *further* evidence will lead non-extreme opinions ever closer together is irrelevant here. We each have a prior distribution, we observe a given amount of evidence, and we each obtain a posterior distribution. If our posterior distributions are still divergent, the sub-jectivistic approach gives us no way of approaching agreement, or even discussing the disagreement fruitfully with an eye to reducing it.

In short, subjective approaches to probability are failures at the one thing we might expect of a theory that is to make sense of local prob-abilistic induction. They work where there is nothing at issue (i.e., where the prior distributions of participants to an inference are not 'too' dis-similar), and they are impotent in the face of disagreement.

Objectivistic (frequency, propensity) approaches to probability do not fare better. In fact, the one thing that can be obtained on the basis of such approach is a characterization of statistical tests. As was true of the Bayesian approach to statistics, there are local situations to which these considerations apply. If we can agree that a given local problematic situa-tion is one to which the objectivistic techniques apply, then we are in good shape. But if we cannot agree, we are out of luck. The objectivistic con-siderations are not even potentially relevant to the question of accepting singular statements characterizing an experimental situation, and thus there is no way this approach can be called upon to adjudicate disputes of this nature. Again, the objectivistic techniques of local induction work productively when there is nothing at issue; and again, they are impotent in the face of disagreement.

The epistemological interpretation of probability, though it has yet to be put to extensive test, does seem to promise to do better. To begin with, it does supply criteria for picking out those situations in which Bayesian techniques are appropriate and for picking out those situations in which objective techniques are appropriate.[9] Epistemological probability is defined for first order languages only, but we may suppose that these languages include set theory, so that almost anything is sayable in them. Most important, epistemological probability supplies a framework within which it is possible to make a constructive effort to resolve disagreements about the import of evidence in the case of local probabilistic induction. It is important to see how this takes place, at least in outline, in order to see that the 'nothing at issue or impotent to resolve disagreement' charge does not apply here.

The definition of epistemological probability is a long and complicated story, which has been given in detail elsewhere.[10] Here a few remarks will suffice to bring out the aspects that are relevant to our special problem.

(1) Probability is defined relative to a body of knowledge, construed as a set of statements of a first order language that is strong enough to include set theory, and, in particular, the mathematics required for statistical inference. At this juncture we are not concerned about how items get into a body of knowledge, but we may suppose that they can get in by 'observation' in some sense, and perhaps by inductive inferences as well. So long as we are concerned merely with the local induction, the source of our (common) body of knowledge is irrelevant. We agree that there is a certain set of things that we take for granted. Of course we couldn't list them; but, as items become relevant to our discussion, they can be listed.

(2) Every probability is based on a known statement of relative frequency or propensity. Since we are concerned here with local induction, we may suppose that all parties to any controversy accept the same statistical background knowledge.

(3) The third required ingredient of a probability statement is an assertion of randomness. Like probability, randomness is metalinguistic and epistemological. It is construed as a four-termed relation, holding between a term (denoting an object), a term (denoting a reference class), a term (denoting another class), and a set of statements, representing a body of knowledge. The probability of '$a \in b$' is the interval (p, q) just in case there is term 'c' denoting a reference class, such that the randomness relation holds between 'a', 'c' 'b', and our body of knowledge. Probability is defined for equivalence classes of statements, where two statements

belong to the same equivalence class if they are connected by a biconditional in our body of knowledge – i.e., s and t belong to the same equivalence class if the biconditional '$s \leftrightarrow t$' is in our body of knowledge.

Given a body of knowledge, there exists a probability for every statement in the language. Finding that probability, however, may be a nontrivial exercise in logic in virtue of the complexity of the randomness relation. It is here, I think, that the material controversies that arise in local induction originate, and it is through the analysis of this relation that we can provide a framework in which they can be constructively resolved.

Consider a statement s; it gives rise to an equivalence class $[s]$. Now consider the set of all triples of terms $\langle a, b, c \rangle$ such that '$a \in b$' is a member of $[s]$, '$a \in c$' is a member of our body of knowledge, and 'c' is a reference term – a term denoting a (potential) reference class. The problem of randomness is essentially the problem of choosing a reference class. There are two ways in which our knowledge of potential reference classes may conflict: (1) the interval (p', q') that represents our knowledge of the proportion of c' that is b' may *differ* from the interval (p, q) that represents our knowledge of the proportion of c that is b, in the sense that neither interval is included in the other; (2) one interval may be properly included in the other, in which cases we say that the statistical statement mentioning the shorter interval is *stronger than* the statement mentioning the including interval. Suppose that $\langle a, b, c \rangle$ conflicts with $\langle a', b', c' \rangle$ in the first sense. There are two circumstances under which we might still want to maintain that the randomness relation held among the terms a, b, c, and our body of knowledge: (a) If we knew that c was included in c'; this reflects the maxim 'choose the narrowest reference class'. (b) If we knew that c was included in a potential reference class c'' that was a cross product and such that either a subset of c' is the first term of the cross product, or that c includes the cross product of a term and a subset of c' as a special case.[12] This reflects the maxim 'use Bayes' theorem when you can'.

A certain number of our triples of terms $\langle a, b, c \rangle$ are now ruled out; in fact there are no conflicts in sense (1) remaining. For if our knowledge of b and c differs from our knowledge of b' and c', at most one of the two triples $\langle a, b, c \rangle$, $\langle a', b', c' \rangle$ can remain a candidate for determining a probability. The remaining class can then be ordered by 'strength': we want as reference class that class about which we have the most precise knowledge. Of course there will be more than one – in fact there will be an infinite number – of the same maximal strength. But since the corresponding statistical statements all mention the same intervals, we may suppose that all the randomness assertions mentioning triples of this last set are valid.

It is no doubt worth examining a couple of instances in which considerations like those just alluded to abstractly can be relevant to the validity of a local induction.

Suppose we are interested in the proportion of A's that are B's, and we take a large sample of A's and count the number of B's. Suppose that in our sample of n, a proportion r are B's. We may argue as follows (taking the mathematics for granted): Let a sample be ε-representative, if the proportion of B's in it differs by less than ε from the proportion of B's among A's in general. The proportion of ε-representative n-fold subsets of A – the proportion of ε-representative members of A^n – varies according to the proportion of B's among A's. It is 1, of course, if either all or none of the A's are B's; it decreases as the proportion in question differs from 0 or 1, and it achieves a minimum if the proportion of A's that are B's is 0.5. Let the proportion of ε-representative subsets of A in the last mentioned case be p. Then, whatever the proportion of A's that are B's, the proportion of ε-representative subsets of A (containing n items) will lie between p and 1. Suppose the sample we have drawn is denoted by s, and the set of ε-representative samples by e. Then if the randomness relation holds of the triple $\langle s, e, A^n \rangle$ we may conclude that the probability is $(p, 1)$ that $s \in e$. Since probabilities are defined to be equal for statements known to be equivalent, this will also be the probability of the statement: the proportion of A's that are B's lies between $r - \varepsilon$ and $r + \varepsilon$.

It is obvious to everybody that such local inductive arguments do not always go through. It should be a task of a theory of local induction to provide an explanation of precisely when and why they fail, or at least to provide a framework within which two people who share the same background knowledge can come to agreement – or at least work constructively *toward* agreement – about which inductive argument goes through in a particular case. Let us see how the framework I have provided serves this function.

(1) Let p' be greater than p, and suppose it is claimed that the probability of our hypothesis (that the proportion of A's that are B's lies

between $r - \varepsilon$ and $r + \varepsilon$, henceforth abbreviated as H) is $(p', 1)$. If this claim is to be justified, there must be a triple of terms $\langle a, b, c \rangle$ such that '$a\varepsilon b \leftrightarrow H$' is a member of our background knowledge, such that the proportion of c's that are b's is known to lie between p' and 1, such that we know '$a\varepsilon c$', and such that the randomness relation holds among a, b, c, and the set of statements constituting our agreed-upon knowledge. Three examples: (a) Our sample s_n was obtained by a certain method of which we know that it yields representative samples more often than 'random' sampling. Let the set of applications of this method be m. It would be being claimed then that the randomness relations hold among s, m, e, and our body of knowledge. (b) We know in advance of sampling that at least three quarters of the A's are B's. Then the minimum proportion of ε-representative samples among the members of A^n is that corresponding to a proportion of $\frac{3}{4}$, and that is represented by p'. (c) A belongs to a certain class of classes, $\mathscr{A}$, among which a proportion between p' and 1 are characterized by having a proportion of B's lying between $r + \varepsilon$ and $r - \varepsilon$ – let us denote the set of such classes by $\mathbb{R}$. It would be being claimed that the randomness relation holds between $A, \mathscr{A}, \mathbb{R}$, and our body of knowledge.

Rebuttal of the claim that the probability of H is the interval $(p', 1)$ would take the same form in the two cases in which it is possible. Under (b), no rebuttal is possible, since no question of randomness is at issue. If we truly share the same background knowledge, I simply made a mistake when I said that the proportion of ε-representative samples in A^n lay between p and 1: I know more than that. Under (a) and (c) the rebuttal of the claim consists in exhibiting a triple of terms $\langle a, b, c \rangle$ concerning which our statistical knowledge *differs* (in the sense described earlier) from the interval $(p', 1)$, and that the conflict cannot be avoided by either of the two gambits mentioned on page 202. Thus, for example, in the case of claim (a) we might argue that the sample was obtained under circumstances C – let the set of samples so obtained be denoted by c – and that the proportion of such samples that are ε-representative lies between q and q', where q is less than p' and q' less than 1; and (we would also have to argue) there is no subset of c matching the proportion to which we know that our sample belongs in which the relevant proportion is $(p', 1)$, and there is no cross product set m', such that a subset of c is the first term in the cross product, or such that the second term of m'

includes the cross product of a term and a subset of c. An example of this circumstance will be given shortly.

The rebuttal of the rebuttal would consist of showing that in fact the conflict can be avoided by one of the gambits mentioned on page 204. At every stage there is a constructive procedure to follow: the *onus probandi* is on a determinate one of the claimants, and can be satisfied – so far as that stage is concerned – by a finite amount of argument.

(2) Let q be less than p, and q' be less than 1, and suppose it is claimed that the probability of H is (q, q'). There must exist, then, a triple of terms $\langle a, b, c \rangle$ such that '$a \epsilon b \leftrightarrow H$' is a member of our background knowledge, such that the proportion of c's that are b's is known to lie in the interval (q, q'), such that we know '$a \epsilon c$', and such that the randomness relation holds between a, b, c, and our background knowledge. Two examples: (d) we already know the proportion of A's that are B's; suppose that proportion is r^*. Then, according as r^* lies in the interval $(r - \varepsilon, r + \varepsilon)$ or not, we *know* whether the sample is representative or not, and the probability is $(0, 0)$ if it is not. We have, in this case, that s is a random member of $\{s\}$ with respect to being ε-representative, and if the difference between r^* and r is greater than ε, we know that the proportion of ε-representative members of $\{s\}$ is exactly 0. Since $\{s\}$ is a subset – and known to be a subset – of A^n, the conflict is resolved in favor of $\{s\}$. (Note that while we always know that s is a member of $\{s\}$, and that $\{s\}$ is a subset of A^n, this knowledge has a bearing on the probability in question only when we know whether s is ε-representative or not; when we lack this knowledge we know only that the proportion of members of $\{s\}$ that belong to e lies between 0 and 1, and this knowledge doesn't *differ* in our sense from any other statistical knowledge.) (e) we know that A is a member of a set of sets $\mathscr{A}$, among which various proportions of B's are distributed according to one of a set D of distributions. Our sample s is ε-representative just in case the pair $\langle A, s \rangle$, which belongs to the set of pairs consisting of populations and samples from them, say $\mathscr{A}\mathfrak{s}$, has the property that s is ε-representative of A, or belongs, say, to $\mathscr{E}$. Given a member of D, we may compute the frequency with which the pair belongs to those pairs in which the second member is ε-representative of the first by straight forward Bayesian techniques. Given the set D, we may therefore compute a maximum and a minimum frequency for this to occur. It is claimed that $\langle A, s \rangle$, $\mathscr{A}\mathfrak{s}$, $\mathscr{E}$, and our body of knowledge satisfy the

randomness relation. In particular although s, A_n, and e conflict with $\langle A, s \rangle$, $\mathscr{A}\mathit{o}$, and $\mathscr{E}$, the conflict is resolved in favor of the latter by the following argument: There is a subset of A^n – namely those exhibiting the frequency r – which is a special case of a subset of a cross product – namely $\mathscr{A}\mathit{o}$ itself – in which the relevant frequency is just the same as it is in $\mathscr{A}\mathit{o}$.

A concrete example may help to illuminate the last abstract example. Suppose we are interested in the mean weight θ of a strain of laboratory animals – say, white rats. We might suppose it well known that the variance in any of a related group of strains is 4, and thus that it is in this one. The difference between the mean and the sample average in a sample of one will therefore be distributed normally with mean 0 and variance 4. Let a be the sample of one, B the set of samples of size one, and C be the set of samples in which the sample mean does not differ by more than 1.96 standard deviations from the population mean θ; 95 percent of the members of B are also members of C. Thus *if a* is a random member of B with respect to belonging to C, relative to what we know, *then* the probability for us, having observed a weight of 75, that θ lies between 71.08 and 78.92, is exactly 0.95

Now we probably, from previous experience, have some idea of the distribution of the means among the strains in the subspecies to which the strain in question belongs. The extent of that knowledge determines whether or not the randomness condition is met. Suppose we know merely that the means are distributed roughly normally among the strains, with an overall mean of between 20 and 100, and a variance somewhere between 9 and 16. The conditional distribution for θ, after our observation, will then be one in the set of normal distributions:[13]

$$\left\{ N\left(\frac{75/4 + m/\sigma^2}{1/4 + 1/\sigma^2}, \frac{1}{1/4 + 1/\sigma^2} \right) : 20 \leqslant m \leqslant 100 \wedge 9 \leqslant \sigma^2 \leqslant 16 \right\}.$$

The range of the measures for the interval (71.08, 78.92) will thus have a maximum at $m = 75$, $\sigma = 9$, and a minimum when $m = 20$ and $\sigma = 9$. The lower measure is 0, and the upper is 0.982. Let P be the population, A be the set of strains. Then if the pair, $\langle a, P \rangle$ is a random member of the set $\{\langle x, y \rangle : x \in y \wedge y \in A \wedge wt(x) = 75\} = S$ with respect to membership in the set of pairs $\langle x, y \rangle$ such that y has a mean (θ) in the interval (71.08, 78.92)

– let the set of such sets be R – relative to what we know, then the probability that θ lies in that interval is (0, 0.982). Under the circumstances we have described, the former assertion of randomness is true, the latter false. The frequency of R among members of $P \times A$ does not *differ* from the frequency of C among members of B.

Now suppose our knowledge of the subspecies is a little more precise. We might know that the distribution of means among the members of A is roughly normal with a mean between 50 and 60, and a variance of between 9 and 16. The set of conditional distributions will then be:

$$\left\{ N\left(\frac{75/4 + m/\sigma^2}{1/4 + 1/\sigma^2}, \ \frac{1}{1/4 + 1/\sigma^2}\right) : 50 \leqslant m \leqslant 60 \wedge 9 \leqslant \sigma^2 \leqslant 16 \right\}.$$

The range of the measures for the interval (71.08, 78.92) is now from 0.012 to 0.68. This range does *not* overlap (include) the previous measure of 0.95, and so indeed the two frequencies are known to differ. Now we shall show that nevertheless $\langle a, P \rangle$ is a random member of S, but a is not a random member of B with respect to C.

Of the gambits mentioned on page 204, the first – choose the narrowest reference class – is no help. Neither of the sets at issue is known to be a subset of the other, nor is there any way of transforming the problem so that this will be the case. The complicated gambit, (b), is the relevant one. There is indeed a potential reference class – namely S itself – which is a cross product, and which is such that the competing class – namely B – is known to have a subset – namely, the set of one-membered samples having a mean weight of 75 – such that the cross product of P and this set is known to be included in S as a special case. We are thus directed to use Bayes' theorem.

The probability, therefore, that θ lies between 71.08 and 78.92 is (0.012, 0.68). The shortest interval whose minimum probability is 0.95 turns out to be roughly (63.7, 75.6).

Finally, if we suppose that we know that the distribution of mean weights among the subspecies A is normal with mean 55 and variance 10, the unique posterior distribution for θ is normal with mean 69.29 and variance 2.86. The object $\langle a, P \rangle$ is a random member of S, and the probability is exactly 0.95 that θ lies between 63.68 and 74.90.

Note that in this last example we have an illustration of the manner in which the epistemological interpretation of probability provides a precise criterion for when our background knowledge is extensive enough that Bayesian procedures properly take over from confidence techniques.

IV

Techniques of local induction can guide collective inquiry, just in case the participants share the same body of knowledge, or the same opinions, at least in relevant respects. To be sure, this condition is vacuous for Robinson Crusoe, at least before the appearance of Friday. But even Robinson Crusoe, to the extent that he wishes his inquiries to be controlled and objective, should on occasion step outside himself and play the part of a critic of Robinson Crusoe. Barring such exceptional cases, however, we want the theory of local induction to provide a framework in within which to criticize particular inductive inferences; we want the theory to be capable of leading a group of scientists to agreement about what the evidence warrants in a particular case. (This is not to say that it will force them to agreement about their beliefs: it is perfectly possible that for someone to say, "Yes, the evidence does indicate H, but I still feel that there is another and deeper explanation".) Theories of local induction, as we have seen, can do this only if the scientists in question share the relevant parts of a body of knowledge, or (in the subjectivistic case) the relevant degrees of belief. What is the scope of the word 'relevant' here? It is well known that inductive logic is peculiar in the sense that if a set of premises inductively yields a conclusion, an enlarged set of premises may not yield that conclusion. Thus it is difficult to limit the scope of relevant knowledge. We thus seem pushed, even in the application of local induction to well circumscribed problems, to consider relatively global matters.

Furthermore, we should be able to achieve some sort of agreement about what the evidence warrants, even if we don't start with the same body of knowledge or the same opinions. We can only be assured of being able to do this if we can apply the techniques of local induction to an ever wider set of circumstances: if, given my background knowledge, you agree that the evidence supports H, but you lack a certain item in my background knowledge, I should be able, using the *same techniques* of local induction, to convince you that the evidence warrants the inclusion of that item in your background knowledge, too. What we require, in short, is that the context of local induction be arbitrarily expandable. But if a theory of local induction has this property, it is already almost a global theory. If my fellow scientist protests ignorance of (almost) everything, and my theory of local induction is arbitrarily expandable, I should be able to fill in his background knowledge to the point where he will agree that the evidence supports *H*. (Note that this does not mean that I have to teach him *everything* I know; but it means that I have to teach him any *particular* thing I know, and that without presupposing a common body of background knowledge.) The conclusion we draw is that the application of local induction is either dependent on the solution of those very problems that constitute the problem of global induction, or else its scope is so limited as to be uninteresting.

Of course it may be that the scope of local inductive logic is severely limited – perhaps even limited to those whose epistemological states are extremely similar. I know of no general arguments to this effect. There are arguments stated within a point of view regarding local induction: thus the subjectivist Bayesian will give limits within which the opinions of participants in an inference must fall in order that they end up in close agreement after observing the evidence. There are general arguments concerning the formation of theories: if you don't accept the same paradigms as I, you will formulate a different hypothesis to account for a given phenomenon. But we are here concerned with the most elementary and straightforward statistical kinds of inductive inference; we leave to one side the question of the case made for the paradigm-dependence of scientific hypothesizing.

To provide a formulation of this problem neutral with respect to the approach to local induction, let us refer to the background *state* of the investigator, rather than to his background knowledge or his prior distribution of belief. In this framework the problem is this: Given a certain background state S, given certain evidence E, a theory of local induction yields an inductive result R (which may be the acceptance of a statement, or the adoption of a new degree of belief in a statement or set of statements). Suppose two scientific workers have different background states S and S'. Given the same evidence E, they will not necessarily obtain the same inductive result. Let the difference between S and S' be represented by D. We suppose that the difference D does not represent merely a difference in data or evidence, for under ordinary circumstances I will not question your data, unless I have some theoretical or general grounds for doubting it. D thus represents inductive results that one worker has obtained and the other not. If local induction is going to resolve the issue between the two workers as to whether or not the inductive result R is yielded by the evidence E, it must first resolve the issue of the difference D between S and S', by leading to the inclusion of the inductive results of D in both S and S', on the basis of a common background state S^0 that both workers can put themselves in. But if we can do this in general, we may simply suppose that the background state S^0 contains no inductive results at all, and we will have at hand a solution to the global problem of induction. A successful local theory would seem to imply a global theory.

V

To what extent can the theories of local induction discussed previously be expanded to take account of differences in background state?

Let us consider first the subjectivistic approach. If our prior probabilities are the same, before we obtain the evidence E, then our probabilities conditional on E will be the same. This is the situation in which local induction applies without any problems. Now suppose our prior probabilities are not the same. They may differ within certain limits without leading our conditional probabilities to be widely different; but suppose they differ more than that? We are stymied. If we could put our finger on a difference in evidence which would account for the difference in prior probabilities, we could get started. But (a) even if we could cite such evidence, its efficacy would depend on our beliefs prior to that evidence being in close enough conformity so that the evidence will constrain us to have similar beliefs, and (b) it need not be the case that we can cite evidence: the prior beliefs relevant to the inductive conclusion may just be prior beliefs – not formed by induction and relative to evidence, but just bare-facedly prior. A logical view would resolve the latter difficulty – but only for those who accept the same purely *a priori* logical distribution of probability over sentences of the language. As is well known, there are no compelling reasons for accepting one logical measure function rather than another.

Now it may be that as we bracket more and more empirical evidence, we will come more and more to share the same degrees of belief. If so, then the subjectivistic approach will yield a solution to the global problem. It would be an unsatisfactory solution, however, since it depends on what, from a logical point of view, must be regarded as sheer coincidence. Is it possible that we could find an evolutionary justification for the primordial predilections to believe? Perhaps; but we are now far off in a speculative realm that has so far been relatively unexplored.

How about objectivistic approaches? Whether or not a certain objectivistic statistical technique can be applied in a certain situation is a matter of what we know or don't know. Can we assume that X is distributed normally with known variance and unknown mean? Should we suppose that X is normally distributed with both mean and variance unknown? Should we suppose only that it has a distribution of a certain type? Answers to these questions depend on accepting the results of prior inductions. If we know the variance of X, for example, and that it is normally distributed, student's test, appropriate for the case of an unknown variance, is just plain wrong. It is true that we can 'back up' to a certain extent. If A claims that the variance is known, and B claims that it is not, it is possible to review the evidence which supports A's claim; we can test for the variance. And if what is at issue is whether or not the distribution of X is normal at all, we can test for normality, or review the evidence on which a claim of normality is made. How far can we go in this way? It seems unlikely that we can go very far. There is no provision, in ordinary objectivistic theories, for the acceptance of singular statements – for that matter, there is no way in which probabilities can be assigned to such statements. There is no mechanism for the acceptance of extreme statistical hypotheses assigning parameters of 1 or 0 to probability measures. There is no mechanism for accepting universal generalizations. These are required for the characterization of the testing situation. Such statements must be accepted if we are to be able to know that a certain kind of test is appropriate to a certain situation. We need background knowledge which is not itself of the form of testable statistical hypotheses. Furthermore most objectivistic writers make a sharp distinction between accepting a hypothesis (which is never regarded as legitimate) and failing to reject a hypotheses (which is the appropriate outcome of a statistical test). If this distinction is maintained, then there is obviously no way in which the background knowledge required for the application of statistical tests can itself be obtained by means of statistical tests.

The epistemological approach can take account of differences in background state only with the help of a rule of acceptance. But in this case a rule of acceptance does not seem unattainable. Suggestions have been made in *The Logical Foundations of Statistical Inference*. We suppose that in a given context there is a degree of probability that constitutes practical certainty. The body of knowledge relative to which we judge probabilities in that context is a set of statements, each of which is acceptable (with that degree of practical certainty), relative to a body of knowledge of some higher degree of practical certainty, and to every degree of practical certainty between it and the given degree of practical certainty. If two people share the same body of knowledge, then as outlined earlier, they will be able to come to agreement, eventually concerning probabilities, and this includes the probabilities of inductively founded statements.

Suppose they do not share the same body of knowledge of a given degree of practical certainty. There are two possibilities: first, that they fail to share the same data; second, that the bodies of statements they regard as acceptable on the basis of that data differ. With respect to the first possibility, the epistemological interpretation of probability offers no help; but this is not surprising: if two investigators do not share each other's data, there can be no way for them to come to agreement. If I insist that your data are cooked or falsified, I will not accept the inductive conclusions you draw from that data. The second possibility cannot occur, for the following reasons: Let A and B share the same data (we suppose that data may be accepted directly into bodies of knowledge of degrees of practical certainty less than one; if we count as data only things of which we are "absolutely" sure, the argument is simplified). Let us assume that the bodies of practical certainties for A and B differ significantly – by which I mean that they differ with respect to those statements that come up in the attempt to adjudicate the probability statement at issue. That is, assume there is a degree of practical certainty such that the body of statements acceptable at that degree of practical certainty for A differs from the body of statements acceptable at that degree of practical certainty for B. Then there is a highest such degree of practical certainty, say p. The bodies of knowledge of A and B of any degree of practical certainty higher than p will then agree in their contents. But then probabilities for A and B, relative to these higher level bodies of knowledge will agree; and since a statement will be acceptable in a lower body of knowledge depending on its content (independent of anybody's body of knowledge), and its probability (which will be the same for both A and B), the contents of the body of knowledge of level p for both A and B will agree, contrary to our hypothesis.

Now of course two scientists A and B do not have identical bodies of knowledge: scientist A knows he had scrambled eggs for breakfast, and scientist B is ignorant of this fact. There are a vast number of bits of data – and induced propositions – that are part of A's body of knowledge that are not part of B's body of knowledge. What is required of a global approach to induction is not that A and B should be forced into epistemological conformity in *all* respects, but (a) that, assuming they have agreed to share data, without which science is impossible, they should be forced into agreement in all *relevant* respects – i.e., in all respects that have a bearing on the arguments concerning the probability of the inductive conclusion at issue, and (b) that there should be no *a priori* constraints as to what is potentially relevant. It may turn out that what scientist A had for breakfast *is* a bit of relevant data; but if it is relevant, its relevance will come out in the dialectic regarding randomness, and, as a piece of data, it will come to be shared. Nor does this approach eliminate the role of intuitions and good guesses. Choice of a local research program – how shall we go about finding out about X – is often a matter of lucky guesses. Within a general paradigm, it may be possible to find patterns of discovery, to provide arguments supporting one program of research as against another. But within a shared language and shared data, it should not be an open question whether or not a given inductive conclusion is probable, or whether it is acceptable at a given level of practical certainty.

VI. CONCLUSIONS

Philosophical work that has oriented itself toward problems of local induction has shed little light on inductive inference. The premises required for demonstrative induction have been phrased with excessive generality by philosophers (e.g., 'The set S includes a necessary and sufficient condition for E'), with the result that there are no real life instances of demonstrative induction in which very much more specific and limited (and powerful) premises are not acceptable. The search for generality here has led away from realism and from relevance. With respect to probabilistic induction, philosophical treatments have suffered from even more serious defects. To the extent that local probabilistic inductions can be put into the frameworks provided by various philosophical approaches, they presuppose the very fundamental agreement that they should be capable of engendering: they require that opinions not differ too much (subjectivism) or that background knowledge not be at issue (objectivism). To the extent that these conditions are not satisfied – as they will not be in controversial instances of inductive argument – the local philosophical theories must be capable of being broadened. A theory which was capable of being broadened in all respects and without limit would be a global theory. Thus a local theory will be either redundant and unnecessary (when everybody is agreed that the inductive conclusion follows from the evidence for it, 'conclusion', 'follows', and 'evidence' being construed broadly enough to include the case that the 'conclusion' is a certain degree of belief in the hypotheses, the 'evidence' is a set of degrees of belief in other statements, and the 'following' consists in adherence to coherence and a principle of belief change) or it will be impotent to resolve disagreements (when our prior degrees of belief differ too much, or when our background knowledge differs to the extent that the objectivistic test appropriate to your circumstances differs from the objectivistic test appropriate to mine).

Different local theories are capable in different degrees and different ways of being broadened in the face of controversy or disagreement. To the extent that they are capable of that broadening, they approach global theories. Levi, though the earliest proponent of local approaches to induction, himself provides an approach that is relatively capable of arbitrary broadening. (His set of 'answers' to the 'problem' can be indefinitely expanded.) This suggests that the ultimate desideratum is, after all, a global theory of induction. Furthermore the global theories that have been proposed have contributed as much or more to the solution of local problems as those theories that are avowedly local. Thus the approach of Reichenbach, developed further by Salmon, has helped to focus attention on the problem of the choice of a reference class, which is fundamental and controversial in a wide range of local problems. The approach of Carnap, originally global in intent, helped to focus attention on the role of languages – again something highly relevant to the discussion of local problems of induction.

A theory of global induction seems to be required in order to avoid the dilemma that faces theories of local induction: irrelevance or impotence. But a plausible theory of global induction must be tied up intimately with an epistemology. There are a host of problems here, of course; but the artificial and technical ones contemplated by theorists of local induction are not among them.

University of Rochester

NOTES

[1] Isaac Levi, *Gambling with Truth*, Knopf, New York, 1967, p. 6.

[2] *Ibid.*, p. 4.

[3] K. Lehrer, 'Induction: A Consistent Gamble', *Nous* 3 (1969), 285–297; 'Induction, Reason, and Consistency', *British Journal for the Philosophy of Science* 21 (1970), 103–114; and elsewhere.

[4] I. Niiniluoto and R. Tuomela, *Theoretical Concepts and Hypothetico-Inductive Inference*, Reidel, Dordrecht, 1973, p. 18.

[5] It was discussed in my old paper 'Demonstrative Induction', *Philosophy and Phenomenological Research* 21 (1960), 80–92; a similar example is used by Niiniluoto and Tuomela, *op. cit.*, p. 240.

[6] R. Hilpinen and J. Hintikka, 'Knowledge, Acceptance, and Inductive Logic', in Hintikka and Suppes (eds.), *Aspects of Inductive Logic*, North Holland, Amsterdam, 1966, pp. 1–20.

[7] Levi, *op. cit.*

[8] Lehrer, *op. cit.*

[9] H. Kyburg, *The Logical Foundations of Statistical Inference*, Reidel, Dordrecht, 1974, Chapter 15.

[10] *Ibid.*

[11] This is perhaps not intuitively clear. For more detail see pages 244 and 366 of the work cited in Note 9.

[12] This possibly cryptic characterization will perhaps be elucidated by an example to be considered shortly; the circumstances in question are characterized explicitly and in detail in the work cited in Note 9.

[13] See, e.g., D. V. Lindley, *Introduction to Probability and Statistics*, Vol. 2, Cambridge University Press, Cambridge, 1965, p. 2.

RADU J. BOGDAN

HUME AND THE PROBLEM OF LOCAL INDUCTION

This paper is an epistemological attempt to outline the problem of local induction relative to the classical framework set forth by Hume. The reconstruction of the latter will be rather unexegetically simplified in the light of the former. Hume's approach to factual knowledge and induction was to a very large extent motivated and influenced by his opposition to, and criticism of, rationalism. I will try to project the problem of local induction against this background. I will claim that, his nuances and sometimes surprising qualifications notwithstanding (particularly in the Appendix to (1739)), Hume's approach to knowledge was atheoretical; and that it is this feature, perhaps more than anything else, which shaped his understanding of induction and left a profound mark on subsequent developments in the philosophy of induction. In contrast with it, I will argue that any epistemologically sound treatment of induction should be theoretical, or relative to a theoretical context of knowledge, and that methodologically this favors a relevance approach to local induction.

I. MEDIUM AND REASONING

In his account of induction Hume was primarily concerned with the logical link or connection that, if available, would make induction a perfectly rational or demonstrative kind of inference. He referred to this link or connection as 'medium'. Hume's quest for such a medium was conceptually associated with his quest for the proper demonstrative reasoning underlying inductive inference[1]. Together, the medium and the reasoning were supposed to explain, rationalize, and justify induction:

(A)   ... It must be acknowledged that there is here a consequence drawn by the mind, that there is a certain step taken, a process of thought, and an inference which wants to be explained. These two propositions are far from being the same: *I have found that such an object has always been attended with such an effect* and *I foresee that other objects which are in appearance similar will be attended with similar effects.* I shall allow, if you please, that the one proposition may justly be inferred from the

R. J. Bogdan (ed.), Local Induction, 217–234. All rights reserved.
Copyright © 1976 by D. Reidel Publishing Company, Dordrecht-Holland

other; I know, in fact, that it always is inferred. But if you insist that
the inference is made by a chain of reasoning, I desire you to produce
that reasoning. The connection between these propositions is not intuitive. There is required a medium which may enable the mind to draw
such an inference, if indeed it be drawn by reasoning and argument.
What that medium is, I must confess, passes my comprehension; and
it is incumbent on those to produce it who assert that it really exists and
is the origin of all our conclusions concerning matter of fact (Hume
(1748), p. 34).

This is, basically, Hume's *logical* problem of induction in terms of
medium and reasoning. (Note in the above quotation the dual role the
medium is supposed to play, i.e., logical, enabling the mind to draw an
inference, and empirical, being the origin of all our conclusions concerning matter of fact.) As formulated in (A), it seems to be restricted
to the empirical knowledge. But Hume was also aware of an ontological
dimension of the problem of induction. Acknowledging the traditional
distinction between (in his terms) 'secret powers' and 'sensible qualities',
Hume thought that the inference from the latter to the former faces the
same logical problem in terms of medium and reasoning:

(B)   Should it be said that, from a number of uniform experiments, we *infer*
            a connection between sensible qualities and the secret powers; this, I
            must confess, seems the same difficulty couched in different terms. The
            question still recurs, on what process of argument this *inference* is
            founded? Where is the medium, the interposing ideas, which join
            propositions so very wide of each other? (*op. cit.*, pp. 36–7)[2].

Hume then remarks: "Here, then, is our natural state of ignorance
with regard to the powers and influence of all objects"; and asks: "How
is this remedied by experience?" (*loc. cit.*) The answer should be referred
to (A). As we shall see, there is here a circularity which cannot be disposed
of on Hume's own grounds. For (A) and (B) face the same difficulty *only
if* induction is allowed to depend on causation, and conversely, *and* both
induction and causation are accounted for in an atheoretical manner.

Let us note for the moment that in both (A) and (B) the inference involved needs an argumentative justification *via* a demonstrative reasoning.
Many philosophers have observed that Hume's requirement that induction should be demonstratively justified is a nonsense since definitionally
it is not supposed to be deduction. It seems to me that Hume's requirement is only a trivial consequence of a deeper one. For Hume's main
concern was to find a reasoning which is both demonstrative and factual.

The key to such a reasoning is the notion of medium. It is the medium which can make any inference demonstrative. And the key to the notion of medium is causation. For, as I will try to show in the next section, what Hume seemed to mean by demonstrative and factual reasoning is nowadays captured, to a very large extent, by the notion of causal explanation.

II. CAUSATION

Hume's problem of causation is very closely related to that of induction. For him, factual knowledge is based on causal relations, and the latter are based on inductive inference from experience[3]. On the other hand, it appears that inductive inference cannot be demonstratively justified because a basic ingredient of causation, i.e. the notion of necessary connection, cannot be empirically established. The only inference able to do it, i.e. induction, was found wanting.

It is important to note that this circularity does not appear in Hume's *own* solution of induction and causation. It is rather an essential aspect built into his critical attempt to demolish rationalism. The notion of causation Hume was attacking was supposed (by most rationalists before him) to operate in an ordered and deterministic universe open to reason or intellect either intuitively, a priori, or demonstratively. For Hume, I believe, *these* notions of causation and reason had to stand or fall together. And he obviously looked for the latter course. In the process, he suggested such (empirically unattainable) standards for inductive inference as to fit the rationalist requirements for both causation and reason, i.e. factual and demonstrative reasoning. These causation-*cum*-reason standards come very close to our notion of causal explanation. The notion of medium Hume (while granting the rationalist's assumptions) was 'expecting' to play both an ontological and logical role, thus making induction both factual and demonstrative, actually has its place in our scheme of causal explanation. Let me amplify these remarks.

Many of Hume's arguments against rationalism seem to suggest that a positive (i.e. along rationalist lines) solution to the problem of causation would lead to a positive (i.e. logical) solution to the problem of induction. For Hume's quest for a medium, in a logical sense, was mirroring his quest for a necessary connection between causes and effects, in an ontological sense. In the scheme of causal explanation, what Hume took for 'causes' are 'initial conditions', while his 'effects' are 'predictions' or 'explananda'. On this reading, both 'causes' and 'effects' satisfy the other two requirements for causation, namely (physical) contiguity and (temporal) succession. The only missing link (and third requirement to be satisfied) is the notion of necessary connection or law. The medium *qua* law statement would logically connect the statements about initial conditions and facts to be explained, respectively, thus providing the required demonstrative reasoning applied to matters of fact.

Besides the general plausibility of this construal there are some specific reasons to hold it. For one thing, the scheme of causal explanation can logically handle problems of explanation proper, prediction, and retrodiction, with which Hume was equally concerned when discussing induction. For another, Hume was perfectly aware that, if justifiably established, a medium (*qua* law statement) between causes (*qua* initial conditions) and effects (*qua* explananda, or predictanda) will not only allow for a demonstrative reasoning ("by which they become proofs of each other's existence", (1748), p. 76)[4], but will do so by providing a *sufficient condition* for establishing the conclusion of such a reasoning. And this is exactly what is logically expected from the law statement in a causal explanation. (It is useful to note, in this respect, the multiple role played by the notion of medium if the rationalist account would have worked: as a necessary connection or law, it is an ontologically sufficient reason; as a piece of empirical knowledge, it constitutes sufficient evidence; and finally, as a logical 'middle statement', it provides a sufficient condition. Anticipating again, it was Hume's mistake to assume that these roles should be accounted for in the same way.) Finally, after deciding that empirical knowledge degenerates into probability and that consequently the problems of induction and causation should be treated at this new level, Hume defined probability itself in terms of causation as being "nothing real in itself, and properly speaking, is merely the negation of a cause, its influence on the mind is contrary to that of causation" ((1739), p. 125)[5]. Probability, too, appears as a poor substitute for causal certainty.

III. ATHEORETICITY

This is roughly an essential part of the rationalist background from which, and against which, Hume mounted his criticism of induction and causation. The weakest spot in this background seems to be the notion of necessary connection and its inferential counterpart, the notion of medium. Hume's criticism of 'necessary connection' and arguments against the availability of a corresponding 'medium', as fundamental prerequisites of his own approach to knowledge and induction, are mostly responsible for his ultimate conclusion (and answer to the sequence of questions quoted in Note 3) that: "all inferences from experience ... are effects of custom, not of reasoning" ((1748), p. 43). Motivated by his antirationalistic stance, Hume began and ended with an entirely atheoretical approach to factual knowledge. He granted rationalism a logical approach to the latter, and found it deficient. And rightly so. But then he ended in a purely experiential approach, and mistakenly so. For neither pure logic nor pure experience can alone, or even taken together, account for factual knowledge and induction. Theoretical knowledge is required as a medium.

However, it seems to me that Hume's atheoretical view of induction has strong preinductive roots, which both his criticism of rationalism and his own solutions tended to reinforce. So we have a systematic framework of atheoreticity which I shall briefly examine in the next subsections.

1. Let us begin with a logical source of Hume's atheoretical view of knowledge and induction, namely his notion of resemblance or similarity. Hume's account of factual knowledge rests on three relations: resemblance, contiguity, and causation. The former has absolute functional priority[6]. Like many philosophers of induction, Hume failed to see the *relativity* of similarity, and its implications.

Things are similar only in certain respects, in so far as the perception and knowledge of similarity are concerned. A point of view or perspective are required in order to identify those respects in which things are similar and the degree to which they are so. Furthermore, things that are similar from one point of view may be dissimilar from another. Similarity, and by implication repetition and classical induction (if any), do presuppose a certain point of view or perspective. In other words, induction is perspective-dependent because, antecedently, so is similarity[7]. The very notion of the uniformity of nature, on which the philosophy of induction rests so heavily, is also relative and perspective-dependent. For one can justifiably ask: in what respects, and relative to what point of view or theory, is nature uniform? And in what respects is the uniformity itself looked at? Is it constancy, or continuity, or lawful evolution, or what? To speak of a neutral, perspective-free uniformity of nature is a very uninformative and entropic way of conceiving of, and looking at, nature.

But neither similarity nor uniformity of nature are perspective-dependent just because we happen to entertain some point of view. A certain problem, area of interest, or cognitive concern are also required in order to direct our intellectual curiosity toward *certain* (classes of) things, and to actuate the relevant criteria of similarity. In theoretically organized knowledge, the resulting process will very often be an inquiry. (We already have here the basic ingredients of local induction *qua* inquiry-oriented and theory-dependent induction. As such, however, it is no longer the notion of induction Hume dealt with. But the important modification involved here stems, I think, from a direct criticism of Hume's atheoretical view.)

Similarity is not only relative in the above sense. It also has a conceptual solidarity with the notion of kind, and by implication with that of *classification* or conceptual scheme (see Quine (1969)). No matter how we will eventually account for the source of primary classifications (as 'inborn expectations' (Popper) or 'prior spacing of qualities' (Quine) or results of actions (Piaget)), and their conceptual relationships to the most sophisticated scientific taxonomies, it seems plausible to hold that both natural and scientific induction operate relative to prior standards of similarity built into, or directly traceable to, relevant classifications. In any mature science, the basic parameters of such classifications will belong to the network of theoretical entities and laws governing their behavior and connections.

His theory of instinct and propensities of mind notwithstanding, Hume saw no natural basis for relative similarity and perspective-dependent knowledge. Nor did he allow any theoretical basis, such as hypothetical or conjectural thinking, for positing classifications:

Every link in the chain [of argument of causes and effects] would ... hang upon another; but there would not be any thing fixed to one end of it, capable of sustaining the whole; and consequently there would be no belief nor evidence. And this actually is the case with all *hypothetical* arguments, or reasonings upon a supposition; there being in them, neither any present impression, nor belief of a real existence (Hume (1739), p. 83).

We note here another important source of Hume's atheoretical view of knowledge and induction, namely his copy theory of (passive) mind with its two basic components, pertaining to impressions and ideas, and causal stimulation of knowledge, respectively.

2. Let us begin with Hume's theory of impressions and ideas. For our purpose, it can be summarized in two sentences: "All ideas are derived from, and represent, impressions", and "... reason alone can never give rise to any original idea" ((1739), pp. 161, and 157, respectively). It follows that hypothetical thinking or theorizing has no perceptual counterpart and is empirically unjustifiable. No idea transcending available impressions can have ontological import. Nor are impressions themselves a reliable guide beyond their own coherence:

> ... Our impressions are causes of our ideas ... [But] as to those *impressions*, which arise from the *senses*, their ultimate cause is, in my opinion, perfectly inexplicable by human reason, and it will always be impossible to decide with certainty, whether they arise immediately from the object, or are produced by the creative power of the mind, or are derived from the author of our being. Nor is such a question any way material to our present purpose. We may draw inferences from the coherence of our perceptions, whether they be true or false; whether they represent nature justly, or be mere illusions of the senses (Hume (1739), pp. 5, and 84, respectively).

This is a most remarkable admission of what, in my opinion, Hume's epistemology stands for: namely, a coordinated retreat from both ontology and theoretical knowledge. There is here a cognitive causal impasse which mirrors that in which, as shown above, induction and causation were inextricably caught. In other words, there is a strong parallel between Hume's initial projection of causal explanation as (according to rationalism) a structural ideal of factual knowledge, whose empirical elusiveness shaped his final construal of causation and induction, on the one hand, and his causal account of cognition itself, on the other hand. The latter will also face the same problem as the former. A medium and a reasoning are required to inferentially connect ideas to impressions, and impressions to senses, and senses to their sources of causal stimulation, whatever these may be. If, at the structural level, the sciences of nature were found causally and inductively imperfect, so were at a cognitive or even psychological level the sciences of man. And let us remember that it was Hume's aim to base the former on the latter. No wonder that the coherence of our perceptions provides the final arbitration since both structurally, and cognitively, we face, according to Hume, the same insoluble problems of induction and causation. (This is a most unfortunate confusion which does not allow for the distinction between what knowledge is ontologically about; what is evidence for such knowledge; and the cognitive processes themselves.)

An important motivation of the above parallel lies in what Hume's causal account of cognition and atheoretical approach to knowledge rely on: i.e. the *passivity* of human knowledge. There is no relative similarity because knowledge has no essentially active powers of discrimination. There is no idea unless there is a sensory stimulation. In general, in order to function both psychologically and theoretically this knowledge has to be constantly stimulated or caused in a chain-like manner, and justified by causally previous, and eventually external, sources. In other words, the resulting foundationalism has to account causally-*cum*-externally for almost each cognitive move, to look for the impression corresponding to an idea, and for its cause, and for the cause of this cause, and so on (see Hume (1739), pp. 82–84). In order to be justified (again with respect to both knowledge claims, evidence for them, and underlying cognitive processes) such an account will need a medium and a demonstrative reasoning connecting the source or cause (whatever its place in the chain) and the result or effect, as well as a principle of uniformity of nature as a 'supermedium' for all possible connections. And these too should be accounted and justified in the same way. The structure of justification, expected to take over what a passive knowledge cannot initiate and accomplish, breaks down under an insuperable logical pressure.

3. Another important stronghold of Hume's atheoretical epistemology, related to the previous ones, is his notion of necessity. At one point Hume was writing: "Perhaps it will appear in the end that the necessary connection depends on the inference, instead of the inference's depending on the necessary connection" (1739), p. 88). Actually, due to the circularity in which (when projected against the rationalist background) causation and induction are involved, Hume had it both ways. On the one hand, according to his copy theory of mind and causal account of cognition, the idea of necessary connection has no sensory counterpart and therefore no experiential import. But it also has no traceable cause and therefore no ontological import. Hence, according to our previous analysis of

causal explanation, there is no medium and no demonstrative reasoning. On the other hand, there is no inference capable of establishing the idea of necessary connection unless it can be demonstratively justified – and it cannot. Perhaps Hume should have rather said that in the end it will appear that the idea of necessary connection depends on the idea of necessary connection!

But such a result follows not only from inferential, evidential, and cognitional failures. It also follows from a self-defeating definition of necessity, as far as factual knowledge is concerned. For it was, I think, in the spirit of Hume's antirationalistic strategy to load the concept of demonstrative reasoning and of causation-based-on-necessary connection with the *same* notion of necessity, and to oppose it to both *his* weaker concepts of induction and causation-based-on-constant-conjunction. Let us consider the following fragment:

> ... No inference from cause to effect amounts to a demonstration. Of which there is this evident proof. The mind can always *conceive* any effect to follow from any cause, and indeed any event to follow upon another; whatever we *conceive* is possible, at least in a metaphysical sense; but wherever a demonstration takes place the contrary is impossible and implies a contradiction. There is no demonstration, therefore, for any conjunction of cause and effect (Hume (1740), p. 188).

What Hume had here in mind is logical necessity and logical conceivability. One can logically conceive any event following from any other event (because they are logically independent) but he cannot do so *theoretically*. And this makes all the difference in the world of knowledge. Causation is relative to theoretical, not logical, conceivability; and so is the idea of necessary connection. Hume was oftentimes praised for having proved that the future is logically independent of the past and that effects are logically independent of causes. I think that this is a most trivial result unless one is expecting to find logical necessity and dependence in nature. If one entertains such an expectation, then Hume's arguments are most convincing. But for one who does not, it is not surprising that Hume could not find any logically necessary connection in nature, nor any impression of it in experience; just as it is not surprising that no inference from experience can establish such a connection.

The conceptual status of Hume's notion of medium is now more clear. We have first assumed with Hume that the notion of logical medium has an ontological and empirical counterpart, and that the latter, if available, would solve the problem of causation and induction. Then Hume showed that they are not available either inferentially or cognitively. Therefore there is no medium imposing logical restrictions on causal relations. As a result: "the mind can always *conceive* any effect to follow from any cause". Now take the latter claim as the basic premise or assumption of the whole edifice and look back: the inevitable conclusion would be that Hume was after the wrong target from the very beginning. This was the point of the previous paragraph. On either reading, however, there is an invariant mistake: i.e. the atheoretical approach to factual knowledge. For neither pure logic nor pure experience can account for the latter; and neither can deliver the genuine medium of factual knowledge: the ontological claims made by a scientific theory.

Let us note Hume's similar approach to the notion of the uniformity of nature:

> ... there can be no *demonstrative* arguments to prove *that those instances, of which we have had no experience, resemble those, of which we have had experience.* We can at least conceive a change in the course of nature; which sufficiently prove that such a change is not absolutely impossible ((1739), p. 89).

And again, I will claim, a 'change in the course of nature' is a question of theoretical, not of logical, conceivability, just as causation is. In either case, logical conceivability is irrelevant, and therefore cannot adequately be used as an argument against induction.

The problem of induction is not logical but theoretical, or relative to a theoretical context. Experience does not face logical reason, and its powers of conceivability, but theoretical reason. It is the latter, not the former, which can provide the genuine medium Hume was looking for. We have to go beyond Hume's way of thinking to find it. And going beyond Hume means in this context reinstating theoretical knowledge as the proper frame of reference for induction. It is relative to this frame of reference that the problem of local induction acquires its legitimate status.

IV. THE PROBLEM OF LOCAL INDUCTION

If my criticism of Hume's reconstruction of the problem of induction is valid, then there are important consequences which can provide an alternative reconstruction as a possible framework for local induction.

In this concluding section I will explore some of these consequences, and take them as a basis for understanding local induction.

1. If induction is made relative to a theory, and *its* evidential environment, we are conceptually able to use the well-known distinction between the ontological *claim* made by a theoretical statement and the *evidence* relevant to it (see Hempel (1965), pp. 350–351, and 379). Against the Humean framework, this distinction allows for the separation of induction from causation, necessary connection, and uniformity of nature – and in general from any ontological function directly attributed to, or expected from, induction.

According to this distinction, the statement of a causal law makes an ontological claim within a certain theory. Individual and empirically unique events will count as instances, and hence as relevant evidence, only in virtue of such claims. More often than not, it is the claim which directs our quest for such instances, which imputes similarity to them, and finds them empirically relevant and informative. Furthermore, the statement of a causal law refers to a *physically* necessary connection claimed to hold between certain kinds or classes of entities to which a certain theory is preinductively committed. The same applies to any nomological law, in general, or to a statistical law. What may change is the form of the ontological claim, reflecting new theoretical classifications and new physical connections between, or dispositions of, the entities considered. On the other hand, we may consider empirical generalizations or laws for which, relative to a determinate theory, there are both instantial relevant evidence (as above) and theoretical evidence in the form of higher level (or theoretical) laws. There is no Humean induction involved here since the instantial empirical evidence is a function of claim and theory, whereas the theoretical evidence is usually introduced as a more general premise.

The notion of uniformity of nature can itself be freed from its experiential connotations and its inductive functions, and regarded as an ontological claim *sui generis* made by a certain theory. Although it does not seem to be the exclusive or primary object of study of most scientific theories, many of the latter have built-in clauses and laws to this effect [8]. It is, however, plausible to assume that in various ontological respects uniformity of nature has or may become the proper object of some theories of evolution (in physics, cosmology, biology, etc.). In such a case, the latter will function as theoretical evidence for the ontological claims made in the other branches of science. But any claim concerning uniformity of nature would be ontologically partial – i.e. pertaining to different aspects, such as constancy, continuity, or lawful evolution, and even different macro-, and micro-cosmic regions of what we call generically 'nature' – and theoretically conjectural and fallible. And this is as it should be once we are prepared to admit that uniformity of nature is a theoretically-dependent class of ontological claims on a par with many others. Thus both causation and uniformity of nature are structural patterns of what we theorize and claim nature is, quite independent of the empirical evidence that might be used relative to the theories and claims themselves. To assume otherwise is to prejudge the results of knowledge in order to keep it working, and working safely. As I have tried to show, only atheoretical knowledge needs such (eventually self-defeating) safeguards.

2. If evidence and claim are distinct, what is the relation between them? If the former pertains to justification and the latter pertains to truth, what relation holds between justification and truth? Many philosophers, both inductivist and noninductivist, believe that the relation is a *logical* one, and reconstruct it in such a way as to make evidence part of claim, and justification (or confirmation, or corroboration) part of truth, even if the latter is only tentatively assumed. For the inductivists in search of a probabilistic formalization, partial inclusion and its measure become the formal standards for dealing with inductive justification. This reconstruction is largely responsible for conceiving of the latter as being inferential and/or measurable within an interval bound by logical truth and contradiction, respectively. But this seems to be just the other side of Hume's atheoreticity. For we are again contemplating knowledge moving between logical reason and pure experience. Theories are just transitional mediums, almost eliminable once the partial inclusion between empirical evidence and a hypothesis is established and the evidence is logically distributed over the domain of logical truth or maximal probability with respect to the hypothesis considered. The strength of the evidence is a function of this distribution, while the very criteria of distribution belong to logical conceivability.

My basic objection to this approach, no matter how probability itself is interpreted, is that logical conceivability of the relationship between evidence and claim is not the same with the theoretical conceivability of that relationship, particularly when the former also incorporates certain linguistic assumptions. We still do not know very well how theoretical conceivability works, but we do know some of the formal and linguistic paradoxes, riddles, and difficulties generated by the attempt to account for induction in terms of logical-*cum*-linguistic conceivability. Using Hume's terminology, the objection is that the quest for a demonstrative reasoning (in this case, logic plus probability calculus) is not independent of the medium which makes such a reasoning possible. And the medium is a certain theory and its methodological machinery. Scientists do conceive formal models (both logical and probabilistic) and apply them to experimental data but the models themselves are conceived within and in terms of a certain theory. They are not just invariant syntactical structures. Furthermore, and very importantly, the application of such models to empirical data, the feedback relevance of the latter to the former, and the theory they belong to, depend on further theoretical evidence and methodological assumptions (see Rosenkrantz (1971)).

A rather similar objection applies to the deductivist account of the relationships between empirical evidence and the claims made by a theory. For just as there is no pure experience prior to the establishment of a theory and directly available for probabilistic treatment, there is no such experience available for theory application, testing, and deductive subsumption either. In both cases, as either input or output, we have to consider methodological assumptions, various pieces of theoretical evidence, etc., which will define the experience in question as being relevant. I do not see how a purely deductive methodology can account for that. There is again a complex theoretical-*cum*-methodological medium which makes the deductive inference from claim to evidence possible, if at all. But it is at the level of this medium that the relationship between evidence and claim is first established.

My point can now be formulated at a higher level of philosophical generality. Any attempt to account for the relation between evidence and claim, or justification and truth, as being a logical, probabilistic, or in general an invariant relation, will face the objection, which I find very plausible, that a reality as posited by a theory is not logically 'transparent' in relation to experience. A reality is posited or claimed to be so-and-so by a theory, and experience is selected and brought to bear on such ontological claims or posits. But there is no logical or invariant relationship between these two acts, and no basis for either inductive or deductive inference unless we have a perfect algorithm for knowledge to function as a formal medium between what we experience and what we claim there is beyond experience. But there is no such algorithm, and this is essentially why our knowledge is hypothetical, theoretically-dependent and problem-oriented, and why its justification is contextual. If a formal medium is expected to relate conceptually homogeneous entities, a piece of empirical evidence cannot share a purely formal relationship with a theoretical law, unless the latter is just an amalgamation of the former (which seems unacceptable) or there are nonlogical assumptions and procedures in the first place governing the relationships between evidence and ontological claims. Which brings us to the very problem of local induction.

A most plausible relation between evidence and claim seems to be that of *relevance*. Basically it amounts to bringing evidence to bear on what we claim the world is, and evaluating the difference it makes with respect to such claims. The first step is mostly methodological, while the second is evaluational. The criteria of evaluation will belong to a rational strategy of local justification. The discussion of these criteria goes beyond the scope of this paper, but there are several relevant approaches to this problem which the reader might be interested in: Levi (1967), and (this volume); Lehrer (1971), (1973), and (this volume); Niiniluoto and Tuomela (1973); Hintikka and Hilpinen (1966); Salmon and Greeno in Salmon et al. (1971).

3. It is now obvious that the way in which evidence may be inductively relevant to an ontological claim is relative to a theory. The problem of local induction will then be the following: under what conditions and according to what criteria can a certain evidence be inductively relevant to an ontological claim made by a certain theory? Granting that in any scientific inquiry a theory (or a set of hypotheses) makes certain ontological claims, (i) how and according to what criteria does that theory define and select the relevant empirical evidence in a nonarbitrary manner and relative to alternative theories and auxiliary sciences (as theoretical evidence), as well as methodological assumptions, etc.? and (ii) how and

according to what criteria does such a theory get from that evidence reliable feedbacks that would improve, refute, or change it – and in general constitute a rational basis for decisions and evaluations concerning its ontological claims, and alternative ones?

It should be noted that the problem of local induction is not only a problem of what is relevant to what, and what follows from this, but also a problem of when this is so. Most advocates and critics of inductive justification assume an invariant structure of justification. But local justification is not only relative to certain contexts of inquiry but also to certain *stages* of inquiry. And this applies to both the selection of relevant evidence, and the evaluation and choice of hypotheses or theories. For example, there is more supporting theoretical evidence, there are easier methodological ways of bringing relevant empirical evidence to bear on ontological claims, and there is less competition among the latter when an indisputable paradigm dominates a certain field of research than when several theories compete for dominance. But the distinction is by no means absolute. Open inquiries involving evidential and theoretical competition may also occur within a dominant paradigm, for no such paradigm is ever so perfect as to be entirely closed and free of internal dissent. Moreover, there is a simple fact of cognitive life which also militates against an absolute distinction: in both cases, as considered above, knowledge is functionally stimulated by problems or difficulties or discrepancies to be solved. In other words, such problems are functional invariants of knowledge no matter what the stage of an inquiry is; nevertheless, the characteristics of the latter will define the problems as open or closed, depending on the availability of effective procedures of solution, and this will influence the structure of justification [9].

As a methodological aspect of local induction, bringing evidence to bear on the ontological claims made by a theory is associated with other methodological criteria and procedures such as correspondence rules, criteria of selection of a reference class and of application to single cases, empirical interpretation of objective probabilities, criteria of significance and decisiveness of empirical tests, application of formal calculi and models to experimental data, etc. The suggestion is that an adequate understanding of the problem of local induction cannot come alone but as a part of a broader methodological enterprise.

Many philosophers seem worried that a local approach to induction may become a new refuge for subjectivism. Their question is: How can different contexts of inquiry, different bodies of evidence which are theoretically selected, promote objectivity and agreement? I think that such worries still reflect atheoretical views of knowledge according to which only logical reasoning and incorrigible experience can constitute sources of objectivity and agreement. It is rather theories which should be taken, even if hypothetically, as starting points for objectivity and agreement. Scientists seem to agree more easily on what a theory claims than on which empirical evidence is certain, if any. Initial empirical evidence is irrelevant if considered outside a theoretical context, i.e. as experience we happen to have. Within a theoretical context, on the other hand, any body of empirical evidence brought to bear on a new theory is also accountable in terms of previous or alternative theories *plus* interpretative or auxiliary theories. This is, I think, a basic fact of scientific methodology as emphasized by the historico-critical reconstruction of science. Thus, there is no relevant empirical evidence without theoretical evidence. It is in the confrontation of ontological claims that our hope for objectivity lies.

What a theoretical approach to knowledge and a local approach to inductive relevance bring about is a sort of *gestalt* switch: that is, seeing knowledge as a move from theories to other theories *via* experience instead of seeing it as a move from one body of experience to another *via* theories. Experience is called upon to 'explain' such inter-theoretical moves – and this is an essential function of local induction. But experience no longer enjoys the absolute status of a prime mover and of an independent and incorrigible source of objectivity, agreement, and cognitive safety – unless we revive Hume and his policy of minimal involvement: "As long as we confine our speculations to *the appearances* of objects of our senses, without entering into disquisitions concerning their real nature and operations, we are safe from all difficulties, and can never be embarrassed by any question" (Hume (1739), p. 638).[10]

Stanford University

NOTES

[1] To my knowledge, the logical problem of induction in terms of medium and demonstrative reasoning is much more explicitly stated in Hume (1748) than in his (1739).

2 To see (as Hume did) in (A) and (B) the same inference facing the same logical problem, the difference being only terminological, is an indirect anticipation of some subsequent developments in the philosophy of knowledge and induction such as Carnap's notion of (theoretically free) qualified instance confirmation or the instrumentalist account of theories and scientific systematizations.

3 "When it is asked, *What is the nature of all our reasonings concerning matters of fact?* the proper answer seems to be that they are founded on the relation of cause and effect. When again it is asked, *What is the foundation of all our reasonings and conclusions concerning that relation?* it may be replied in one word, *experience.* But if we still carry on our sifting humor, and ask, *What is the foundation of all conclusions from experience?* this implies a new question ... (Hume (1748), p. 32).

4 Where, let us remember, *proofs* are exactly "those arguments which are derived from the relation of cause and effect and which are entirely free from doubt and uncertainty" (Hume (1739), p. 124).

5 In this context, Hume paid serious attention to the (probabilistic) problem of causal contrariety, a remarkable anticipation of some aspects of statistical explanation and ambiguity: see (Hume (1739), Bk. I, Part III, Sec. XI–XIII); Hempel (1965) Ch. 12; and Salmon (1971). Interestingly enough, a good deal of the recent controversies on this topic is still revolving around the inferential-*cum*-causal standards to be attributed to, or denied of, statistical explanation.

6 "[Resemblance] is a relation without which no philosophical relation can exist; since no objects will admit of comparison..." (Hume (1739), p. 14). The ideas of connection, cause, necessity, and secret power will also depend on that of resemblance: see ((1739), pp. 164–5; and (1748), pp. 76–79).

7 See Popper ((1959), Appendix *X, §1; (1963), Ch. 1, Sec. IV) for a critical discussion of Hume's theory of induction from this angle. See also Quine (1969).

8 This seems to be the case with representational theories (concerned with the structure and mechanism of the material systems involved) rather than with black-box or phenomenological theories (concerned with the external behavior of such systems): see Bunge (1964). Significantly, the latter rely much more on experience than on hypothetical theorizing, and seem to need a principle of uniformity of nature as a *ceteris paribus* clause designed to take care of the box' inside. It seems to me that Hume's epistemology as well as many varieties of empiricism take as models such phenomenological theories.

9 This applies to the evaluation and selection of evidence (see Lehrer (1971), (1973); Schick (1970); Hilpinen (1970)) as well as of hypotheses (see Levi (1967); Hintikka (1967); Hempel (1965), Ch. 2; Niiniluoto and Tumela (1973); Niiniluoto (this volume)).

10 I am most grateful to Stephen Pink, Robert Howarth, and Joseph Rotondo for their critical comments and suggestions.

BIBLIOGRAPHY

Bunge, M.: 1964. 'Phenomenological Theories', in M. Bunge (ed.), *The Critical Approach to Science and Philosophy*, The Free Press, N.Y., pp. 234–254.

Hempel, C. G.: 1965, *Aspects of Scientific Explanation*, The Free Press, N.Y.

Hilpinen, R.: 1970, 'On the Information Provided by Observations', in J. Hintikka and P. Suppes (eds.), *Information and Inference*, D. Reidel, Dordrecht-Holland, pp. 97–122.

Hintikka, J. and Hilpinen, R.: 1966, 'Knowledge, Acceptance, and Inductive Logic', in J. Hintikka and P. Suppes (eds.), *Aspects of Inductive Logic*, North-Holland, Amsterdam, pp. 1–20.

Hintikka, J.: 1967, 'The Varieties of Information and Scientific Explanation', in van Rootselaar and Staal (eds.), *Logic, Methodology, and Philosophy of Science III*, North-Holland, Amsterdam, 1968, pp. 311–331.

Hume, D.: 1739, *A Treatise of Human Nature* (ed. by L. A. Selby-Bigge), At the Clarendon Press, Oxford, 1888 (1973).

Hume, D.: 1740, *An Abstract of a Treatise of Human Nature*, in Hume, *An Inquiry Concerning Human Understanding* (ed. by C. W. Hendel), Bobbs-Merrill, Indianapolis, 1965.

Hume, D.: 1748, *An Enquiry Concerning Human Understanding* (ed. by L. A. Selby-Bigge), At the Clarendon Press, Oxford, 1927.

Lehrer, K.: 1971, 'Induction and Conceptual Change', *Synthese* **23**, pp. 206–225.

Lehrer, K.: 1973, 'Evidence, Meaning, and Conceptual Change', in G. Pearce and P. Maynard (eds.), *Conceptual Change*, D. Reidel, Dordrecht-Holland, pp. 94–122.

Levi, I.: 1967, *Gambling with Truth*, Knopf, N.Y.

Niiniluoto, I., and Tuomela, R.: 1973, *Theoretical Concepts and Hypothetico-Inductive Inference*, D. Reidel, Dordrecht-Holland.

Popper, K. R.: 1959, *The Logic of Scientific Discovery*, Harper & Row, N.Y., 1968.

Popper, K. R.: 1963, *Conjectures and Refutations*, Harper & Row, N.Y., 1965.

Quine, W. V.: 1969, *Ontological Relativity and Other Essays*, Columbia U.P., N.Y., pp. 115–138.

Rosenkrantz, R.: 1971, 'Inductivism and Probabilism', *Synthese* **23**, pp. 167–205.

Salmon, W. C. *et al.*: 1971, *Statistical Explanation and Statistical Relevance*, University of Pittsburgh Press, Pittsburgh.

Schick, F.: 1970, 'Three Logics of Belief', in M. Swain (ed.), *Induction, Acceptance, and Rational Belief*, D. Reidel, Dordrecht-Holland, pp. 6–26.

L. JONATHAN COHEN

A CONSPECTUS OF THE NEO-CLASSICAL
THEORY OF INDUCTION

The classical tradition in inductive logic is the tradition of Bacon (1620), Herschel (1833), Whewell (1840) and Mill (1843). Admittedly these authors did not always distinguish as well as they should have done between the study of criteria for the evaluation of scientific hypotheses and the study of procedures for scientific discovery. But in bk. III, ch. IX, Sect. 6 of the third edition of his (1843) Mill came to see clearly that the former, rather than the latter, was the primary domain of inductive logic. During the past decade I have tried to make some substantial further developments in this philosophical tradition so as to bring out, in a fairly systematic way, the inherent richness of its basic ideas. But my writings on the subject have been scattered over quite a number of books and periodicals (which are listed in the bibliography below), and while the theory has been growing some earlier statements of it have been modified or corrected. Moreover the need to explore numerous points of detail has sometimes made it rather difficult to maintain a proper perspective. So I offer here a conspectus of the current state of the theory. But a conspectus is necessarily just an outline. Those who wish to evaluate or criticise the theory must supplement their reading of the outline by a reading of the much fuller formulations and arguments that I offer elsewhere and particularly in (1970a) and (1976a).

1.

There are two key ideas for the neo-classical philosophy of induction. One bears on the nature of inductive evidence, the other on what this evidence is evidence for. The first idea (1.1.–1.273. below) is manifest in Bacon's tables of presence and absence, in Mill's methods of agreement and difference, and in Whewell's principle of consilience. It was put most succinctly by Herschel (1833, p. 155) when he remarked that experiments

become more valuable, and their results clearer, in proportion as they possess this quality (of agreeing exactly in all their circumstances but one), since the question put to nature becomes thereby more pointed, and its answer more decisive.

R. J. Bogdan (ed.), Local Induction, 235–262. *All rights reserved.*
Copyright © 1976 *by D. Reidel Publishing Company, Dordrecht-Holland*

This idea is represented in the present theory by what I call 'the method of relevant variables'. The second key idea (2–2.325. below) is that grades of inductive reliability, evidence for which is evaluated by the method of relevant variables, are steps on a staircase that mounts towards necessary truth – or at least towards the statement of a law. Whewell (1840) occasionally hinted at this idea (e.g. part I, bk. III, ch. VIII) but did not develop its implications. Its role in the present theory is based on arguments to establish that the logical syntax of statements about grade of inductive reliability is analogous to that of statements about necessity in the standard interpretation of the Lewis–Barcan calculus S4, together with arguments to establish that the latter system (and not, say, S2 or S5) gives a correct account of a familiar notion of logical necessity. These two key ideas also generate some important considerations about the justifiability of induction (in 3–3.6. below), about the relationship between the theory of induction and the theory of probability (in 4–4.382.), and about certain general problems in epistemology (in 5–5.5.). And in keeping with the classical tradition of Bacon, Herschel, Whewell and Mill, at least as much importance is attached to showing that the method of relevant variables is in actual use among non-philosophers as is attached to developing the details of the method's structure and exploring its implications.

1.1. In experimental science the reliability of a hypothesis is often identified with its capacity to resist falsification under cumulatively more and more challenging conditions, and its grade of reliability may be assessed in the light of correspondingly appropriate observations. A conspicuous paradigm is provided by von Frisch (1950) in his work on the colour sensitivity of bees, as analysed in my (1973a). For a particular field of experimental study this method of relevant variables may be reconstructed, in summary form, as follows (1.11–1.193.).

1.11. Let us suppose that all hypotheses in the field are initially formulated as generalised conditionals, out of a certain set c of first-order predicates (which may be, severally, monadic or dyadic or ... or n-adic for any finite n), plus truth-functional connectives, quantifiers and individual-symbols (unbound occurrences of which are understood to function as individual constants).

1.12. A relevant circumstance for these c-category generalisations is defined as a characteristic which is not describable in the terminology of the category and which is such that its presence in a particular type of situation suffices to exclude at least one generalisation belonging to the category from holding good for that type of situation. A relevant (physical) variable for the category is defined as any set of mutually incompatible relevant circumstances for it: these circumstances are its variants. A common series of relevant variables for all c-category generalisations is then defined as a suitably (cf. my (1976a)) trimmed and ordered set of non-exhaustive relevant variables. For example, the variable oral/intravenous administration might be first in the common series for hypotheses about the results of treating patients with antibiotics. A series of relevant variables, $v_1, v_2, ..., v_n$ for any particular c-category generalisation is the common series for the category, prefaced by the variable (v_1) that has as its sole member the characteristic described by the antecedent of the generalisation.

1.13. The inductive reliability of a c-category generalisation is graded by a monadic support-function $s_c[\ ] \gtreqless i$ which maps propositions on to ordinal numbers $0 \leqslant i \leqslant n$. Specifically $s_c[U] \geqslant 1$ iff U resists falsification when no variant of any of its relevant variables except v_1 is present; $s_c[U] \geqslant 2$ iff U resists falsification when any possible combination of variants of v_1 and v_2, and no variant of any other of its relevant variables, is present; $s_c[U] \geqslant 3$ iff U resists falsification when this is true for v_1, v_2 and v_3; and so on. $s_c[U] < i$ iff U is falsified by some combination of variants of $v_1, v_2, ..., v_i$ when no variant of any other relevant variable is present.

1.14. In 1.13. I gave the *truth-conditions* for statements of the form $s_c[\ ] \gtreqless i$. In order to give the *justification-conditions* for such statements – i.e. the method of assessing inductive reliability – it is necessary to suppose a series of canonical tests $t_1, t_2, ..., t_n$. In a performance of t_1 an instance of the antecedent condition of U is to be studied when no variant of any other relevant variable is present; in t_2 instances are to be studied for each variant of v_2 when no variant of any other relevant variable is present; in t_3 for each possible combination of v_2 and v_3 when no variant of any other relevant variable is present; and so on up to t_n. If E reports that there was some occasion on which these conditions held good for t_i (or for $t_{i+1}, ...$ or t_n) and the consequent of U was satisfied, then the truth of E normally entitles us to assert $s_c[U] \geqslant i$. If E reports a case of $t_1, t_2, ...$ or t_i in which the consequent of U was not satisfied, then the truth of

E normally entitles us to assert $s_c[U] < i$. Thus evidential support for a hypothesis is evidence that grades its reliability as greater than zero.

1.15. If from a true E it turns out that we should be entitled in this way to infer both $s_c[U] \geqslant i$ and also $s_c[U] < i$, both entitlements lapse. A hidden variable must have been operating in the test-performances reported by E, and our assumed common series of relevant variables for c-category generalisations needs to be amended accordingly. There will then be a different series of canonical tests for c-category generalisations.

1.16. A dyadic support-function $s_c[U, E] \gtreqless i$ grades what reliability we should be entitled to infer for U if we were given the truth of E. Hence $s_c[U, E] < i$ will be true not only if E reports a failure under tests $t_1, t_2, \ldots$ or t_i, but also if E does not report a performance of any of the tests $t_i, t_{i+1}, \ldots$ or t_n on U; whereas $s_c[U] < i$ can be inferred from E only if E reports a failure under tests $t_1, t_2, \ldots$ or t_i.

1.17. For appropriately richer languages this method of grading reliability, and evaluating evidential support, may be extended to hypotheses about causal connections or quantitative correlations. This is done by assigning to v_1 an appropriate range of variants instead of just one variant, and constructing correspondingly more complex tests, as in my (1970a, p. 73ff.). Thus, while tests $t_2, t_3, \ldots, t_n$ correspond to Mill's method of agreement, test t_1 comes to correspond to his method of difference (in the case of causal hypotheses) and to his method of concomitant variations (in the case of correlational ones).

1.18. The method of relevant variables may also be extended to grading systems of scientific theory that set out to explain a certain specific variety of quantitative correlations in the way that, say, classical mechanics set out to explain planetary movements, the acceleration of falling bodies, etc. Each relevant variable for a scientific theory is the set of combinations of circumstances constituting t_n for one of the correlational generalisations that the theory is to explain. And in order to ensure that such a consilience, as Whewell called it, is not achieved trivially – by a mere conjunction of the generalisations to be explained – the method of relevant variables requires that the theory graded should predict new and hitherto unknown correlations: cf. my (1970a), p. 83ff., and (1975c), *ad fin.* What Lakatos (1970, p. 118) calls an empirically progressive problem-shift represents an increase of inductive support in these terms, while a degenerating problem-shift does not: cf. my (1971).

1.181. Note that at no point in the method of relevant variables is there any restriction to monadic or non-quantitative predicates. To publish an inductive logic for scientific theory which turns out to be confined to monadic, qualitative predicates, as in Pietarinen (1972) or Niiniluoto and Tuomela (1973), is like publishing a textbook of mechanical transportation which turns out to apply only to bicycles. The excuse will no doubt be offered that a more powerful logic may one day be built on the same fundamental principles. But it might be more realistic to take the present impasse in the Carnap-Hintikka programme as evidence that these fundamental principles are mistaken. One has to choose, perhaps, between a non-quantitative grading of inductive support, like the neo-classical one, which applies quite adequately to hypotheses about measurable quantities, and a method of grading which affects to achieve a quantitative measure of inductive support but at the cost of being unable itself to measure support for hypotheses about quantities.

1.19. A further extension of the method, for any particular category of generalisations, is achieved when the terminology from which these generalisations may be constructed is taken to include not only the initial set of predicates, but also the terminology describing variants of relevant variables, plus a term predicating the absence of any variant of a relevant variable other than v_1, a term predicating this for v_1 and v_2, and so on up to $v_1, v_2,...$ and v_n. It is then possible to construct modified versions of a generalisation so that if in its unmodified version it fails test t_i the modified version is not exposed to falsification by this test, as in my (1970a, p. 142ff.): cf. (1962, p. 311ff.). Whatever the evidence, there will be some version that has the desired grade of support.

1.191. For example, if the generalisation

$$(x)(Rx \to Sx)$$

is falsified just by variant V_2^1 in test t_2, at least 2nd grade support is ensured for the version

$$(x)((Rx \ \& \ V_2^2x) \to Sx)$$

where V_2^2 is another variant of v_2.

1.192. Under these circumstances there will be the same grade of support also for the version

$$(x)((Rx \ \& \ T_1x) \to Sx)$$

where '$T_1 x$' signifies 'x is not characterised[1] by any variant of a relevant variable other than v_1'.

1.193. When this process of inductive modification, as in 1.192., is applied to a correlational generalisation, or to the description of its domain, we can achieve that kind of hypothesis which tells us what would happen under ideal conditions, such as a hypothesis about movement in a frictionless medium, – a hypothesis that appears to correct the observed facts, rather than bow to them. It is also clear why,[2] under sufficiently well-understood and well-controlled laboratory conditions, a single instance of a phenomenon can establish a universal generalisation about it: the generalisation that can be established is one that is maximally qualified by the method of 1.191.

1.2. The method of relevant variables has several consequential features that are worth noticing (1.21.–1.273.).

1.21. Different lists of relevant variables determine different modes of evaluation for support-functions. Hence, if generalisations belong to different categories, they can be described as having the same grade of support only if this is 0th or nth grade.

1.22. A statement of the form $s_c[U, E] \geqslant i$, or 'If E is true, then $s_c[U] \geqslant i$', is empirically corrigible. Such a statement is in fact a consequence of the empirically corrigible hypothesis that the common series of relevant variables for category c generalisations is $v_2, v_3, ..., v_n$. The latter hypothesis is a generalisation about category c generalisations and so it may be learned to have greater and greater reliability as it is found to apply to wider and wider varieties of such generalisations without being refuted by the discovery of, say, a hidden variable as in 1.15. Thus the basis for assessments of evidential support is itself subject to higher-level inductive assessments (cf. my (1968)). But there is a way to avoid being forced into founding one's lower-level assessments on an inadequately supported basis. If a higher-level hypothesis about a common series of relevant variables is suitably restricted in respect of the category of generalisations to which it is to apply, it may be taken (in view of 1.19.–1.193. above) to have full support.

1.23. In revising such a higher-level hypothesis in the light of experience we are free either to maintain unaltered the basic category of predicates and just alter the common list of relevant variables, or to alter the former and not the latter, or to alter both. It is certainly not the case, however, that a hypothesis should never be projected unless its terms (or co-extensive terms, or disjunctive terms of which the hypothesis' terms are disjuncts) have already occurred in successfully projected hypotheses. If the founders of modern chemistry had read, and agreed with, Goodman (1954, p. 87ff.) – contrast Schwartz *et al.* (1970) –, chemists might still be talking about earth, air, fire and water rather than about hydrogen, lithium, beryllium, boron, etc. Instead what must be insisted on is just that a vocabulary for hypothesis-construction should pay its way by being fruitful of hypotheses that turn out to be well-supported when they are tested against an appropriate series of relevant variables.

1.24. It is conceivable that in reality there are an infinite number of relevant variables for this or that particular category of generalisations. But, in order to ensure specificity we must assume finiteness whenever we hypothesise a list of relevant variables. We lose nothing by this assumption since the hypothesis can always be corrected, and the list extended, if experience teaches us to do so.

1.25. From the terminology used to formulate the generalised conditionals of a particular category (plus appropriate individual constants) very many other types of proposition may be constructed. But it is the generalisations – or rather a subset of them that are properly testable, as in my (1970a, p. 96ff.) – which are the primary beneficiaries of evidential support in accordance with the method of relevant variables. Propositions of other types enjoy what support they do in virtue of their logical relationships with these generalisations. For example, arguments may be deployed to show the validity of the following principles (1.251.–1.255.).

1.251. For any E, H and I, if I is a consequence of H according to some non-contingent assumptions, such as laws of logic or mathematics, and I is not a modified version (as described in 1.19–1.193. above), then $s[I] \geqslant s[H]$ and $s[I, E] \geqslant s[H, E]$. Note that, when one employs the method of relevant variables, these principles are fully compatible with ascribing inductively assessed support to general theories with infinite domains. On the other hand, M. Hesse (1974, pp. 142–150), in order to reconcile such a principle with her view that evidential support for H is measured by the difference between H's prior and posterior mathematical probabilities, is driven to suppose that the grading of evidential support can be concerned only with theories about finite sets of individuals. But she does not (and presumably cannot) give any historical grounds for supposing that the normal scientist has ever felt himself thus restricted in his assessments.

1.252. For any E, F and H, if F is a consequence of E according to some non-contingent assumptions, then $s[H, E] \geqslant s[H, F]$.

1.253. For any E, H and I, if $s[H] \geqslant s[I]$, then $s[H \& I] = s[I]$, and if $s[H, E] \geqslant s[I, E]$, then $s[H \& I, E] = s[I, E]$.

1.254. For any E and H, if $s[H] > 0$, $s[-H] = 0$, and, if $s[H, E] > 0$ and $s[-E] < n$, then $s[-H, E] = 0$.

1.255. For any E, U and P, if P is just a substitution-instance of a first-order generalisation U, then $s[U] = s[P]$ and $s[U, E] = s[P, E]$. Different arguments for this principle are given in my (1970a) and (1973a), respectively.

1.26. Since ith grade evidential support for a generalisation U is evidence that grades U's reliability correspondingly, and ith grade reliability is the capacity to resist falsification under test t_i (1.13.–1.14.), it follows that to report the results of a certain experiment as giving ith grade evidential support to U is to imply the replicability of these results, whatever they may be. Indeed, whether H be a generalisation or not (cf. 1.25. above), to claim that certain results afford inductive evidence for, or against, H is to imply the replicability of these results. The title to imply this replicability derives ultimately from the higher-level hypothesis (cf. 1.22.) that just $v_2, v_3, \ldots, v_n$ are capable of *causing* the falsification of such a generalisation. What causes, or fails to cause, a certain result in one case, will do so again and again if all the relevant circumstances are the same.

1.261. So the neo-classical analysis of inductive reasoning mirrors the standard assumption about replicability in experimental science. If you accept that assumption, you must regard any attempt at the actual replication of an experiment as being merely a check up on its evidential validity. More evidence of the same favourable (or unfavourable) kind as before must not be supposed to provide an increase (or decrease) of evidential support, as in so-called 'enumerative induction': cf. my (1970a, p. 106ff.) and (1972).

1.262. Now in the system of Carnap (1971), with the addition of Reichenbach's convergence axiom, it can be proved (pp. 161–165) that $c(h_1, e \& h_2) > c(h_1, e)$ when h_1 is a singular proposition, e is consistent with h_1, and h_2 reports another instance of the same attribute as h_1

reports. But the discovery of this proof, though welcomed by Carnap, was in fact a heavy blow to his claim to be explicating natural-scientific modes of discourse about evidential support. The discovery showed that this 'principle of instantial relevance', as Carnap called it, is rather deep-rooted in his system; and the harsh truth is that functions for which the principle is provable are cut off therewith from assessing the support given by experimental evidence in any context in which an assumption of replicability operates.

1.27. If support is assessed by the method of relevant variables, we can have an E that implies, without contradiction, both $s[U] \geqslant i$ and $s[U] < j$, where $j > i$. That is because test t_j manipulates more relevant variables than t_i but no variant of any of the additional variables is present in t_i.[3] It follows that we can have $s[U, E] > 0$ even when E reports some evidence contradicting U. More specifically, – cf. my (1973c) – we can quite consistently suppose a theory to be well-supported even when we know of anomalies that conflict with it, rather as Lakatos (1970, p. 137ff.) allows that a scientific research programme may progress 'in an ocean of anomalies', and (1971, p. 114) that 'most important theories are born refuted'.

1.271. This is quite an essential feature for any analysis of inductive reasoning that is to apply unrestrictedly to generalisations over fairly extensive domains, like scientific theories, since even when replicable kinds of counter-evidence within these domains are unknown it is often obviously rash to suppose them non-existent. Scientists often detach, from a report of existing evidence, the conclusion that theory U_2 is decidedly more reliable than theory U_1 – i.e. $s[U_2] > s[U_1]$ – without having to imply thereby that U_2 is totally true, for a certain domain, and has no counter-instances in time or space. And for that it is essential to be able to suppose that even if replicable counter-evidence to U_2 became known it would not refute the view that $s[U_2] > 0$.

1.272. Otherwise one is caught in a situation like Popper's. Because his concept of 'corroboration' assigned zero-level corroboration for any properly falsified theory, as in his (1959, p. 268), Popper had to develop a supplementary concept, the concept of 'verisimilitude', as in his (1963, p. 233ff.), in order to be able to arbitrate between two theories which may both be false. But it is very difficult to develop such a concept satisfactorily, as Miller (1974) has shown.

1.273. Moreover, if we were to allow U's falsity to imply U's failure to attain even 1st grade reliability, then by contraposition we should have to allow that where U has even 1st grade reliability U is true. So there would no longer be any need for more than one grade of support.

2.

The logical syntax of monadic and dyadic inductive support-functions is controlled by certain principles, of which examples were given in 1.251.–1.255. All such principles may be axiomatised into a system of modal logic that generalises on the Lewis–Barcan system S4. The details of the system have been progressively developed and revised in my (1966a), (1970a) and (1976a). But the key idea is to introduce, for any given category of generalisations, a set of modal operators $\Box^1, \Box^2, \ldots$ $\Box^n$ to represent grades of reliability, as well as the operator $\Box^d$ that is to represent logical truth, where $d > n$. $\Box^0 H$ may be defined as an abbreviation for $\Box^1 Hv - \Box^1 H$. Among the axioms we have $\Box^i H \rightarrow H$ where $i \geqslant n$; $\Box^j H \rightarrow \Box^i H$ where $j > i$; and, quite generally, $\Box^i(H \rightarrow I) \rightarrow$ $\rightarrow \Box^i(\Box^i H \rightarrow \Box^i I)$. $s[H] \geqslant i$ is defined as $\Box^i H$; $s[H, E] \geqslant i$ as $\Box^n(E \rightarrow \Box^i H); s[H] \leqslant i$ as $- \Box^j H$, and $s[H, E] \leqslant i$ as $- \Box^n(E \rightarrow \Box^j H)$, where $i = j - 1$; and so on. A large number of theorems may then be proved about the logical syntax of inductive support, without generating any paradoxes.

2.1. This system is not offered as an axiomatisation of a philosopher's intuitions about inductive support, since these are of dubious validity; cf. my (1975b). It is founded instead on arguments from the actual judgements of reputable experimentalists like von Frisch (1950) (as in 1.1.) and the scientists discussed by Lakatos (1970) (as in 1.18.).

2.2. To make sense of the fact that this syntax treats grades of inductive reliability as steps on a staircase that mounts towards necessary truth one can think of the matter, in the following way, as a scale of invariance. In accordance with the method of relevant variables and the way of modifying generalisations described in 1.19., a generalisation U_2 has a higher grade of inductive reliability than a generalisation U_1 iff U_2 remains true under more kinds of inductively relevant replacement of its non-logical terms than does U_1. If U_2 remains true under all such replacements, then $s[U_2] \geqslant n$, and U_2 states what is often called a law of nature –

a law that, within its own specified domain, no factors of any kind can prevent from operating. And a proposition is logically true if it remains true under all uniform and grammatical replacements of its non-logical terms (including those terms that specify, implicitly or explicitly, its domain of discourse). It might also seem worthwhile to describe this scale of necessity in terms of truth in wider and wider circles of so-called 'possible worlds'. But the point does not depend on the intelligibility of that kind of description, which is controversial.[4] Nor is it any philosophical help here to construct a formal semantics for the logistic system mentioned in 2 above. The philosophical claim is not that some formal model for a certain calculus exhibits a certain characteristic, but that a concept of inductive support, which is actually used in experimental science, exhibits it.

2.3. Many writers on inductive logic have confined its application to evaluations of support for natural-scientific hypotheses and their surrogates. But the method of relevant variables, and the logical relations it generates, have a much wider application. So any monadic or dyadic function which has a syntax like that described in 2 above for inductive *support* will here be termed an inductive function. And the following are some important varieties of inductive function (2.31.–2.325.).

2.31. The different versions of a generalisation that were described in 1.19.–1.193. may be graded in their order of simplicity. A generalisation becomes less simple as it gets qualified in regard to more relevant variables. So we can conceive of a simplification-function that grades evidentially permissible simplification. Where $s'[H, E]$ is such a function, it determines the maximum grade of simplicity E allows a version of H to have if this version of H is fully supported by E. There is an equivalence between simplification-functions and support-functions, such that for any H, E and $i, s'[H, E] = i$ iff $s[H, E] = i$. But the two functions have different roles to play in our reasoning. This is particularly noticeable in regard to questions about acceptability. The simplification-function directs attention to the question of how much simplicity has to be sacrificed by a man who wants to accept some hypothesis about a problem but will only accept one that he believes to be fully supported, while the support-function directs attention to the question whether the available grade of evidential support is high enough for some less demanding threshold of acceptance.

2.32. Inductive functors can take a non-standard interpretation in regard to the permissible types of filler for their argument-places, as well as in regard to the type of grading (cf. 2.31.) that they signify, because the method of relevant variables can be applied not only to the hypotheses of experimental science, as in 1.1.–1.193. above, but also to hypotheses about other subject-matters. One can compare the way in which Lakatos's theory of research-programmes is applicable not only to scientific enquiry (as in 1.18. above), but also, as he saw (cf. his (1971)), to philosophy and historiography.

2.321. For any particular branch of a legal system that appeals to precedents it is possible to determine a common list of relevant variables – sets of circumstances that under certain conditions constitute exceptions to general rules of the form

> For any persons x and y, if x has R to y, then x has a good cause of action against y (i.e. if x sues y, x ought to win)

or of a similar but more complex form. On the basis of such a list a rule of law can be graded by an inductive function in the light of recognised precedents: cf. my (1970a, p. 155ff.).

2.3211. It is sometimes said that legal reasoning from judicial precedent cannot be inductive, because it often proceeds by analogy from single precedents to single conclusions and not by generalisation from a multiplicity of instances. But in well-understood circumstances (*i.e.* when the *ratio decidendi* of the previous case is clear) neo-classical induction permits argument from a single instance, as in 1.19. above; and if all relevant circumstances are specified it does not matter whether the conclusion is general or singular, since a first-order generalisation has just the same grade of inductive support as each of its substitution-instances, as in 1.255 – the natural justice of *similia similibus*.

2.3212. It is sometimes said that legal reasoning is often concerned more to evaluate the strength, or width of application, of an unquestionably valid rule, than to evaluate the level of support that exists for a rule of dubious validity. But an inductive simplification-function (cf. 2.31. above) achieves the former just as well as an inductive support-function achieves the latter.

2.322. If we suppose an office for conscience, or moral sensitivity, that is analogous to the function of judicial precedent or scientific experiment, we can apply inductive reasoning to general issues in each of the various fields of ethical discussion such as those of personal, commercial, professional, or international morality.

2.3221. The principle of 1.255., which is closely connected with causal uniformity in experimental science (cf. 1.26.), and with natural justice in legal reasoning from precedent (cf. 2.3211. *ad fin.*), appears now as the principle of ethical universalisability.

2.3222. Within certain limits it is also possible to use an inductive function to grade the moral wrongness of an act, so far as this varies with the width of application of the rule forbidding it: cf. my (1970a, p. 175ff.).

2.323. Inductive functions have various applications in connection with language.

2.3231. They can be used to appraise the grade of support that exists for grammatical hypotheses, whether elementary or theoretical: cf. my (1970a, p. 177).

2.3232. They can be used (analogously to 2.3221.) for grading grammaticalness, as in my (1965), in a way that does not relate it to mathematical probability (cf. 4.1.–4.112. below).

2.3233. They can be used, as in my (1970b) and (1973d), to elucidate how a child could learn the syntax of its language in a relatively short space of time, without having any innate mechanisms that are concerned solely with syntax-learning. A child's rejection of degenerate data in syntax-learning is precisely analogous to an experimental scientists' construction of idealised, fact-correcting generalisations as in 1.19.

2.3234. Like things should normally be named or described alike (cf. 1.255. and 2.3221.). But some aspects of resemblance are relevant to naming and some are not. Semantical categories may therefore be viewed as the relevant variables for evaluating hypotheses about how to name or describe things.

2.32341. Such a hypothesis, however, may need qualification, as in 1.19., in respect of many semantical categories. A child's word-learning is often a matter of progress in learning these qualifications. It is not that any large non-person is named a motorcar, for example, but that any artefact that is terrestrial, mobile, carries passengers, etc. etc. is named a motorcar.

2.32342. Also the need to order any list of inductively relevant variables is reflected in the fact that certain features are normally much more central than others to the meaning of a descriptive term. It is when one of these restrictive features is ignored in adult speech that we have a typical case of metaphor. Young children use metaphors without knowing that they are doing so: cf. my (1970c).

2.324. Inductive functions can be used, as in my (1970a), to appraise certain forms of non-demonstrative reasoning in mathematics which have been discussed by Polya (1954a) and (1954b) and by Lakatos (1964).

2.325. The argument in support of a philosophical theory may also sometimes have an inductive structure. For example, a philosophical theory of probability needs to be able to account for several types of probability-judgement that are, at least superficially, very different from one another, and its success in doing so is tested, as in 1.18., by its ability not merely to put a Procrustean gloss on what we already know but also to predict some hitherto unnoticed type of truth. By this standard it is of little value to claim that the underlying unity of the different familiar types of probability-judgement consists in their identity of mathematical structure or in their being interconnected by family-resemblances: cf. my (1975a). Similarly the whole neo-classical theory of induction may be viewed as a 'major research programme' in the study of synthetically based inductive reasoning, as desiderated by Lakatos (1974, p. 261).

3.

There is an old puzzle about the justifiability of induction that goes back to Hume (1739). The puzzle is about how there can ever be any reasonable argument from a statement of observed fact to a statement of as yet unobserved fact, since the former cannot logically imply the latter. A neo-classical analysis of induction produces a reformulation and resolution of the puzzle, as in my (1970a, p. 183ff.).

3.1. The puzzle is often formulated either as the problem whether (and, if so, how) an adequate reconstruction of inductive reasoning can operate from non-controversial premises according to solely formal-logical or mathematical criteria of validity, or as the problem whether (and, if so, how) inductive procedures can be presented in a form in which they are an inherently reasonable policy of action. In both cases the attempt is to reduce inductive reasoning to some other form of inference – deductive or practical, respectively – and both attempts have been notoriously unsuccessful. Induction shows no signs of being open either to validation, as it has been called, or to vindication.

3.2. It has also often been claimed that, since induction sets up its own standards of justification, it is not itself open to further justification by reference to other standards. But Hume's puzzle is about just this question – whether inductive inferences are properly regarded as modes of *reasoning* and inductive criteria are properly regarded as standards of *justification*.

3.3. So the heart of Hume's puzzle is concerned not with validation or vindication but with ratification – with how to find a philosophical ratification for the ordinary, vulgar convention, whereby terms like '... is a reason for expecting ...' are taken to be assignable, or their assignments gradable, in accordance with inductive criteria.

3.4. The solution of the puzzle lies in accepting the terms in which Hume posed it. Hume's paradigm of rationality for the inference from one statement Ra to another Sa was when Ra logically implies Sa. So what Hume regarded as rationality is just the limiting case – the most extreme form – of a connection between the antecedent and consequent of a generalisation $(x)(Rx \to Sx)$ which we have already seen to exist in increasingly closer and closer forms: cf. 2.2. above. If such a generalisation is logically true, it is maximally reliable, on a scale of reliability which is constituted by the increasing reasonableness of inferring from its antecedent to its consequent in particular cases. It is not that what we have to ratify is a popular belief that all inferences from the observed to the unobserved are equally reasonable. Rather, the belief to be ratified is that there are gradations of reasonableness whereby some such inferences are more reasonable than others. So it is natural to treat deducibility as a limiting case here, both in view of the scale of invariance described in 2.2., and also in view of the fact that, in accordance with 2., all the necessary-but-not-sufficient conditions of a proposition's having such-or-such a grade of inductive support or reliability are also necessary-but-not-sufficient conditions of its being logically true. It would be about as absurd to restrict the title of 'reasoning' to valid deductions and exclude less reliable inferences, as it would be to restrict the title of 'number' to integers and exclude $\frac{3}{4}$, $\sqrt{2}$, $\sqrt{-1}$, and so on.

3.5. What has been discussed in 3.4. is the rationality of inference from the antecedent to the consequent of an inductively supported generalisation. So perhaps it will be asked why the method of relevant variables should be regarded as a justifiable way to evaluate the grade of reliability, or reasonableness, of a contingent generalisation. But the question asked is too general. What should be asked instead is why some particular criteria of evaluation for a support-function are reasonable ones in relation to generalisations of a certain category. And the answer to that lies in the reliability or reasonableness of the higher-level hypothesis which specifies a certain common list of relevant variables for all generalisations of that category. There is no need to assert, like Mill (1843, bk. III, ch. XXI, Sect. 4), that belief in the uniformity of nature is inductively justifiable, or, like Keynes (1921, p. 260), that belief in the limited independent variety of nature is inductively justifiable, or to suppose, like Lakatos (1974, p. 260–1), that inductive reasoning must be based on a synthetic fundamental principle that is irrefutable and purely conjectural. Such excessively general hypotheses are quite untestable, and are therefore incapable of acquiring inductive support from observable evidence. What are needed instead are inductively testable hypotheses about which variables are relevant in each particular field of enquiry; yet higher-level inductively testable hypotheses about the relevant variables for testing *those* hypotheses; and so on, just so far as the challenge to reasonableness proceeds. If the general method of relevant variables does deserve to be called a reasonable procedure, it is merely because this is *a* method of grading reasonableness in the sense of 3.3. (even if there are other such methods).

3.6. Hume's puzzle was about the rationality of inductive reasoning in regard to observable facts of nature. But the ratification of induction that has been sketched in 3.4.–3.5. applies, *mutatis mutandis*, to inductive reasoning about precedent-based legality, about conscience-based morality, and any other subject-matter like those mentioned in 2.32.–2.325. So this ratificatory resolution of Hume's puzzle tells us nothing special about nature, and treats the puzzle as being just one form of a more general problem. If there is a special problem of induction about nature, any solution it has must lie primarily in some scientifically discoverable fact about nature, rather than in some philosophically recognisable fact about the structure of inductive reasoning.

4.

It is fruitful to ask the question: what has inductive reasoning to do with probability? But the question has a deep-seated ambiguity, depending on the type of definition provided for probability.

4.1. According to one definition, accepted in my (1966b), (1970a) and (1973a), a probability-function is one that satisfies the axioms of the mathematical calculus of chance, when this calculus is set up as a purely formal system, like the system described by Popper (1938). The question posed in 4. may then be reformulated as: can the logical syntax of inductive reasoning that is generated by the method of relevant variables (as in 1.1.–1.27.) be mapped on to the mathematical calculus of chance? It is obvious, in view of 1.253., 1.254. and 1.27., that monadic and dyadic inductive functions do not have the same logical syntax as monadic and dyadic probability-functions, respectively. But perhaps dyadic inductive functions have the same logical syntax as some function of probability-functions, like the difference, say, or the ratio, between a prior and a posterior probability?

4.11. There are at least two quite distinct arguments to show that this is not so. However, both arguments begin by showing that for any H and E, $p[E, H]$, $p[H]$ and $p[E]$ are the only probabilities with which we need to concern ourselves in relation to $s[H, E]$, wherever, as normally, $1 > p[E] > 0$.

4.111. The first argument, which is too lengthy to repeat here, then proceeds by showing that if a support-function that satisfies 1.253. has to be a function of $p[E, H], p[H]$ and $p[E]$, it is subject to unviably restrictive conditions. The argument was first formulated in my (1966b), and its formulation has been revised successively in my (1970a), (1973a) and (1976a). The criticisms that have been made of its earlier formulations do not touch its hard core, and the version offered in my (1976a) is immune to all these criticisms.

4.112. A second, much shorter and simpler argument is based on 1.27. Consider two logically independent first-order generalisations U and U', such that $p[U] = p[U']$. Let E report, as it might well do, an unfavourable result of test t_i on U, plus a favourable result of test t_i on U' and an unfavourable result of t_j on U', where $j > i$. We shall then have $s[U', E] > s[U, E]$, as in 1.27., while $p[E, U] = p[E, U'] = 0$ (because E contradicts both U and U') and $p[U]=p[U']$. So we have a by no means abnormal type of case in which the grades of inductive support for two generalisations are different while each pair of mathematical probabilities concerned are identical. Therefore the former cannot be a function of the latter.[5]

4.113. Note that, while the argument mentioned in 4.111. established that the logical syntax described in 2. cannot be mapped on to the mathematical calculus of chance, the argument set out in 4.112. establishes that if support-functions did have a logical syntax that could be mapped on to that calculus it would not be possible to evaluate them in accordance with the method of relevant variables. In the former argument it is primarily the logical syntax of inductive functors which is invoked to show that grade of inductive support cannot be a function of mathematical probabilities, in the latter it is primarily their semantics.

4.2. The arguments of 4.111. and 4.112. assume a formalist definition of probability. But no philosophical theory of probability can rest content with such a definition because it is insufficiently explanatory. It does not elucidate why there should be the familiar diversity of proposed criteria of probability – relative frequency, degree of belief, logical relation, etc. – each satisfying the axioms of the mathematical calculus in its own way and each with some substantial utility or interest of its own. For this purpose we need an umbrella semantics for the term 'probability' that generates, in some readily intelligible fashion, the familiar variety of more specific criteria and has its validity confirmed by the fact that it also generates some other criterion, which has not been noticed previously by philosophers but has a demonstrable role to play in human judgements (as in 2.325. above). If we had such a theory we could interpret the question 'What has inductive reasoning to do with probability?' in a different sense than 4.1. An appropriate theory of probability is developed in my (1975a) and (1976a), and may be sketched, in rather bare outline, as follows (4.21.–4.254.). The question posed in 4. then receives, in 4.3., a different answer from that of 4.1.

4.21. In any artificial language, or system of thought, an item S is provable from R iff there is a primitive or derived rule that licenses the immediate derivation of S from R. Let us speak of such a rule as being inferentially sound, in an interpreted system L, if and only if the conclusion of any derivation which it licenses is true whenever the premise or premisses are. Then the hypothesis to be considered is that to grade the probability of S on R, according to the criterion of L, is to talk qualitatively, comparatively, ordinally or quantitatively about the degree of inferential soundness of the rule that would entitle one to derive S from R in L.

4.22. If we do not confine our attention to the rather limited variety of derivation-rules that is normally discussed in metamathematical proof-theory, we can ask at least three important questions about the rules of any deductive system. Are they general, setting up proof-schemas, or singular, legitimating inferences only from one specific item to another? Are they necessary, like laws of logic or arithmetic, or contingent, like rules read off from some well-established scientific theory? Are they extensional, as in number theory, or non-extensional, as in the logic of indirect discourse?

4.23. All the main philosophically familiar criteria of probability can be categorised within a matrix determined by the three binary questions of 4.22., whether we regard ourselves as *discovering* the meanings of natural-language sentences about probability, or with *inventing* meanings for the formulas of the mathematical calculus of chance.

4.231. If probability-functions correspond to general rules of inference their arguments are predicables, and when they correspond also to rules that are extensional and necessary, the probability can be graded by the corresponding rule's success-ratio. That is, the probability can be assessed *a priori* by a ratio between class-membership sizes, as typically for games of chance.

4.232. When probability-functions correspond to rules of inference that are general, extensional and contingent, the probability is normally an empirically estimatable relative frequency.

4.233. If probabilities are ever natural propensities, and therefore predicatable distributively of individual objects (like sub-atomic particles), they are determined by criteria corresponding to rules of inference that are general, non-extensional and contingent (like causal laws).

4.234. If probabilities are measured by range-overlap (as in Carnap (1950)) or by some other logical relation between propositions, they are determined by criteria corresponding to rules of inference that are singular and necessary.

4.235. If probabilities are degrees of belief, as measured by acceptable betting-odds within a coherent betting policy, they are determined by criteria corresponding to rules of inference that are singular, non-extensional and contingent.

4.24. But a fourth important question, also, arises about any set of inference-rules: is the deductive system to which they belong complete or not? So what kind of criteria for degree of inferential soundness would be analogous to a set of derivation-rules that ensured a system's completeness as to provability?

4.241. Where the system contains negation and is consistent as well as complete, a closed wff S is provable from the axioms R iff not-S is not provable. The analogous thesis for probability will be $p[S, R] = 1 \rightarrow$ $\rightarrow p[-S, R] = 0$, on the assumption that R is consistent and that all values for the probability-functor are in the closed interval $[0, 1]$.

4.25. It follows that in the probabilistic analogue of an *in*complete system, the complementational principle for negation (as in 4.241) will *not* operate. There will be some R and some S such that $p[S, R] = 0 =$ $= p[-S, R]$. But has such a non-Pascalian probability any role to play in human judgement? There are reasons (4.251.–4.254.) for supposing that it has.

4.251. What Keynes (1921) called evidential weight was in fact a measure of inferential soundness in an incomplete system. One argument, on his view, has more weight than another if based on a greater amount of relevant evidence, while it has more probability if the balance in its favour, of what evidence there is, is greater than the balance in favour of the argument with which we compare it. But we can in fact evaluate the probability of S on R in accordance with the weight of R if what we grade, where on balance R favours S rather than not-S, is the amount of relevant evidence stated by R. By such a criterion, if R is consistent and $p[S, R] > 0$, $p[-S, R] = 0$, but the converse will not hold: i.e. the negation principle will be non-complementational.

4.252. There are at least six reasons (4.2521.–4.2527.), which are discussed at length in my (1976a), for thinking that the concept of probability invoked by the two main standards for proofs of fact in Anglo-American lawcourts does not conform to the principles of the mathematical calculus of chance. It should be borne in mind that, while in criminal cases Anglo-American law requires the prosecutor to prove his charge at a level of probability which puts the matter beyond reasonable doubt, in civil actions he is only required to prove his case on the balance of probability. It is not correct to claim, as does Suppes (1970, p. 8), that "it is characteristic of legal analysis, as well as of classical physics, not to be satisfied with open-ended probabilistic results". But the probabilities with which we are here concerned are those invoked in calibrating the strength of the connection between admissible evidence and proposed conclusion, not those which sometimes (when sworn by experts) form part of the evidence. It should also be emphasised that we are concerned throughout with proofs of fact, not with reasoning about matters of law as in 2.321.–2.3212. above.

4.2521. A multiplication principle for the probability of conjunctions would impose intolerable constraints on a plaintiff in a civil suit who needed to establish several different points for his case to succeed.

4.2522. Such a principle would make it much easier than judges normally allow for a proof to be successful by inference upon inference – from R to S and from R & S to T and therefore from R to T.

4.2523. A complementational principle for negation enables a man to win a civil suit (by a proof on the balance of probability) even when there is a substantial probability – perhaps 40% – that the defendant is in the right.

4.2524. The reason for thinking a man's guilt has not been proved beyond reasonable doubt is normally a loop-hole in the evidence, rather than an insufficiently high mathematical probability.

4.2525. None of the familiar criteria for mathematical probability are applicable to the proof of individual facts about the past.

4.2526. If testimonial corroboration, or the convergence of circumstantial evidence, is to be measured by a mathematical probability, the conclusion must have a greater-than-zero prior probability: cf. my (1976b). But this is unacceptable in the adversary procedure of Anglo-American law. Proofs may not invoke facts that are not before the court and therefore not open to challenge.

4.253. It is clear that when the balance of evidence before an Anglo-American court favours a certain conclusion, the strength with which that conclusion is established is taken to depend on the extent to which the evidence before the court constitutes all the relevant facts. The probability concerned is thus analogous to inferability in an incomplete system – which explains why it does not conform to the principles of the mathematical calculus of chance.

4.254. Nor is this just a technical, legal concept of probability, since jurors are required to use the same standards of probability in deciding cases within the juryroom, as they use in their everyday affairs outside.

4.3. We are now justified (by 4.21.–4.27.) in interpreting the question 'What has inductive reasoning to do with probability?' in another sense than that adopted in 4.1. The concept of probability is not now to have a syntactic definition, in terms of the mathematical calculus of chance, but a semantical one, as a gradation of inferential soundness in accordance with 4.21.–4.25. Consequently it can now be said, in answer to the question, that there is a type of probability, analogous to inferability in an incomplete system, which may be reconstructed within the neo-classical theory of inductive reasoning. The details are given in my (1976a), and may be summarised as follows (4.31.–4.382.).

4.31. The inductive probability of Ra on Sa is equal to the inductive reliability of (grade of monadic support for) the generalisation that for any x, x's being R is a cause, or a sign, of its being S.

4.32. In any field of enquiry, like human conduct, for which it is difficult to discover simple and precise laws, inductive probabilities can be increased if we are prepared to qualify their covering generalisations, as in 1.19. above. So in such a field Sa may well be more probable on the evidence of Ra & V_1a & V_2a than on that of Ra alone.

4.33. The evidence reported by the second argument of a two-place inductive probability-function $p_I[S, R]$, should not be confused with the experimental evidence reported by the second argument of a two-place inductive support-function $s[H, E]$, as in 1.16. Here R abbreviates Ra, and S abbreviates Sa.

4.34. Correspondingly the logical syntax of dyadic inductive probability-functors is not wholly the same as that of dyadic inductive support-functors, though its principles are all derivable from the same axioms. The following examples may be mentioned (4.341.–4.344.).

4.341. The conjunction principle is analogous to 1.253.

4.342. The negation principle is analogous to 1.254., in conformity with 4.25. and 4.251.

4.343. But, where R contradicts S, we have $p_I[S, R] = 0$, which is not analogous with 1.27., though there is, of course, a corresponding principle for mathematical probability.

4.344. Also we have a principle of symmetry. Assessments of inductive probability are invariant under all uniform transformations of their references to individuals. An analogous principle is to be found in Carnap (1950) p. 483ff.

4.345. If we define a monadic, or prior, inductive probability function $p_I[S]$ as $p_I[S, T]$ where T is tautological, it can be proved that $p_I[S] = s[S]$.

4.35. It can be shown that an analysis of juridical proofs of fact in terms of inductive probability obviates quite smoothly each of the difficulties (cf. 4.2521.–4.2526) that are encountered by an analysis in terms of mathematical probability.

4.36. The concept of inductive probability has several other important uses, e.g. 4.361.–4.363. [6]

4.361. It can be used to grade the strength of historical explanations that tacitly invoke covering laws.

4.362. It can be used to determine a threshold for justifiable belief, or rational acceptance, in a way that avoids the well-known difficulties encountered by mathematical probability in this connection. After all, we are hardly prudent to adopt an acceptance-criterion that does not concern itself at all with how much of the relevant facts we know; and, if a mathematicist criterion concerns itself with this issue by requiring us always to know *all* the relevant facts, it is not coming to grips with the actual nature of the problem in everyday life.

4.363. It can be used to grade the strength of a disposition, where analysis in terms of mathematical proability conflicts with commonplace assumptions about the compounding of dispositions.

4.37. For every statement about $p_I[S, R] < n$ there is a corresponding statement about the certainty, if R is true and all other relevant factors are favourable, that S is true. But the strength of the proviso about the other relevant factors here varies inversely with the grade of probability involved, as in 1.192. and 2.31.

4.38. Though it undoubtedly satisfies some objectives for which no concept of mathematical probability is readily suited, the concept of inductive probability suffers from certain comparative disadvantages (4.381.–4.382.), which it inherits from the parent concept of inductive support.

4.381. Inductive probability-functions map pairs of propositions on to integers (cf. 1.13.), not real numbers, because inductive probability is non-additive.

4.382. Cross-field comparisons of inductive probability are heavily restricted, as in 1.21.

5.

From an overall consideration of the neo-classical theory of inductive reasoning certain general, epistemological conclusions emerge.

5.1. The inductive reliability of a generalisation depends on its resistance to falsification by variants of relevant variables. But this resistance may be understood in one or other of two quite different ways – 5.11. and 5.12.: cf. my (1976a).

5.11. According to a realist interpretation the resistance may exist even if never manifested. This is because of facts about physical causation, say, or because judicial precedents declare pre-existing customary law, as in Blackstone (1765).

5.12. According to a nominalist interpretation the resistance exists only so far as it is manifested. This jibes with a truth-functional analysis of causal laws, and with the view advocated by Austin (1832) that judicial precedents create legal rules.

5.2. Juries have accepted that guilt has been proved beyond reasonable doubt in innumerable cases. In terms of an inductivist analysis this means that the inductive probabilities in question stem from generalisations that are (implicitly) taken to be fully supported. So an inductivist analysis of the standards of proof in Anglo-American courts is not compatible with a philosophy of science which declares inductive knowledge to be unobtainable and all science to be guesswork: cf. my (1974b). Of course it may very rarely happen that any scientist accepts a generalisation that is in fact fully reliable, and it may be still rarer that he has obtained the right evidence to show this. But if he has done both things then he knows the truth of the generalisation, even if, because he does not know that he has done both things – perhaps he does not know that his list of relevant variables is complete –, he does not know that he knows the truth of the generalisation.

5.3. Thus the thesis that if a man knows, then he knows that he knows, is inapplicable to inductive knowledge, where at each level reasonable belief is at least possible even when there is inadequate evidence for fully supported belief about the appropriate series of relevant variables. The thesis that if a man knows, then he knows that he knows, stems instead from a foundationalist epistemology which postulates a single sure basis (Cartesian clear and distinct ideas, say, or Carnapian sense data) for all knowledge and a single sure criterion for evaluating the relation of scientific theories to this basis (deducibility or analytic confirmation-assessments, respectively).

5.4. Because the pretensions of foundationalism are easily exposed it tends to provoke a sceptical reaction (Hume against the seventeenth-century rationalists; Kuhn, Feyerabend, Toulmin, etc. against the twentieth-century logical empiricists). But neo-classical inductivism is immune to these sceptical challenges: cf. 3.–3.6. and 5.2. above, and my (1973b) and (1975c).

5.5. The need to invoke empirical criteria for the validity of a fundamental explanation in natural science is forced on us by the inductive desire to make our explanations as unified and comprehensive as possible – to make the same fundamental theory cover both observable and unobservable events. For then any theory involving unobservable events has to have empirically testable consequences. So it is inductive logic that, at bottom, justifies empiricism in the epistemology of scientific theory, not empiricism that needs to justify the use of inductive logic or to demonstrate the rationality of inductive reasoning: cf. my (1974a).

The Queen's College, Oxford

NOTES

[1] It is not necessary to say 'affected' here, instead of 'characterised', as I did in (1970a) pp. 54, 56, 147, 148 etc.

[2] Mill (1843, bk. III, ch. III, sect. 3 *ad fin.*) attached great importance to the solution of this problem. He wrote: "Why is a single instance, in some cases, sufficient for a complete induction, while in others myriads of concurring instances, without a single exception known or presumed, go such a very little way towards establishing an universal proposition? Whoever can answer this question knows more of the philosophy of logic than the wisest of the ancients, and has solved the problem of Induction". Perhaps Mill exaggerated. But it is a pity that so many modern writers on inductive logic have ignored the problem: cf. in particular Carnap's 'principle of instantial relevance', which is discussed in 1.26. below.

[3] This could not be said of the method of relevant variables as it was described in (1970a), since there the series of relevant variables for a particular category of generalisations was mistakenly allowed to include exhaustive variables. What I now require (cf. 1.12. above and (1976a)) is that for each relevant variable in the series there must be some logically possible circumstance that is incompatible with any of its variants.

Typically this is a non-relevant circumstance for the category, like perhaps still air is for meteorological predictions where the relevant variable is wind-direction.

[4] Lewis (1973, p. 84ff.), has tried to give a realist, as distinct from a linguistic, account of possible worlds. But his account rests quite specifically (p. 85) on the assumption that 'world' is a sortal concept, and this is an important part of what is disputed. A world here is not a continent, or a galaxy, but a possible totality of facts. So to take a world for an *entity* of a certain *sort*, as Lewis does, is like taking what Kant called a regulative idea for what he called a constitutive one – a typically metaphysical manoeuvre. Indeed, for one kind of determinist (e.g. Spinoza) the idea of alternative possible worlds is self-contradictory. A purely linguistic account of possible worlds, which treats sentences about them as material-mode equivalents for formal-mode sentences about state-descriptions, escapes these difficulties. But such a semantics achieves no more than the mapping of one formal language onto another.

[5] In my (1970a, p. 50) I wrongly assumed that this argument could be side-stepped by tying $s[U, E]$ not to $p[U, E]$ but to $p[P, E]$, where P is a substitution-instance of U that does not refer to any individual referred to by E, as in Carnap (1950)'s theory of instance-confirmation. That escape-route is in fact blocked by the fact that in accordance with 1.252. and 1.255. we can have $s[P, E] > 0$, when E contradicts P, even if E refers to the individual or individuals referred to by P: a well-supported prediction may turn out false.

[6] Strictly speaking there is both a strong and a weak concept of inductive probability, depending on whether or not the nature of the evidence is crucial to the probability. Strong inductive probability is as defined in 4.31 and applies as in 4.35 and 4.361. The weak inductive probability of Ra on Sa is equal to the inductive reliability of the generalisation that for any x, if x is R, then x is S. Weak inductive probability is appropriate in 4.362 and 3.463.

BIBLIOGRAPHY

Austin, John: 1832; *The Province of Jurisprudence Determined.*

Bacon, Francis: 1620; *Novum Organum.*

Blackstone, William: 1765; *Commentaries on the Laws of England*, Vol. I.

Carnap, Rudolf: 1950; *Logical Foundations of Probability*, Routledge and Kegan Paul, London.

Carnap, Rudolf: 1971, 'A Basic System of Inductive Logic', in R. Carnap and R. C. Jeffrey (ed.), *Studies in Inductive Logic and Probability*, University of California Press, Berkeley, pp. 33–165.

Cohen, L. Jonathan: 1962, *The Diversity of Meaning*, Methuen, 2nd ed., London, 1966.

Cohen, L. Jonathan: 1965, 'On a Concept of Degree of Grammaticalness', *Logique et Analyse* 8, 141–153.

Cohen, L. Jonathan: 1966a, 'A Logic of Evidential Support', *British Journal for Philosophy of Science* 17, 21–43 and 105–126.

Cohen, L. Jonathan: 1966b, 'What has Confirmation to Do With Probabilities', *Mind* 75, 463–481.

Cohen, L. Jonathan: 1968, 'An Argument that Confirmation Functors for Consilience are Empirical Hypotheses', in I. Lakatos (ed.), *The Problem of Inductive Logic*, North-Holland, Amsterdam, pp. 247–250.

Cohen, L. Jonathan: 1970a, *The Implications of Induction*, Methuen, London. Paperback, with corrections, 1973.

Cohen, L. Jonathan: 1970b, 'Some Applications of Inductive Logic to the Theory of Language', *American Philosophical Quarterly* 7, 299–310.

Cohen, L. Jonathan: 1970c, with Avishai Margalit: 'The Role of Inductive Reasoning in the Interpretation of Metaphor', *Synthese* 21, 468–487; reprinted with corrections in *Semantics of Natural Language* (ed. by D. Davidson and G. Harman), Reidel, Dordrecht-Holland, 1972, pp. 722–740.

Cohen, L. Jonathan: 1971, 'The Inductive Logic of Progressive Problem-Shifts', *Revue Internationale de Philosophie*, No. 95–96, 1971, pp. 62–77.

Cohen, L. Jonathan: 1972, 'A Reply to Swinburne', *Mind* 81, 249–50.

Cohen, L. Jonathan: 1973a, 'A Note on Inductive Logic', *Journal of Philosophy* 70, 27–40.

Cohen, L. Jonathan: 1973b, 'Is the Progress of Science Evolutionary' (review of S. Toulmin, *Human Understanding*, Vol. I, 1972), *British Journal for Philosophy of Science* 24, 41–61.

Cohen, L. Jonathan: 1973c, 'The Paradox of Anomaly', in R. J. Bogdan and I. Niiniluoto (eds.), *Logic, Language and Probability*, Reidel, Dordrecht, pp. 78–82.

Cohen, L. Jonathan: 1973d, 'Is Contemporary Linguistics Value-free?', *Social Science Information, Information sur les Sciences Sociales* 12, 53–64.

Cohen, L. Jonathan: 1974a, 'Why Should the Science of Nature Be Empirical?' (lecture to Royal Institute of Philosophy, London, October, 1974). Forthcoming in 1974/5 volume of *Royal Institute of Philosophy Lectures*, Macmillan, London.

Cohen, L. Jonathan: 1974b, 'Guessing', *Proceedings of the Aristotelian Society* 74, 189–210.

Cohen, L. Jonathan: 1975a, 'Probability – the One and the Many', Annual Philosophical Lecture to the British Academy, 1975, Oxford University Press, London. To be reprinted in *Proceedings of the British Academy* 61, 1977.

Cohen, L. Jonathan: 1975b, 'How Empirical is Contemporary Logical Empiricism?'. Forthcoming in *Philosophia*. Also forthcoming in *Memorial Volume for Yehoshua Bar-Hillel* (ed. by A. Kasher), Reidel, Dordrecht-Holland.

Cohen, L. Jonathan: 1975c, 'The Progress of Science' (paper read to Italian Society for Philosophy of Science at Chiavari, September, 1974). Italian translation forthcoming in *Il concetto di progresso nella scienza* (ed. by E. Agazzi), Feltrinelli, Milan.

Cohen, L. Jonathan: 1976a, *The Probable and the Provable*, forthcoming.

Cohen, L. Jonathan: 1976b, 'How Can One Testimony Corroborate Another?', Forthcoming in *Memorial Volume for Imre Lakatos* (ed. by R. S. Cohen, P. Feyerabend, and M. W. Wartofsky), in *Boston Studies in the Philosophy of Science*, Reidel, Dordrecht-Holland.

Goodman, Nelson: 1954, *Fact, Fiction and Forecast*, Athlone Press, London.

Herschel, J. F. W.: 1833, *A Preliminary Discourse on the Study of Natural Philosophy*.

Hesse, Mary: 1974, *The Structure of Scientific Inference*, Macmillan, London.

Hume, David: 1739, *A Treatise of Human Nature*.

Keynes, J. M.: 1921, *A Treatise on Probability*, Macmillan, London.

Lakatos, Imre: 1964, 'Proofs and Refutations', *Brit. Jour. Phil. Sci.* XIV, 1–25, 120–139, 221–245, and 296–342.

Lakatos, Imre: 1970, 'Falsificationism and the Methodology of Scientific Research Programmes', in I. Lakatos and A. E. Musgrave (eds.), *Criticism and the Growth of Knowledge*, Cambridge U.P., Cambridge, pp. 91–195.

262   L. JONATHAN COHEN

Lakatos, Imre: 1971, 'History of Science and Its Rational Reconstructions', in R. C.
    Buck and R. S. Cohen (eds.), *Boston Studies in the Philosophy of Science*, Vol. VIII,
    Reidel, Dordrecht, pp. 91–136.
Lakatos, Imre: 1974, 'Popper on Demarcation and Induction', in P. A. Schilpp (ed.),
    The Philosophy of Karl Popper, Open Court, La Salle, pp. 241–273.
Lewis, David: 1973, *Counterfactuals*, Blackwell, Oxford.
Mill, J. S.: 1843, *A System of Logic*.
Miller, David: 1974, 'Popper's Qualitative Theory of Verisimilitude' and 'On the
    Comparison of False Theories by Their Bases', *Brit. Jour. Phil. Sci.* XXV, 160–188.
Niiniluoto, I. and Tuomela, R.: 1973, *Theoretical Concepts and Hypothetico-Inductive
    Inference*, Reidel, Dordrecht-Holland.
Pietarinen, J.: 1972, *Lawlikeness, Analogy and Inductive Logic*, North Holland,
    Amsterdam.
Polya, G.: 1954a, *Patterns of Plausible Inference*, Princeton U.P., Princeton.
Polya, G.: 1954b, *Induction and Analogy in Mathematics*, Princeton U.P., Princeton.
Popper, K. R.: 1938, 'A Set of Independent Axioms for Probability', *Mind* xlvii,
    275–277. (reprinted, with further developments, in (1959), pp. 318–358).
Popper, K. R.: 1959, *The Logic of Scientific Discovery*, Hutchinson, London.
Popper, K. R.: 1963, *Conjectures and Refutations*, Routledge and Kegan Paul, London.
Schwartz, R., Scheffler, I., and Goodman, N.: 1970, 'An Improvement in the Theory
    of Projectibility', *Journal of Philosophy* lxvii, 605–608.
Suppes, P.: 1970, *A Probabilistic Theory of Causality*, North-Holland, Amsterdam.
Von Frisch, K.: 1950, *Bees, Their Vision, Chemical Senses, and Language*, Cornell U.P.,
    Ithaca.
Whewell, William: 1840, *The Philosophy of the Inductive Sciences*.

ILKKA NIINILUOTO

INQUIRIES, PROBLEMS, AND QUESTIONS: REMARKS ON LOCAL INDUCTION

1. INTRODUCTION

Isaac Levi opens his *Gambling with Truth* (1967) with a distinction between global and local justification of beliefs. A 'globalist' wishes to justify the totality of beliefs held at a given time. Local justification is less demanding: it arises in the context of specific inquiries, and it appeals to corrigible statements which may later lose their evidential status.

Levi has a Peircean view of inquiry as a process which is hoped to relieve us from agnosticism by leading from doubt to true belief. This process is *local*, Levi argues, in the sense that it is relative to a given problem or question:

> ... justification of conclusions reached from given evidence ought to be made relative not only that evidence but to a characterization of what is to count as a 'relevant answer' to the problem raised. (Levi, 1967, p. 32.)

The relevant answers to a given problem are those "eligible for acceptance as strongest via induction from the evidence", and their set is determined by an 'ultimate partition' (*op. cit.*, pp. 32–35).

> Ultimate partitions have been introduced to take cognizance of the obvious fact that investigators do not entertain as relevant answers to their questions all those sentences whose truth values are undecided by the evidence available to them. Whether for any given context there is a system of criteria for determining the ultimate partition that ought to be used is a difficult question; it cannot be answered here. To attempt to answer this question would be like trying to decide whether there is some logic of questions that determines what is the right question to ask on any given occasion. (*Op. cit.*, pp. 36–37.)

Thus, Levi contends that investigators are always committed to some ultimate partition, even if he is not willing to propose any criteria for selecting them.

Levi's allusion to a 'logic of questions', in the above passage, is very interesting. There exists by now a considerable amount of literature, both by logicians and linguists, about the 'erotetic logic' and the grammar of

R. J. Bogdan (ed.), Local Induction, 263–296. All rights reserved.
Copyright © 1976 by D. Reidel Publishing Company, Dordrecht-Holland

interrogatives.[1] On the other hand, several philosophers have suggested that the methods of science should be analyzed as procedures for problem-solving.[2] In this context, 'problem' does not refer only to practical decision problems but also to cognitive problems which require new information for their solution. As Levi himself has made clear, the idea of science as problem-solving is attractive also to a 'cognitivist' who regards science as a systematic attempt to look for informative truths. Scientists request information by questions, and the results of their inquiry can be viewed as answers to questions that have been asked in scientific problem-situations.

If the idea of science as a problem-solving activity is not a mere metaphor, it is reasonable to expect that the logic of questions will prove to be a valuable tool for a philosopher of science.[3] For example, scientific explanations can be viewed as answers to why-questions (and how-questions).[4] It is the crucial idea of the experimental method in science that, instead of passively observing the course of events, the investigator puts *questions to nature* through experiments.[5] From this viewpoint, the logic of explanation and the logic of experimentation are branches of a general logic of questions.

One of the most central topics in the logic of questions has been the problem of *answerhood*, that is, the problem what counts as a potential answer to a given question. This problem is generally different from the problem of 'questionhood', that is, the problem what one should ask in the first place on a given occasion. Levi is right, of course, in doubting the existence of a logic of questions which would effectively determine what is the right question to ask on any given occasion. The art of making questions is an essential aspect of the heuristics of science, and there seems to be no reason to suppose that it could be reduced to a number of mechanical rules. This by no means implies that the making or the designing of questions in science is arbitrary; rather, it is *guided* by the desired information and by the background knowledge of the questioner. For example, even if the logic of questions does not determine what to ask in a given problem-situation, it may still show the pragmatical inappropriateness of some questions.[6]

It is sometimes said that a well-formulated question already provides a half of the answer. What is really meant by this saying, I think, is that the problem of answerhood has an almost trivial solution for a class of 'well-formulated' questions which itself contains a list from which all potential answers can be effectively generated. As soon as we know a question of this type, we also know its potential answers. It is precisely these 'closed' questions that Levi has in mind: the function of his 'ultimate partitions' is just to provide us with the lists of basic answers. Levi's thesis that induction is always relative to an ultimate partition thus amounts to the claim that the results of induction are always answers to 'closed' questions.

But Levi does not only claim that inductive acceptance is always relative to an ultimate partition: he argues further that requirements of consistency and deductive closure and that measures of informational content should be made relative to ultimate partitions (*op. cit.*, pp. 37–38, 70–71). Some philosophers have agreed with Levi that the acceptability of hypotheses should depend on the size of the corresponding ultimate partition (see Schick (1970), pp. 18–19), while some others have argued that Levi's analysis of inductive acceptance is *too sensitive* to the choice of ultimate partition (see Hacking (1967); Hilpinen (1968), (1972)).

The principal aim of this paper is to embed the theory of inductive acceptance in a systematic view of the question-making and the answer-providing procedures in science, and thereby to show that Levi's acceptance rule is oversensitive to the choice of ultimate partition. The 'abstract' theory of ultimate partitions (or P-sets, as I call them) is developed in Section 2, and it is interpreted in terms of 'problems' in Section 5. Section 3 reviews some elements of the logic of questions, and Section 4 discusses the role of 'open' questions in science. The analysis of problems in Section 5 gives the background for the treatment of inductive acceptance in Section 6. The argument developed in Section 6 largely agrees with Hilpinen (1968), and it restates a part of his criticism of Levi's rule. I hope that my way of defending this criticism has some independent interest and applications beyond the scope of this paper. And I also hope that my remarks serve to clarify at least some aspects of the general question: In what sense, and to what extent, is induction *local*?

2. P-SETS

Throughout this section it will be assumed that logically equivalent sentences have been identified. In other words, it is supposed that all the languages to be considered are provided with an underlying logic, and that sentences are identified with (or replaced by) equivalence classes determined by the relation of logical equivalence. Thus, if h and k are sentences (i.e., equivalence classes), we may write simply

(1) $h = k$,

instead of

(2) $\vdash h \equiv k$.

The symbol 'T' is used for tautology (i.e., for the class of logically true sentences) and '$\perp$' for contradiction.

DEFINITION 1. A non-empty set $B = \{h_i \mid i \in I\}$ is a P-set, if I is finite and

(a) $\vdash \bigvee_{i \in I} h_i$,

(b) $\vdash \sim (h_i \,\&\, h_j)$ for all $i \neq j$,

(c)   not $\vdash \sim h_i$ for all $i \in I$.

Conditions (a), (b), and (c) can be written, according to our convention (1), in the following form:

(a') $\bigvee_{i \in I} h_i = T$,

(b') $(h_i \,\&\, h_j) = \perp$ for all $i \neq j$,

(c') $h_i \neq \perp$ for all $i \in I$.

A P-set is thus a non-empty finite set of mutually exclusive non-contradictory sentences precisely one of which is true. The set which contains only the tautology T is trivially a P-set.

DEFINITION 2. $\mathbb{O} = \{T\}$ is the *trivial P-set*.

The disjunctive closure of a P-set B is defined to be the set of all finite disjunctions of members of B.

DEFINITION 3. If $B = \{h_i \mid i \in I\}$ is a P-set, then the *disjunctive closure* of B is the set

$$D(B) = \{ \bigvee_{i \in I(h)} h_i \mid \emptyset \neq I(h) \subseteq I \}.$$

By Definition 3, each sentence h in $D(B)$ can be represented in the form

$$(3) \qquad h = \bigvee_{i \in I\,(h)} h_i.$$

where $h_i \in B$. Disjunction (3) is called the *normal form* of h relative to P-set B (or the B-normal form of h), and $I(h)$ is the *index set* of sentence h relative to B. The operation D has the following properties.

THEOREM 1. Let B and B^* be P-sets. Then

(a) $\qquad B \subseteq D(B),$

(b) $\qquad \mathbb{O} \subseteq D(B),$

(c) $\qquad D(\mathbb{O}) = \mathbb{O},$

(d) $\qquad D(D(B)) = D(B),$

(e) $\qquad B \subseteq D(B^*) \quad \text{iff} \quad D(B) \subseteq D(B^*).$

This theorem is an immediate consequence of Definitions 1–3.

The inclusion relation $\subseteq$ between the disjunctive closures of P-sets induces a natural (partial) ordering $\leqslant$ of the class of P-sets. By Theorem 1 (e), we know that $D(B) \subseteq D(B^*)$ if and only if $B \subseteq D(B^*)$. When this condition holds, B^* is said to be at least as *fine* as B; this is denoted by $B \leqslant B^*$. If $B \leqslant B^*$ and $B \neq B^*$, B^* is said to be *finer* than B.

DEFINITION 4. Let B and B^* be P-sets. Then

$$B \leqslant B^* \quad \text{iff} \quad B \subseteq D(B^*).$$

Let $B = \{h_i \mid i \in I\}$ and $B^* = \{h_j^* \mid j \in J\}$. By Definition 4, B^* is at least as fine as B if and only if each h_i in B can be expressed as a finite disjunction of sentences h_j^* in B^*. In other words, $B \leqslant B^*$ if and only if for all $i \in I$ there is a set $J(i) \subseteq J$ such that

$$(4) \qquad h_i = \bigvee_{j \in J\,(i)} h_j^*.$$

Disjunction (4) is the *representation* of $h_i \in B$ relative to the finer P-set B^*, and $J(i)$ is the *index set* of h_i relative to B^*. Note that if $B \leqslant B^*$, then the index sets $\{J(i) \mid i \in I\}$ constitute a partition of J into mutually exclusive and jointly exhaustive classes.

The relation $\leqslant$ has the following properties.

THEOREM 2. Let B, B^*, and B^{**} be P-sets. Then

(a) $\mathbb{O} \leqslant B$,

(b) $B \leqslant B$,

(c)   if $B \leqslant B^*$ and $B^* \leqslant B^{**}$, then $B \leqslant B^{**}$,

(d)   if $B \leqslant B^*$ and $B^* \leqslant B$, then $B = B^*$.

Proof. (a) follows from Theorem 1(b). Further, (b) and (c) follow, via Theorem 1(e), from the reflexivity and the transitivity of the inclusion relation $\subseteq$. Finally, from the assumptions of (d) it follows that $J(i) = \{i\}$ for all $i \in I$, i.e., that each sentence h_i in B is logically equivalent to some sentence of B^*. By our convention, this means that B and B^* are identical.

If $\mathscr{B}_L$ is the class of all P-sets in a language L, then by Theorem 2 relation $\leqslant$ is a *partial ordering* of $\mathscr{B}_L$ with the trivial P-set $\mathbb{O}$ as the minimal element. As $\leqslant$ is not connected in class $\mathscr{B}_L$, relation $\leqslant$ is not a linear ordering of $\mathscr{B}_L$.

The class of P-sets does not possess an ordinary set-theoretical structure, since the union and the intersection of distinct non-trivial P-sets is not a P-set. Still, there is a natural way of defining the least upper bound (sup) and the greatest lower bound (inf) of two P-sets.

DEFINITION 5. If $B = \{h_i \mid i \in I\}$ and $B^* = \{h^* \mid j \in J\}$ are P-sets, then the *combination* of B and B^* is the set

$$B \oplus B^* = \{h_i \ \& \ h^* \mid i \in I, j \in J\} - \{\bot\}.$$

It is easy to see that the combination $B \oplus B^*$ of two P-sets B and B^* is always a P-set. The operation $\oplus$ has the following properties.

THEOREM 3. Let B and B^* be P-sets. Then

(a) $B \oplus B^* = B^* \oplus B$,

(b) $B \oplus \mathbb{O} = \mathbb{O} \oplus B = B$,

(c) $B \leqslant B \oplus B^*$; $B^* \leqslant B \oplus B^*$,

(d) $B \leqslant B^*$ iff $B \oplus B^* = B^*$,

(e) $B \oplus B^* = \sup\{B, B^*\}$.

Proof. (a) holds, since $h_i \ \& \ h_j^* = h_j^* \ \& \ h_i$. Further, (b) holds, since

h_i & $T=T$ & $h_i=h_i$. To prove (c), it is sufficient to note that

$$h_i = \bigvee_{j \in J} (h_i \& h_j^*) \quad \text{for all} \quad i \in I.$$

To prove (d), assume first that $B \leqslant B^*$. By Formula (4), we have for all $i \in I$

$$
\begin{aligned}
h_i \& h_j^* &= h_j^* \quad \text{for all} \quad j \in J(i), \\
&= \perp \quad \text{for all} \quad j \in J - J(i).
\end{aligned}
$$

Moreover, if $j \in J(i)$, we have for all $k \in I, k \neq i$,

$$h_k \& h_j^* = \perp.$$

Therefore, $B \oplus B^*$ contains precisely the sentences of B^*. Assume then that $B \oplus B^* = B^*$. Now, if $h_i \& h_j^* \neq \perp$, then $h_i \& h_j^* = h_j^*$. The latter condition implies that

$$(5) \qquad \vdash h_j^* \supset h_i.$$

Formula (5) cannot hold for any $h_k \in B$, where $k \neq i$; for otherwise h_j^* would be a contradiction. Therefore, sentence h_i will be the same as the disjunction of all $h_j^* \in B^*$ which satisfy (5). Hence, $B \leqslant B^*$.

To prove (e), assume that $B \leqslant C$ and $B^* \leqslant C$. By Definition 4, we have $B \subseteq D(C)$ and $B^* \subseteq D(C)$. Now, if $(h_i \& h_j^*) \in B \oplus B^*$, then

$$(6) \qquad h_i \& h_j^* = \bigvee_{k \in K_1} c_k \& \bigvee_{k \in K_2} c_k$$

for some index sets K_1 and K_2, where $c_k \in C$ for all $k \in K_1 \cup K_2$. The right hand side of Formula (6) is an element of $D(C)$. Therefore, $B \oplus B^* \subseteq D(C)$, so that $B \oplus B^* \leqslant C$. Combination $B \oplus B^*$ is thus the least upper bound of B and B^*.

We shall next define a P-set $B \times B^*$ which generates the intersection of the disjunctive closures of P-sets B and B^*, i.e., $D(B \times B^*) = D(B) \cap D(B^*)$.

DEFINITION 6. Let $B = \{h_i \mid i \in I\}$ and $B^* = \{h^* \mid j \in J\}$ be P-sets. Then the *meet* of B and B^* is the set

$$B \times B^* = \{g \in D(B) \mid g = \bigvee_{i \in I(g)} h_i \in D(B^*) \text{ and for all proper}$$
$$\text{subsets } G \text{ of } I(g): \bigvee_{i \in G} h_i \notin D(B^*)\}.$$

By Definition 6, the meet $B \times B^*$ contains precisely those common elements of $D(B)$ and $D(B^*)$ which are 'minimal' with respect to the number of disjuncts in their normal form. This guarantees that $B \times B^*$ is a P-set. In particular, we always have

(7) $B \cap B^* \subseteq B \times B^*.$

If the only common element of $D(B)$ and $D(B^*)$ is tautology T, then P-sets B and B^* are said to be *isolated*. Thus, we have

(8) B and B^* are isolated iff $B \times B^* = \mathbb{O}.$

Note, however, that $B \cap B^* = \emptyset$ does not imply $B \times B^* = \mathbb{O}$.
    The operation $\times$ has the following properties.

THEOREM 4. Let B and B^* be P-sets. Then

(a) $B \times B^* = B^* \times B,$
(b) $B \times \mathbb{O} = \mathbb{O} \times B = \mathbb{O},$
(c) $B \times B^* \leqslant B; \qquad B \times B^* \leqslant B^*,$
(d) $B \leqslant B^*$ iff $B \times B^* = B,$
(e) $B \times B^* = \inf\{B, B^*\}.$

Proof. Let $B = \{h_i \mid i \in I\}$ and $B^* = \{h^* \mid j \in J\}$. If $g \in B \times B^*$, then

$$g = \bigvee_{i \in I(g)} h_i = \bigvee_{j \in J(g)} h_j^*$$

holds for some $I(g) \subseteq I$ and $J(g) \subseteq J$. If further

$$g' = \bigvee_{j \in J(g')} h_j^* \in D(B),$$

where $J(g')$ is a proper subset of $J(g)$, then correspondingly $I(g')$ is a proper subset of $I(g)$, and

$$g' = \bigvee_{i \in I(g')} h_i \in D(B^*).$$

This contradicts Definition 6, and concludes the proof of (a).
    To prove (b), note that by Theorem 1 (b) and (c)

$$D(B) \cap D(\mathbb{O}) = D(B) \cap \mathbb{O} = \mathbb{O}.$$

By Definition 6, $B \times B^* \subseteq D(B)$ and $B \times B^* \subseteq D(B^*)$, which implies (c). (d) is immediate from Definition 6. To prove (e), note first that

$$(9) \qquad D(B) \cap D(B^*) \subseteq D(B \times B^*),$$

since $B \times B^*$ contains the 'minimal' common elements of $D(B)$ and $D(B^*)$ and, hence, its disjunctive closure contains all elements in $D(B) \cap \cap D(B^*)$. If now $C \leqslant B$ and $C \leqslant B^*$, then $C \subseteq D(B) \cap D(B^*) \subseteq D(B \times B^*)$, so that $C \leqslant B \times B^*$. This concludes the proof of Theorem 4.

The converse of the inclusion (9) is valid, since $D(B \times B^*) \subseteq D(B)$ and $D(B \times B^*) \subseteq D(B^*)$ hold by Theorem 4(c). Similarly, Theorem 3(c) implies that $D(B) \subseteq D(B \oplus B^*)$ and $D(B^*) \subseteq D(B \oplus B^*)$. Thus, the following theorem is valid.

THEOREM 5. Let B and B^* be P-sets. Then

(a) $D(B) \cup D(B^*) \subseteq D(B \oplus B^*)$,
(b) $D(B) \cap D(B^*) = D(B \times B^*)$.

The converse of the inclusion (a) generally fails, however.

Theorems 3 and 4 together show that class $\mathscr{B}_L$, provided with the relation $\leqslant$, is a *lattice*, i.e., a partially ordered set such that each pair of its elements has supremum and infimum in $\mathscr{B}_L$.

THEOREM 6. $\langle \mathscr{B}_L, \leqslant \rangle$ is a lattice with meet $\times$ as the infimum operation, combination $\oplus$ as the supremum operation, and the trivial P-set $\mathbb{O}$ as the null element.

P-sets correspond to Levi's 'ultimate partitions' and to what Schick (1970) has called 'inquiry sets'; proper subsets of P-sets correspond to Schick's 'problem sets'. If F is a proper subset of a P-set B, then F itself is not a P-set. By joining into such an F one new element, we can transform it into a P-set.

DEFINITION 7. If $B = \{h_i \mid i \in I\}$ is a P-set and $F = \{h_i \mid i \in I_0\}$ is a nonempty proper subset of B, then the *completion* of F is the set

$$C(F) = F \cup \{\sim (\bigvee_{i \in I_0} h_i)\}.$$

For example, if $B = \{h_1, h_2, ..., h_n\}$ and $F = \{h_1, h_2, ..., h_m\}$, where $1 \leqslant m < n$, then $C(F) = \{h_1, h_2, ..., h_m, \sim(h_1 \vee ... \vee h_m)\}$. By Definition 7, the completion of a proper subset of a P-set is always a P-set. Moreover, the operation C has the following properties.

THEOREM 7. Let B be a P-set and let F_1 and F_2 be proper subsets of B. Then

(a) $C(F_1) \leqslant B$,

(b) $C(F_1) \oplus C(F_2) \leqslant B$,

(c) $F_1 \subseteq F_2$ iff $C(F_1) \leqslant C(F_2)$.

Further, let B and B^* be two P-sets such that $B \cap B^* \neq \emptyset$. Then

(d) $C(B \cap B^*) \leqslant B$; $C(B \cap B^*) \leqslant B^*$,

(e) $C(B \cap B^*) \subseteq D(B) \cap D(B^*)$,

(f) $C(B \cap B^*) \leqslant B \times B^*$.

Proof. (a) and (c) are immediate from Definition 7. Further, (b) follows from (a) by Theorem 3(e). (d) is a special case of (a), (e) follows directly from (d) by Definition 4, and (f) follows from Theorem 4(e).

The theory of P-sets can be relativized to *background knowledge* b in a straightforward manner. A finite non-empty set $B = \{h_i \mid i \in I\}$ is a *P-set relative to b* if and only if

(a″) $b \vdash \bigvee_{i \in I} h_i$,

(b″) $b \vdash \sim(h_i \& h_j)$ for all $i \neq j$,

(c″)   not $b \vdash \sim h_i$ for all $i \in I$.

If B is a P-set, then the *truncation* of B by b is the set

$$B_b = \{h_i \in B \mid \text{not } b \vdash \sim h_i\}.$$

In other words, B_b contains precisely those elements of B which are compatible with b. By definition, B_b is always a P-set relative to b. The trivial P-set is now replaced by the set containing the equivalence class of the deductive consequences of b. Theorems 1–7 remain valid for P-sets which are relativized to the same b.

3. QUESTIONS

Some elements of the logic of questions are introduced in this section. Only those aspects of questions are considered that are relevant to the account of the role of questions in science (cf. the next two sections).[7]

Hintikka (1974) has developed a useful approach to questions regarded as *requests of information*. According to this analysis, a question of the form

(10)   Why (who, what, etc.) p?

is represented by the imperative

(11)   Bring it about that I know why (who, what, etc.) p?

The description of the epistemic state of affairs which follows the imperative operator 'Bring it about that' is called the *desideratum* of the question. The *presupposition* of a question can usually be obtained by first translating its desideratum into a corresponding 'know that'-construction and then dropping the operator 'I know that' from this desideratum.

Questions can be classified according to the logical nature of their potential answers. The potential answers to *propositional questions* are sentences (statements) or series of sentences; the potential answers to *x-questions* are values of a quantified variable. Propositional questions include why-questions ('Why did Brutus kill Caesar?'), how-questions ('How did Caesar die?'), narrative what-questions ('Whatever happened to Baby Jane?'), and disjunctive questions ('Is number 6 odd or even?'). Disjunctive questions are also called *nexus questions*. The general logical form of a nexus question with a finite number of propositional alternatives $h_1, h_2, ..., h_n$ is the following:

(12)   Bring it about that I know whether $h_1, h_2, ..., $ or h_n.

If the presupposition of the whether-question (12) is

$$(13) \qquad \bigvee_{i=1}^{n} h_i \ \& \ \bigwedge_{\substack{i,j=1 \\ i \neq j}}^{n} \sim (h_i \ \& \ h_j),$$

(i.e., $\{h_1, h_2, ..., h_n\}$ is a P-set), then the desideratum of (12) is

$$(14) \qquad \bigvee_{i=1}^{n} \text{I know that } h_i.$$

In this case, (12) is a *unique alternative whether-question*. If the presupposition of (12) contains only the first conjunct of (13), then the desideratum of (12) may be either (14) or

$$(14') \qquad \bigvee_{i=1}^{n} (h_i \supset \text{I know that } h_i).$$

In the former case, (12) is a *non-exhaustive whether-question*, and in the latter case (12) is a *complete list whether-question*. *Yes-or-no-questions* are also included among nexus questions. The question

(15)   Is it the case that p?

can be analysed as a whether-question

(16)   Bring it about that I know whether p or $\sim p$.

While p is the presupposition of the why-question 'Why is it the case that p?', question (15) has only the tautology '$p \lor \sim p$' as its presupposition.

x-questions include who-questions ('Who killed Caesar?'), where-questions ('Where were you born?'), when-questions ('When were you born?'), and many what-questions ('What is the color of your hair?'). These questions are also called *WH-questions*. They can generally be reformulated as which-questions (who = which person, where = which place, when = which time). The potential answers to WH-questions are values of a quantified (individual or predicate) variable. *Which-individual-questions* are of the form

(17) $(?x) F(x),$

where $F(x)$ is an open formula ('Which individual x is such that $F(x)$?'). *Which-property-questions* are of the form

(18) $(?P) P(a),$

where a is an individual constant and P is a predicate ('Which property P is such that $P(a)$?'). Questions (17) and (18) may be either *exhaustive* or *for-instance*-questions. In the former case, the desideratum of question (17) is

(19) $(x)(F(x) \supset \text{I know that } F(x)),$

and in the latter case

(20) $(\exists x)(\text{I know that } F(x))$.

The problem of answerhood for nexus questions has a simple solution: the potential answers to question (12) are the sentences $h_1, h_2, \ldots, h_n$. For WH-questions (17) and (18) the potential answers are singular terms and predicates, respectively, or lists of singular terms and predicates. However, if a is given as an answer to question (17), it has to be required that the questioner *knows who a is* – otherwise 'I know that $F(a)$' does not imply the required desideratum (20).[8]

An important distinction can be made between open and closed questions. This distinction does not concern our ability to find correct answers to questions, i.e., the solvability problems related to questions, but rather the lists of their potential answers. For *closed questions* we are in a position to give an effective method of enumerating or of building the potential answers.[9] Whether-question (12) is a typical example of a closed question. WH-questions may also be closed ('Who among your brothers is responsible for this?', 'What is the number of planets in our solar system?'). Closed WH-questions can be represented by

(21) $(?x \in X)\, F(x)$,

and

(22) $(?P \in P)\, P(a)$,

where the ranges of x and P are restricted to sets X and P, respectively. If sets X and P are finite, questions (21) and (22) reduce to a whether-question of the form (12). On the other hand, a WH-question may be *open* in the sense that no effective restrictions are given for the ranges of variables x and P. It also may happen that there are no effective criteria for determining whether we already have given a complete answer to an exhaustive which-property-question. Narrative what-questions have a similar open character: the series of sentences given as an answer to them can be continued at will without changing the nature of the answer. Further, why-questions are open in the sense that usually they do not have ultimate answers: any answer to a why-question may be subjected to a question of the same logical type, i.e., we may ask why it is the case that the sentences hold which were given as an answer to the original why-question.

4. INQUIRIES

A physicist studying the refraction of light, a chemist studying the chemical constitution of the rare earth metals, a genetists studying the mechanisms of heredity, an ornithologist studying the mating patterns of a species of birds, and a sociologist studying religion as a social phenomenon – all these investigators are engaged in an activity which I shall call an *inquiry*. It is a characteristic feature of inquiries that they are systematic attempts to find answers to *open questions* – such as open what-questions ('What are the basic laws of refraction?', 'What are the mating patterns of cranes like?'), why-questions ('Why does light refract when it is passed from a medium to another with a different density?'), how-questions ('How is the chemical constitution of the rare earth metals related to their chemical properties?'), and exhaustive which-property-questions ('What factors influence the persistence of characters throughout successive generations?'). These questions typically refer to some *domains* ('universes') of objects, events, or processes – such as the classes of the (past, present, and future) cranes, the (pieces of) rare earth metals, and human societies. They also typically give some specification of the *kinds* of attributes that may be primarily employed within the corresponding inquiry. (For example, the sociological study of religion is distinguished from the psychological, philosophical, or theological approaches to religion through its concentration on the social functions of religion and on the relations of religious practice to social structure.)

The open question which characterizes an inquiry may have presuppositions which constitute a part of the general background assumptions shared by those engaged in the same inquiry. An inquiry may be a part ('subinquiry') of another, or it may be absorbed or reduced to another inquiry through a change in the conceptual and theoretical background assumptions. Inquiries in science thus form a dynamical network of developing and partially overlapping 'research programmes' for obtaining certain kind of information about certain kind of objects.

The 'openness' of inquiries means that the set of the potential results of an inquiry is *open-ended*: no answer to the basic question characterizing an inquiry is ultimate or recognizable as ultimate. That this kind of 'openness' is characteristic to science in general has been argued by several philosophers from different traditions. But even if inquiries do not lead to final results, it is still true that they often lead to tentative *partial answers* to the given question. This means that the 'global' inquiry – for example, the attempt to establish all the true sentences (of a certain kind) about a given domain of objects – will be restructured in terms of closed questions. In brief, to get partial results from an inquiry you have to make it local!

A central aspect of the open-ended character of inquiries is the relative freedom that the investigator has in his choice of the conceptual system employed at a particular stage of an inquiry. Even if the nature of an inquiry may put some restrictions to this choice, the investigator is, at least to a certain extent, free to change the conceptual system in the course of the inquiry. This conceptual change is a major dynamical feature of inquiries; it may take place e.g. by introducing new concepts and by adopting new background postulates or theories about them and their relations to the old concepts. Therefore, the choice of a language which represents a conceptual system is a very important way of making inquiries 'local'.

To take a very simple example, let L be a monadic first-order language with set λ of primitive predicates, and let

(23)
$$S_L = \text{the set of all state-descriptions of } L,$$
$$R_L = \text{the set of all structure-descriptions of } L,$$
$$B_L = \text{the set of all constituents of } L.[10]$$

Then S_L, R_L, and B_L are P-sets (cf. Section 2) such that

(24) $$B_L \leqslant R_L \leqslant S_L.$$

Moreover, we have

(25) $$B \leqslant S_L \text{ for all } P\text{-sets } B \text{ in } L.$$

For P-sets S_L and B_L, we have the results:

(26) $$D(S_L) \cup \{\perp\} = \text{the set of all sentences of } L,$$

and

(27) $$D(B_L) = \text{the set of all generalizations of } L.[11]$$

If language L is now interpreted in a domain U, then P-set S_L generates all the potential answers to the question

(28)   Which sentences of L hold true in U?,

and similarly P-set B_L generates all the potential answers to the question

(29)   Which generalizations of L hold true in U?

Questions (28) and (29) are thus closed questions which correspond to certain inquiries relative to domain U.

Let L' be a monadic first-order language such that $\lambda \subseteq \lambda'$, i.e., L is a sublanguage of L'. (L and L' are supposed to have the same individual constants.) Then P-sets $S_{L'}$, and $B_{L'}$, are finer than P-sets S_L and B_L, respectively, i.e.,

(30) $S_L \preccurlyeq S_{L'}$

and

(31) $B_L \preccurlyeq B_{L'}$.

Relations (30) and (31) illustrate the fact that the richer language L' allows us to give a 'finer' description of a domain than the poorer language L. But they also illustrate the open-ended nature of inquiries. Let λ_i, $i = 1$, $2, \ldots$, be sets of predicates such that $\lambda_j \subseteq \lambda_{j+1}$ for all $j = 1, 2, \ldots$, and let $L_i, i = 1, 2, \ldots$, be the corresponding languages. Then we have, for example

(32) $B_{L_1} \preccurlyeq B_{L_2} \preccurlyeq \cdots \preccurlyeq B_{L_i} \preccurlyeq \cdots$

If languages L_i, $i = 1, 2, \ldots$, are interpreted in a same domain U, then sets B_{L_i}, $i = 1, 2, \ldots$, represent sets of increasingly 'fine' potential 'partial results' of an inquiry relative to domain U.

Induction may play different roles within an inquiry. It may sometimes be a part of the reasoning by which the investigator hits upon his partial results.[12] More importantly, the attempts to justify these results – that is, the evaluations of knowledge claims or the considerations of the acceptability of hypotheses – are normally inductive by their nature. The Bayesian theories of inductive inference, including subjectivism à la de Finetti and inductive logic à la Carnap and Hintikka, are *prima facie* concerned with closed questions of the type (28) and (29). In this sense, they are not theories of inquiries, but rather theories of certain kinds of stages within an inquiry.

However, systems of inductive logic (like Hintikka's) have a built-in dynamics which brings them closer to a full-fledged theory of inquiries. A system of inductive logic is primarily intended to assign inductive probabilities to sentences in a fixed language L which is interpreted in a fixed domain U. Inductive probabilities of the form $P(h/e)$, where h and e are sentences in L, depend upon the logical form of hypothesis h, the number and the kinds of individuals in the sample described by evidence e, the set λ of primitive predicates of language L, and parameters reflecting the degree of disorder or irregularity in domain U. In this model, the major 'local' features of induction are the inductive parameters which represent contextual presuppositions of induction. (Carnap's original idea was that, given the domain U, one should choose language L so as to contain precisely those predicates that are sufficient to express all the qualitative attributes exhibited by the members of U (and nothing else), but later he rejected this 'requirement of descriptive completeness'.) The dependence of inductive probabilities upon the choice of language L allows one to study the effects of conceptual enrichment to induction, which suggests that inductive logic can be used as a tool for discussing and analysing important dynamical aspects of inquiries (cf. (32) above).[13]

5. PROBLEMS

It was seen in the previous section that one way of making a global inquiry 'local' is to fix the language L. If L is a finite monadic language, then the elements of S_L are the strongest sentences expressible in the vocabulary of L and the elements of B_L are the strongest generalizations expressible in L. The questions corresponding to P-sets S_L and B_L can be expressed by

(33)   Which element of S_L is true?

and by

(34)   Which element of B_L is true?

Questions (33) and (34) are unique alternative whether-questions of the type (12). More generally, any question of the form

(35)   Which element of B is true?

where B is a P-set, will be called a *problem*. If $B = \{h_i \mid i \in I\}$, then the ultimate *potential answers* to question (35) are the sentences h_i, $i \in I$. Even if
disjunctions of elements of B do not give complete answers to question (35),
they still reduce the uncertainty concerning the correct answer to (35).
It is thus natural to regard elements of $D(B)$, the disjunctive closure of
P-set B, as (potential) *partial answers* to problem (35).

As we shall in this paper exclude questions involving an infinite set of
potential answers, all the closed questions which occur at different stages
of inquiries can be represented by nexus questions of the form (12). It
seems, moreover, that in science the unique alternative whether-questions
are the basic type of questions of the form (12). This means, in effect, that
(with the above qualifications) the theory of closed questions in science
can be reduced to the theory of problems (in the above technical sense).
In the first place, if the investigator is considering a finite set of non-exclusive propositional alternatives, this set can be generated in an obvious
way from a P-set. In the second place, whether-questions involving a set
of alternatives which are not collectively exhaustive can be reduced to
problems by considering the completion of the given set of alternatives
(see Section 2).[14]

Let B and B^* be two P-sets such that $B \leqslant B^*$. The problems corresponding to P-sets B and B^* (cf. question (34)) will also be denoted by B and
B^*, respectively. Now, each answer to problem B can be represented as a
disjunction of answers to problem B^*. In other words, each complete
answer to B is a partial answer to B^*. It is therefore natural to define B
to be a *subproblem* of B^* (and B^* as the extension of B) just in case $B \leqslant B^*$.

For example, if h is an element of a P-set B, then the question

(36)   Is it the case that h?

corresponds to the subproblem $\{h, \sim h\}$ of B. More generally, if F is a
non-empty proper subset of P-set B, then the question

(37)   Which element (if any) of F is true?

corresponds to the subproblem $C(F)$ of B (cf. Theorem 7 in Section 2).
Conversely, corresponding to each subproblem B' of B there is an appro-

priate question such that the elements of B' are its potential answers. In a 'problem-situation' where the ultimate answers are defined by a P-set B the *subproblems* of B are in a natural *one-to-one* correspondence with (whether-) *questions* that may be asked relative to this problem-situation.

It was seen in Section 2 that, mathematically speaking, the class of problems has the structure of a lattice. The ordering relation $\preccurlyeq$ has already been interpreted in terms of subproblems to a given problem (and in terms of questions asked relative to a problem-situation). The *combination* $B \oplus B^*$ of two problems (P-sets) represents now the least problem which contains both B and B^* as subproblems (cf. Theorem 3 in Section 2). By definition, each answer to problem $B \oplus B^*$ entails an answer to problem B *and* an answer to problem B^*. The *meet* $B \times B^*$ of two problems (P-sets) represents the largest common subproblem of B and B^* (cf. Theorem 4 of Section 2). The meet $B \times B^*$ is the largest problem such that all of its

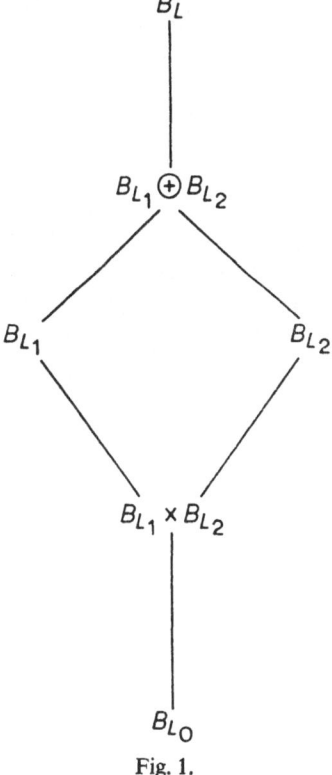

Fig. 1.

partial answers are partial answers to both B and B^*, and conversely (cf. Theorem 5 of Section 2).

To illustrate these notions, let L_1 and L_2 be two monadic languages with $\lambda_1 \cap \lambda_2 \neq \emptyset$. Let L_0 be the monadic language with $\lambda_0 = \lambda_1 \cap \lambda_2$, and let L be the monadic language with $\lambda = \lambda_1 \cup \lambda_2$. Thus L_0 is the *common sublanguage* of L_1 and L_2, and L is the *common extension* of L_1 and L_2. Figure 1 illustrates the relations between the corresponding sets of constituents.

6. THE COMPATIBILITY REQUIREMENT

The notion of knowledge provides a link between the notions of problem and of inductive acceptance. A person X may be said to know that h if and only if h is true and X has evidence e such that h is acceptable on the basis of e (see Hilpinen (1968), p. 28). Here 'h is acceptable on evidence e' roughly means that X is rationally (even if perhaps inductively only) justified in concluding the truth of h from e. Therefore, in order that h_i is a satisfactory answer to a unique alternative whether-question

(38)   Bring it about that I know whether $h_1, h_2, ..., $ or h_n,

there must be some evidence statement e such that h_i is rationally acceptable on e. This idea can, of course, be generalized to other types of whether-questions as well.

Statisticians have largely adopted the view – which Jerzy Neyman has forcefully defended against R. A. Fisher – that the acceptance and the rejection of statistical hypotheses is always relative to the set of the relevant alternative hypotheses. In particular, as Neyman noted in the late 1920's, low probability (relative to evidence) is not sufficient for the rejection of a statistical hypothesis – it should be compared with the probabilities of the rival hypotheses. It is thus very natural to assume, as Levi (1967) in effect does, that acceptance is always relative to a P-set. In our terminology, it may be said that the acceptance and the rejection of hypotheses makes sense only in relation to a problem.

Let B be a P-set. Instead of 'acceptance relative to P-set B' I shall speak of 'B-acceptance'. The following notation will be used:

(39) $Ac_B(h/e)$ iff h is *B-acceptable* on e,

(40) $Re_B(h/e)$ iff h is *B-rejectable* on e,

where B is a P-set relative to evidence e, and h is a member of $D(B)$. It is thus assumed that also the partial answers to problem B may be B-acceptable or B-rejectable. h is said to be B-*plausible* on e if and only if h is not B-rejectable on e. If h is neither B-acceptable on e nor B-rejectable on e, the investigator *suspends judgement* on e (relative to B) in regard to the truth and falsity of h. Thus, we define

(41) $Pl_B(h/e)$ iff   not $Re_B(h/e)$,

(42) $Su_B(h/e)$ iff $Pl_B(h/e)$ and not $Ac_B(h/e)$.

The B-acceptance of h on e is tantamount to the B-rejection of $\sim h$ on e, i.e.,

(43) $Ac_B(h/e)$ iff $Re_B(\sim h/e)$.

It follows that

(44) $Su_B(h/e)$ iff $Su_B(\sim h/e)$,

and

(45) $Pl_B(h/e)$ iff   not $Ac_B(\sim h/e)$.

If h is B-acceptable on e, then it is not B-rejectable on e. This simple consistency principle can be expressed, by (43), as

(46)   If $Ac_B(h/e)$,   then not $Ac_B(\sim h/e)$,

and by (41) as

(47)   If $Ac_B(h/e)$,   then $Pl_B(h/e)$.

It is further assumed that the following conjunction principle is valid:

(48) $Ac_B(h_1 \& h_2/e)$ iff $Ac_B(h_1/e)$ and $Ac_B(h_2/e)$.

By (43), principle (48) can be expressed as

(49) $Re_B(h_1 \vee h_2/e)$ iff $Re_B(h_1/e)$ and $Re_B(h_2/e)$,

so that we also have

(50) $Pl_B(h_1 \vee h_2/e)$ iff $Pl_B(h_1/e)$ or $Pl_B(h_2/e)$.

These principles, especially (49), suggest a simple analysis of the notions of acceptance and rejection. Suppose that we have a criterion for deciding which elements of a P-set B are rejectable on given evidence. By (41), all the non-rejectable elements of B are then B-plausible. By (49), the disjunction of B-rejectable elements of B is likewise rejectable. On the other hand, by (50) we see that a disjunction of elements of B is plausible if and only if at least one of the disjuncts is plausible. We should thus reject only those disjunctions of elements of B which contain only B-rejectable disjuncts. By (42) and (43), this also means that we should suspend judgement in regard to those disjunctions which do not contain all plausible elements of B. The notions of acceptance and rejection relative to a P-set B can thus be defined as follows: If h is in $D(B)$, then

(51) $Ac_B(h/e)$ iff   the B-normal form of h contains *all the plausible* elements of B.

(52) $Re_B(h/e)$ iff   the B-normal form of h contains *no plausible* elements of B.

(53) $Su_B(h/e)$ iff   the B-normal form of h contains *some but not all* plausible elements of B.

By (51), the disjunction of all the plausible elements of B is the *strongest B-acceptable* hypothesis in $D(B)$; all the other B-acceptable hypotheses in $D(B)$ are logically entailed by it.

The decision-theoretic approach leads, in a natural way, to the analysis (51)–(53) of acceptance and rejection, as Levi (1967) and Hilpinen (1968) have shown. Levi's inductive acceptance rule corresponds to this analysis, with the following criterion for the rejection of elements of B:

(54)   Reject $h_i \in B$ on e iff $P(h_i/e) < q/w$,

where $P(h_i/e)$ is the probability of h_i on e, q is parameter in the interval $[0, 1]$ (index of boldness), and w is the number of elements in B. Hilpinen's acceptance rule A_{cont} employs the following criterion:

(55)   Reject $h_i \in B$ on e iff $P(h_i/e) < qP(h_i)$,

where $P(h_i/e)$ and q are as above and $P(h_i)$ is the prior probability of h_i.[15]

The acceptance rule AG_n of Hintikka and Hilpinen (1966) requires that the strongest B-acceptable hypothesis in $D(B)$ is an element of B. Therefore, it satisfies the following principles:

(56) $\quad Ac_B(h_1 vh_2/e) \quad$ iff $\quad Ac_B(h_1/e)$ or $Ac_B(h_2/e)$,
(57) $\quad Re_B(h_1 vh_2/e) \quad$ iff $\quad Re_B(h_1/e)$ or $Re_B(h_2/e)$.

Levi's rule and Hilpinen's rule A_{cont} do not generally satisfy these two principles. (See Hilpinen (1968), pp. 81–84.)

Let B and B^* be two P-sets, and let h be a hypothesis in $D(B)$ and in $D(B^*)$. If h is B-acceptable (B-rejectable), what can be said about its B^*-acceptability (B^*-rejectability)? To obtain more insight to this problem, let us consider its special case, where B^* is obtained from B by 'splitting' one element of B into two mutually exclusive disjuncts. Thus, assume that $B = \{h_1, \ldots, h_n\}$ and $B^* = \{h_1, \ldots, h_{n-1}, h_n^*, h_{n+1}^*\}$, where $h_n = h_n^* \vee h_{n+1}^*$. If plausible elements of a P-set are denoted by • and rejectable elements by ∘, then it is intuitively obvious that the following two cases are possible:

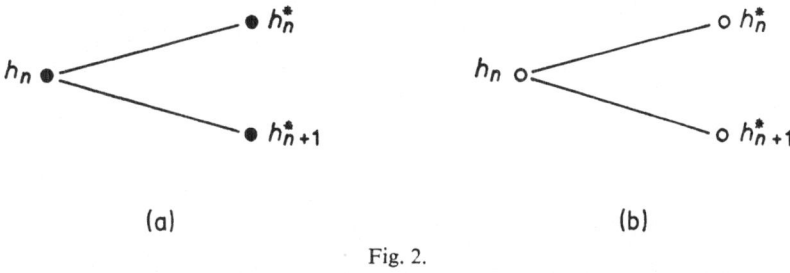

(a) (b)

Fig. 2.

If h_n is B-plausible, it may be possible to 'analyze' it into two parts such that, on this deeper level of analysis, one of them is rejectable. Similarly, if h_n is B-rejectable, it may be possible to analyze it into two parts such that, on this deeper level of analysis, one of them is plausible. Thus, the following two cases seem possible:

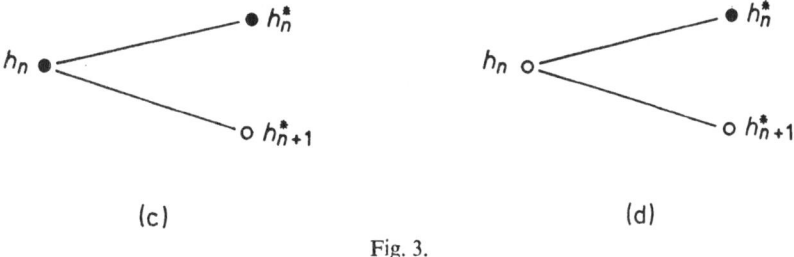

(c) (d)

Fig. 3.

On the other hand, it seems that the following case should be excluded:

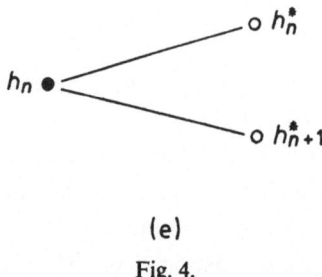

(e)

Fig. 4.

This claim can be supported by noting that if both h_n^* and h_{n+1}^* are B^*-rejectable, then also their disjunction, i.e., h_n, is B^*-rejectable. The basic alternatives incompatible with h_n are the same (i.e., $h_1, \ldots, h_{n-1}$) both relative to B and to B^*. It therefore seems natural to assume that if h_n is B^*-rejectable, it is B-rejectable as well.

The principle of excluding the case (e) (see Figure 4), with the corresponding requirement that h_i $(i < n)$ should not be B-plausible and B^*-rejectable at the same time, is equivalent with the following *Rejection Principle*:

(RP)   Let B and B^* be P-sets relative to e such that $B \leqslant B^*$, and let $h \in B \subseteq D(B^*)$. If $Re_{B^*}(h/e)$, then also $Re_B(h/e)$.

As P-sets correspond to problems, rule RP expresses the principle that any hypothesis which is rejectable as a complete or partial answer to a problem is rejectable as a complete answer to a subproblem of the given problem.

To illustrate the Rejection Principle, suppose that a rich uncle has died and that, relative to the background knowledge, the following three hypotheses constitute a P-set:

h_1 = the uncle died a natural death.
h_2 = the uncle was poisoned by his niece.
h_3 = the death of the uncle was caused by the fact that his niece prevented him from taking his urgently needed medicine.

Suppose that you consider h_1 as the only plausible hypothesis relative to

P-set $B^* = \{h_1, h_2, h_3\}$, so that you B^*-reject h_2 and h_3 on your evidence. If someone asks you, whether the uncle was murdered or not, the relevant P-set is $B = \{h_1, h_2 \vee h_3\}$. Principle (RP) requires now that you B-reject the hypothesis $h_2 \vee h_3$, that is, your answer will be that the uncle was not murdered. (We shall see below that the situation is different if we start with a problem and then consider its extensions. We may at first be willing to reject the hypothesis of murder. But when we come to realize that, relative to our background knowledge, the murder could have happened precisely in two ways, indicated in h_2 and h_3, then we might be willing to suspend judgement in regard to h_3. This corresponds to Figure 3 (d) above.)

By (49), rule RP is equivalent to a principle which generalizes RP to all elements of $D(B)$:

(GRP)   Let B and B^* be P-sets relative to e such that $B \preccurlyeq B^*$, and let $h \in D(B) \subseteq D(B^*)$. If $Re_{B^*}(h/e)$, then also $Re_B(h/e)$.

By (43), rule GRP can be expressed also in terms of acceptance as follows:

(GAP)   Let B and B^* be P-sets relative to e such that $B \preccurlyeq B^*$, and let $h \in D(B) \subseteq D(B^*)$. If $Ac_{B^*}(h/e)$, then $Ac_B(h/e)$.

Rule GAP expresses the principle that any satisfactory (complete or partial) answer to a problem is also a satisfactory answer to those sub-problems of the given problem relative to which it is expressible. As simple corollaries to GAP we obtain:

(58)   Let F be a non-empty subset of a P-set B (relative to e), and let $h \in D(F)$. If $Ac_B(h/e)$, then $Ac_{C(F)}(h/e)$.

(59)   Let B be a P-set relative to e, and let $h \in B$. If $Ac_B(h/e)$, then $Ac_{\{h, \sim h\}}(h/e)$.

(60)   Let B and B^* be two non-isolated P-sets relative to e, and let $h \in D(B) \cap D(B^*)$. If $Ac_B(h/e)$ or $Ac_{B^*}(h/e)$, then $Ac_{B \times B^*}(h/e)$.

As a corollary to GRP, GAP, (41), and (42), we obtain

(61)   Let B and B^* be two P-sets relative to e such that $B \preccurlyeq B^*$. and let $h \in D(B) \subseteq D(B^*)$. If $Su_B(h/e)$, then $Su_{B^*}(h/e)$.

As the case (d) (see Figure 3) is not excluded, the converses of principles GRP and GAP are not valid: when $B \preccurlyeq B^*$, it is possible that a hypothesis

h is B-acceptable (B-rejectable) but still not B^*-acceptable (B^*-rejectable). Thus, there is a sharp asymmetry between the subproblems and the extension of a given problem – or, on the level of language-shifts, between the 'downward' and 'upward' movement in Figure 1. This asymmetry has a natural explanation. In Section 5, problems were characterized as closed questions with ultimate answers. Extensions of a given problem B (i.e., problems B^* such that $B \leqslant B^*$) represent more detailed analyses of the problem-situation than B; the 'ultimate' answers to B are split into disjunctions of new 'ultimate' answers to B^*. The change from B to B^* is thus a 'radical' problem-shift, so that what is acceptable relative to B may fail to be acceptable relative to B^*. On the other hand, the subproblems of a given problem B (i.e., problems B' such that $B' \leqslant B$) correspond to different questions that can be asked in the problem-situation as characterized by B, and answers to these questions are partial answers to the original problem. No 'radical' problem-shift is involved in this change, so that what is acceptable relative to B should be acceptable relative to B' as well.

For example, even if a hypothesis h were rejectable relative to P-set $\{h, \sim h\}$, it may still be reasonable to suspend judgement in regard to h relative to P-set $\{h, g_1, \ldots, g_n\}$ provided that the hypotheses $h, g_1, \ldots,$ and g_n are equally well supported by the available evidence. But if h is a satisfactory (i.e., acceptable) complete answer to the question 'Which of the hypotheses $h, g_1, \ldots, g_n$ is true?', then h is a satisfactory answer also to the question 'Is it the case that h?'.

Let now B and B^* be two non-isolated P-sets, and let $h \in D(B) \cap D(B^*)$. By Theorem 5 of Section 2, we have $h \in D(B \times B^*)$. By Theorem 4 of Section 2, $B \times B^* \leqslant B$. If h is now B-acceptable, then principle (60) implies that h is also $(B \times B^*)$-acceptable. If h is B^*-rejectable as well, rule GRP implies that h is also $(B \times B^*)$-rejectable, i.e., $\sim h$ is $(B \times B^*)$-acceptable. However, this is a contradiction with principle (46). We have thus seen that the Rejection Principle RP implies the following general principle:

(CR)   Let B and B^* be two non-isolated P-sets relative to e, and let $h \in D(B) \cap D(B^*)$. If $Ac_B(h/e)$, then not $Re_{B^*}(h/e)$. Similarly, if $Re_B(h/e)$, then not $Ac_{B^*}(h/e)$.

CR is called the *Compatibility Requirement*: it expresses the principle that on the basis of same evidence, a hypothesis which is acceptable (rejectable) relative a P-set B cannot be rejectable (acceptable) relative to another P-set B^*. By (43), principle (CR) is equivalent to the following requirement:

(CR′)   Let B and B^* be two non-isolated P-sets relative to e, and let $h \in D(B) \cap D(B^*)$. Then it is not the case that $Ac_B(h/e)$ and $Ac_{B^*}(\sim h/e)$, and similarly it is not the case that $Re_B(h/e)$ and $Re_{B^*}(\sim h/e)$.

In other words, CR requires that the answers given to non-isolated problems on the basis of the same evidence should be compatible with each other.

Principle CR allows us to sharpen the remarks made above concerning the relation of a problem B to its extensions B^*. It was seen that a B-acceptable (B-rejectable) hypothesis h may fail to be B^*-acceptable (B^*-rejectable). However, CR shows that to make a B-acceptable (B-rejectable) hypothesis h B^*-rejectable (B^*-acceptable) it is not sufficient merely to make B finer, but new evidence is also required.

Levi's acceptance rule (cf. (54)) does not satisfy the Compatibility Requirement CR unless $q \leqslant \frac{1}{2}$. As Hacking (1967) pointed out, for sufficiently large values of q, Levi's rule gives the following result:

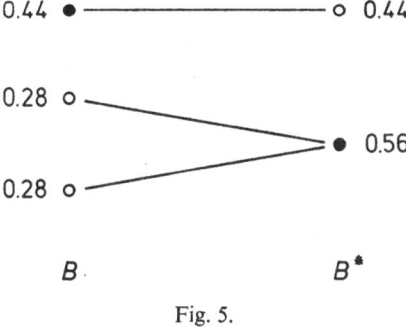

Fig. 5.

In this situation, the hypotheses acceptable relative to B and to B^*, respectively, are incompatible with each other. If the value of q is restricted to those less or equal to $\frac{1}{2}$, the probability of the strongest hypothesis acceptable by Levi's rule is higher than $\frac{1}{2}$, and examples like the one given in Figure 5 are blocked. However, both Hacking (1967) and Hilpinen (1968) argue that "playing with the q value" does not save Levi's rule from troubles arising from its oversensitivity to the choice of the ultimate partition (i.e., P-set). The reason, I suggest, is that Levi's rule, even with the restriction $q \leqslant \frac{1}{2}$, does not satisfy the fundamental Rejection Principle RP. This is shown by the following example, where q may be chosen to be less than $\frac{1}{2}$ but sufficiently close to $\frac{1}{2}$.

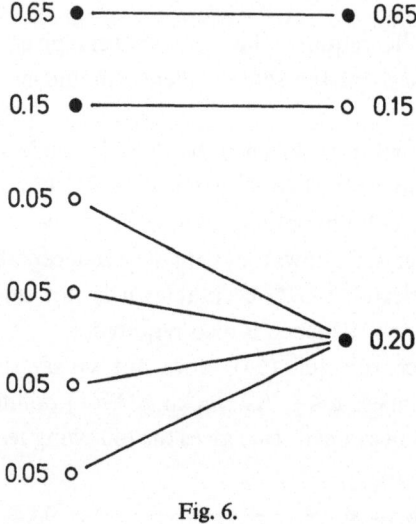

Fig. 6.

As Levi's rule does not generally satisfy the Compatibility Requirement CR, it does not satisfy the principles RP, GRP, and GAP, either.[16] On the other hand, Hilpinen's rule A_{cont} does satisfy all of these principles.[17] (This is also true for Hintikka's and Hilpinen's n-rule AG_n.[18]) The crucial difference between Levi's rule and A_{cont} is the fact that definition (55) is independent on the corresponding P-set, so that RP is automatically satisfied, but (54) depends upon the (size of the) corresponding P-set in such a manner that RP may be violated.

7. CONCLUDING REMARKS: ACCEPTANCE VS SUPPORT

Throughout the preceding section we have assumed that the problems considered in relation to each other are represented by P-sets relative to the same evidence and background assumptions. New observational evidence and changes in the conceptual or the theoretical background assumption may effect 'radical' changes in the problem-situation and in the corresponding sets of acceptable hypotheses (cf. Niiniluoto and Tuomela (1973) for an attempt to deal with problems of this kind). For fixed evidence and background assumptions, a simple Rejection Principle RP was suggested above. This principle allows the acceptability and the rejectability of hypotheses to be, to a reasonable extent, relative to the choice of P-set; but it also implies a Compatibility Requirement CR which contradicts Levi's claim that

... conclusions that are reached relative to different ultimate partitions, like conclusions reached relative to different bodies of evidence, cannot be pooled, and the resulting set of sentences then expected to be consistent and deductively closed. (Levi (1967), p. 38.)

Levi supports this view with an example of two *isolated* P-sets (*op. cit.*, pp. 37–38); this is not evidence for Levi's rule and against CR, however, since CR is concerned only with non-isolated P-sets.

I conclude with a remark concerning the relation between inductive acceptance and inductive support. Let $B = \{h_1, h_2, h_3\}$ be a P-set, and let e_1 and e_2 be evidence statements such that

$$
\begin{array}{ll}
P(e_1/h_1) = 0.90 & P(e_2/h_1) = 0.10 \\
P(e_1/h_2) = 0.02 & P(e_2/h_2) = 0.98 \\
P(e_1/h_3) = 0.01 & P(e_2/h_3) = 0.99 \, .
\end{array}
$$

(62)

Stegmüller ([1973], pp. 93–94) notes that Hacking's (1965) 'law of likelihood' implies, in this situation, that evidence e_2 supports hypothesis h_2 more than evidence e_1 supports hypothesis h_1. Stegmüller finds this result intuitively inadequate, since h_1 is on e_1 "45 times more plausible than its nearest rival" h_2 in B, but h_2 is not even the most plausible hypothesis in B on evidence e_2.

I do not find this argument very convincing. Inductive support (of a hypothesis on the basis of some evidence) is a considerably weaker notion than inductive acceptance. For example, the best supported hypothesis in a P-set need not be rationally acceptable as true. Moreover, it is quite possible that two mutually incompatible hypotheses are both well-supported by the same evidence. In the above example, both h_2 and h_3 are supported by e_2, but only h_1 is supported by e_1. In my intuition, a reasonable theory of acceptance and rejection should imply that (at least when h_1, h_2, and h_3 do not initially differ too much in probability) h_1 is the strongest acceptable hypothesis on e_1, and the disjunction $h_2 \vee h_3$ is the strongest acceptable hypothesis on e_2. This is precisely the result that is given by Levi's rule, Hilpinen's A_{cont}, and (almost all) 'likelihood tests' of Hacking (see Note 15). Thus, one might claim that Stegmüller's remarks about the example (62) are really relevant to the notion of acceptance, but not to the notion of support – the latter need not be relative to the rival hypotheses *in the same way* as the former.

Stegmüller's criticism of Hacking suggests an interesting principle which links the notions of acceptance (or rejection) and support. To make this principle more perspicious, let us modify the likelihoods (62) to

$$
\begin{array}{lll}
 & P(e_1/h_1) = 0.50 & P(e_2/h_1) = 0.30 \\
(63) & P(e_1/h_2) = 0.25 & P(e_2/h_2) = 0.55 \\
 & P(e_1/h_3) = 0.20 & P(e_2/h_3) = 0.95.
\end{array}
$$

If now $1 < \alpha \leqslant 1.7$, then Hacking's likelihood test rejects h_2 and h_3 on e_1, and both h_1 and h_2 on e_2, even if e_2 supports h_2 better than e_1 supports h_1 (by the law of likelihood). Thus, if 'support' is defined by the law of likelihood and if 'rejection' is defined by likelihood tests, then the following principle is not generally valid:

(SP)   Let h_1 and h_2 be elements of a P-set B, and suppose that e_2 supports h_2 more than e_1 supports h_1. If h_1 is the only non-B-rejectable element of B on e_1, then h_2 is not B-rejectable on e_2.

In terms of acceptance, SP can be formulated as follows:

(SP')   Let h_1 and h_2 be elements of a P-set B, and suppose that e_2 supports h_2 more than e_1 supports h_1. If $Ac_B(h_1/e_1)$, then $Pl_B(h_2/e_2)$.

If 'support' is defined in terms of positive relevance (i.e., $P(h/e) > P(h)$) and if 'acceptance' is defined by rule A_{cont}, then SP' is satisfied. The reason is that, in this analysis, each $h \in B$ which is supported by e is also B-plausible on e. But as I do not know any independent evidence for the principle SP, the validity of SP (and of other principles linking support and acceptance) has to be left here as an open problem.

Stegmüller uses his example (62) to argue that comparisons between the support that different hypotheses receive from different evidence statements are not allowed: the comparative notion of support is not a four-place but a three-place relation. However, example (62) does not create a problem for a Bayesian who uses something like the ratio $P(h/e)/P(h)$ or the difference $P(h/e)-P(h)$ as an 'absolute' measure of support – and would therefore be able to define a four-place comparative support-relation. For example, if h_1, h_2, and h_3 have equal initial probabilities, a Bayesian would find in example (62) that h_1 is better supported by e_1 than h_2 is supported by e_2. In this sense, (62) provides also an argument against the 'law of likelihood'. As the ratio $P(h/e)/P(h)$ is equal to $P(e/h)/P(e)$, a Bayesian is prone to take as a measure of support the likelihood $P(e/h)$ considered in relation to the total probability $P(e)$ of evidence e. This probability $P(e)$ is calculable only in relation to a (possibly infinite) P-set $B=\{h_i \mid i \in I\}$ such that the likelihoods $P(e/h_i)$ are defined for all $i \in I$ and the initial probabilities $P(h_i)$ are known for all $i \in I$. (The same remarks concern also the posterior probabilities $P(h/e)$.) The Bayesian approach is therefore applicable only in relation to suitably chosen P-sets or problems in our technical sense. This agrees with the remarks at the end of Section 4, and gives some indication of the sense in which the notion of inductive support may be said to be 'local'.[19]

University of Helsinki

NOTES

[1] For an evaluation of this literature with a comprehensive bibliography, see Hintikka (1974).

[2] See, for example, Popper (1963), Levi (1967), and Tondl (1973). According to Popper, "science progresses from problems to problems", but his scheme does not seem to leave any place for genuine *answers* (cf. Levi (1967), pp. 108–112).

[3] For an interesting attempt to apply ideas from the logic of questions to the philosophy of science, see Tondl (1973), especially pp. 126–153. The role that different kinds of questions play in science is in fact one of the earliest topics in the philosophy of science: in *Analytica Posteriora* II, 1 (89b23) Aristotle makes a distinction between four types of inquiry in terms of four different questions (whether-questions including yes-or-no-questions, why-questions, whether-questions concerning existence, and what-questions). The Stoic logicians and some medieval Aristotelians were interested in the theory of questions, and the methodologies of the Renaissance 'dialecticians' consisted largely of standardized sequences of different kinds of questions.

[4] See, for example, Bromberger (1966) and Hempel (1965), pp. 334–335.

[5] Charles Peirce (1931–35), 5.168, quotes Julius Stöckhardt as saying that an experiment is a question put to nature.

[6] I have attempted to defend the covering law theory of explanation against some of its critics by pointing out that certain kinds of explanation-seeking why-questions are not appropriate relative to certain kinds of background knowledge (see Niiniluoto, 1975a). For example, if you know already that a bird a is a raven and if you want to know why a is black, you should not ask 'Why is this bird black?' but rather 'Why is this raven black?'.

[7] Terminology and distinctions developed by Belnap, Harrah, Katz, Åqvist, and Hintikka are freely used in this section.

[8] Hintikka (1974) shows that a singular term a is a satisfactory (potential) answer to a which-individual question (17) if and only if the questioner knows who (what, etc.) a is. Even if Aristotle did not include questions of type (17) among his list of scientific questions, they still play a considerable role within scientific inquiry. Hintikka's solution to the problem of answerhood has then an interesting corollary concerning a class of questions that scientists ask. To each scientific question there corresponds an indefinite class of potential *questioners*. Therefore, a satisfactory answer to a scientific question of the type (17) should contain, besides a singular term a, enough information so that any questioner (in the scientific community) who receives the answer comes to know who (what, etc.) a is.

[9] Tondl (1973), pp. 128–129, attributes this characterization of closed questions to Giedymin.

[10] For the notion of constituent, see Hintikka (1973). If L is a first-order language with relation symbols and if $B_L{}^d$ denotes the class of the constituents of L with the depth d, then $B_L{}^d$ is a P-set for all $d=0, 1, \ldots$, and

$$B_L{}^0 \leqslant B_L{}^1 \leqslant \cdots \leqslant B_L{}^d \leqslant \cdots.$$

Furthermore,

$$\bigcup\nolimits_{d=0} D(B_L{}^d) = \text{the set of all generalizations of } L.$$

[11] In this context, generalizations are simply quantificational sentences without individual constants.

[12] For the role of induction within the 'context of discovery', see Achinstein (1970).

[13] See Niiniluoto and Tuomela (1973) and Niiniluoto (1975b) for attempts to give an account of the inductive effects of conceptual enrichment.

[14] In contrast with my usage in this paper, Schick (1970) calls 'problems' those sets of basic alternatives which are not collectively exhaustive with respect to the background information. A one-membered problem is a 'question' in Schick's terminology. If $\{h\}$ is a 'question', in this sense, then the complementary proposition $\sim h$ is "of no interest to us" and believing it would "serve no epistemic purpose" (*op. cit.*, p. 19). This kind of 'question' does not correspond to the whether-question 'Is it the case that h?' (cf. '(15) above), but to the whether-question 'Ought I believe h or not?' (*op. cit.*, p. 7). A cognitivist analysis of inductive acceptance presupposes the former alternative: if h is a scientific hypothesis worthy of consideration, then both its truth and its falsity (i.e., the truth of $\sim h$) serve some epistemic purpose. Therefore, I assume that scientifically relevant 'problems' and 'questions' in Schick's sense are all reducible to 'problems' in my sense.

[15] Hacking's (1965) *likelihood tests*, when applied to the situation discussed here, correspond to the above analysis in the sense that a composite (i.e., disjunctive)

hypothesis is B-rejectable on evidence e if and only if each of its disjuncts is B-rejectable on e, and a hypothesis $h_i \in B$ is B-rejectable on e if and only if

$$\frac{P(e/h_j)}{P(e/h_i)} > \alpha(> 1)$$

holds for some $h_j \in B$. If now $h_i \in B$ is B-rejectable on e according to Hilpinen's criterion (55), then there is at least one $h_j \in B$ such that $P(h_j/e) > qP(h_j)$. As now

$$\frac{P(e/h_i)}{P(e)} < q < \frac{P(e/h_j)}{P(e)},$$

we have that $P(e/h_j) > \alpha P(e/h_i)$ for an α which depends upon q. In other words, if a hypothesis is B-rejectable according to Hilpinen's rule A_{cont}, it is also B-rejectable in a likelihood test à la Hacking.

[16] I have shown in Niiniluoto (1971) that Keith Lehrer's rule IR does not generally satisfy the Compatibility Requirement CR (see Lehrer, 1970), so that it does not satisfy principles RP, GRP, and GAP, either.

[17] Principle CR corresponds to Hilpinen's (CA3) and principles GAP and GRP correspond to Hilpinen's (C9.1). (See Hilpinen (1968), pp. 101,110.)

[18] This remark needs some qualifications – see Hilpinen (1968), pp. 83–84, and cf. the results in Niiniluoto (1975b).

[19] The notion of inductive support may be said to be 'local' also in the sense that it is relative to some background knowledge: instead of the two-place relation 'e supports h' we have the relation 'e supports h relative to b'. In a probabilistic theory of induction this relation may be defined by the following equality:

$$P(h/e \ \& \ b) > P(h/b).$$

It follows from this definition that evidence e cannot support a hypothesis h relative to background knowledge b if e is included in, or entailed by, b. In other words, only evidence which is new relative to b can support h relative to b. This result seems to give a reasonable answer to many of the problems discussed in Musgrave, 'Logical versus Historical Theories of Confirmation', *British Journal for the Philosophy of Science* **25** (1974) 1–23. Moreover, this result is independent of the question, whether the probabilities involved are logical or not. As one can make sense of the 'historical' or 'local' dimension of inductive support even within a theory of logical probabilities, Musgrave's contrast between logical and historical theories of confirmation (*ibid.*, pp. 2–3) is less sharp and less accurate than he suggests.

BIBLIOGRAPHY

Achinstein, P.: 1970, 'Inference to Scientific Laws', in R. H. Stuewer (ed.), *Historical and Philosophical Perspectives in Science, Minnesota Studies in the Philosophy of Science*, Vol. V, University of Minnesota Press, Minneapolis.

Bromberger, S.: 1966, 'Why-Questions', in R. G. Colodny (ed.), *Mind and Cosmos*, Pittsburgh University Press, Pittsburgh.

Hacking, I.: 1965, *Logic of Statistical Inference*, Cambridge University Press, Cambridge.

Hacking, I.: 1967, 'Review of Isaac Levi, *Gambling with Truth*', *Synthese* **17**, 444–448.

Hempel, C. G.: 1965, *Aspects of Scientific Explanation*, The Free Press, New York.

Hilpinen, R.: 1968, *Rules of Acceptance and Inductive Logic* (*Acta Philosophica Fennica* 21), North-Holland, Amsterdam.

Hilpinen, R.: 1972, 'Decision-Theoretic Approaches to Rules of Acceptance', in R. E. Olson and A. M. Paul (eds.), *Contemporary Philosophy in Scandinavia*, The John Hopkins Press, Baltimore.

Hintikka, K. J.: 1973, *Logic, Language-Games, and Information: Kantian Themes in the Philosophy of Logic*, Oxford University Press, Oxford.

Hintikka, K. J.: 1974, 'Questions about Questions', in M. K. Munitz and P. Ungar (eds.), *Semantics and Philosophy*, New York University Press, New York.

Hintikka, K. J. and Hilpinen, R.: 1966, 'Knowledge, Acceptance, and Inductive Logic', in K. J. Hintikka and P. Suppes (eds.), *Aspects of Inductive Logic*, North-Holland, Amsterdam.

Lehrer, K.: 1970, 'Induction, Reason, and Consistency', *British Journal for the Philosophy of Science* 21, 103–114.

Leyi, I.: 1967, *Gambling with Truth: An Essay on Induction and the Aims of Science*, Alfred A. Knopf, New York.

Niiniluoto, I.: 1971, 'Can We Accept Lehrer's Inductive Rule?', *Ajatus* 33, 254–265.

Niiniluoto, I.: 1975a, 'Inductive Explanation, Propensity, and Action', in J. Manninen and R. Tuomela (eds.), *Essays on Explanation and Understanding*, D. Reidel, Dordrecht-Holland and Boston-U.S.A.

Niiniluoto, I.: 1975b, 'Inductive Logic and Theoretical Concepts', forthcoming in *The Proceedings of the Conference on Formal Methods in the Methodology of Empirical Sciences* (held in Warsaw, Poland, in June, 1974), D. Reidel, Dordrecht-Holland and Boston-U.S.A.

Niiniluoto, I. and Tuomela, R.: 1973, *Theoretical Concepts and Hypothetico-Inductive Inference*, D. Reidel, Dordrecht-Holland and Boston-U.S.A.

Peirce, C. S.: 1931–35, *Collected Papers*, Vol. 1–5 (ed. by C. Hartshorne and P. Weiss), Harvard University Press, Cambridge, Mass.

Popper, K. R.: 1963, *Conjectures and Refutations*, Routledge and Kegan Paul Ltd., London. (3rd revised ed., 1969.)

Schick, F.: 1970, 'Three Logics of Belief', in M. Swain (ed.), *Induction, Acceptance, and Rational Belief*, D. Reidel, Dordrecht-Holland.

Stegmüller, W.: 1973, *Personelle und Statistische Wahrscheinlichkeit. Probleme und Resultate der Wissenschaftstheorie und Analytischen Philosophie*, Band IV, Teil III, Springer-Verlag, Berlin, Heidelberg, and New York.

Tondl, L.: 1973, *Scientific Procedures*, D. Reidel, Dordrecht-Holland and Boston-U.S.A.

HÅKAN TÖRNEBOHM

ON PIECEMEAL KNOWLEDGE-FORMATION

I. AN OUTLINE OF PIECEMEAL KNOWLEDGE-FORMATION

Piecemeal knowledge-formation (to be distinguished from the production of syntheses of knowledge) may be described as a sequence of acts as follows:

(1) A cognitive problem is raised, i.e. a problem, the solution of which deserves to be confirmed.

(2) A set of hypotheses $\{h, h', h'', \ldots\}$ is framed each one of which could be accepted as a solution of the problem, provided that it deserves to be confirmed.

(3) A testing plan for $\{h, h', h'', \ldots\}$ is designed. This plan consists of subplans for subsets of $\{h, h', h'', \ldots\}$. For the sake of simplicity I will confine my attention to a single hypothesis h in the set $\{h, h', h'', \ldots\}$.

(4) A testing plan for h is implemented. As a result a body e of information is collected. e will be employed as evidence pro or con h.

(5) A verdict about h is passed in the light of the evidence e. 'h is (strongly, moderately or weakly) supported by e', or 'h is (strongly, moderately or weakly) undermined by e'.

(6) A decision is taken to opt for one of the following alternatives:

(a) to reject h;

(b) to modify h into another hypothesis h', such that favorable evidence for h is also favorable for h' and unfavorable evidence for h is not unfavorable for h';

(c) to procure further evidence; or

(d) to accept h.

(7) In the case that the alternative (6d) is opted for, a researcher proceeds to argue on behalf of h before a court of 'authorized legitimizers' or 'umpires'.

(8) The hypothesis h is confirmed if the umpires also accept h.

Remarks. 1. Confirmation as here conceived is analogous to an act of promulgation in the sphere of legislation. 2. An hypothesis which is con-

R. J. Bogdan (ed.), Local Induction, 297–318. All rights reserved.
Copyright © 1976 by D. Reidel Publishing Company, Dordrecht-Holland

firmed in this sense might or might not deserve to have the status of knowledge.

II. PROBLEMS CONCERNING PIECEMEAL KNOWLEDGE-FORMATION

The sequence of moves leading to a confirmed hypothesis suggests a number of problems. These problems will be dealt with in this essay. I refer to the ordinal numbers which were used in Section I as labels.

Ad 3. How should testing plans be designed? What kind of evidence should researchers look for when they want to test an hypothesis?

Ad 4. How should the relation(s) between an hypothesis and a body of evidence be construed (a) in the case of favorable and (b) in the case of unfavorable evidence? What is the sense of assigning qualifications such as 'weakly', 'moderately' and 'strongly'?

Ad 5. How should a researcher opt for one of the alternatives open to him in the light of a verdict on the bearing of evidence upon an hypothesis?

Ad 7. Under what conditions does a confirmed hypothesis deserve to be confirmed?

A final problem which I will also deal with is this one: What long run trends are to be expected in a field of research in which no attempts are made to produce syntheses of knowledge (=narratives as in history or theories as in physics) upon sets of confirmed hypotheses?

In order to deal with these problems in a systematic fashion a formalism *KF* (for *k*nowledge-*f*ormation) will be presented in the following section.

III. A FORMALISM *KF* (FOR KNOWLEDGE-FORMATION)

Introduction

This formalism will be presented in two sections, KF_0 and KF_1. KF_0 caters for the notion of positive evidential support but not for the relation between evidence e and an hypothesis h, in the case that e undermines h.

KF_1 which is identical with KF caters for all relations which may obtain between a body of evidence and an hypothesis.

KF_0 is based on a set S_0 of propositions which are individually and mutually consistent. It contains two mappings, C and M.

C maps S_0 on the set of real numbers. C is defined by means of a set of postulates in such a manner that C may be interpreted as a measure of the content of a proposition.

M maps ordered pairs of propositions in S_0 on the set of real numbers. M is defined by means of a postulate which in conjunction with the C-postulates ensures that $M(p, q)$ (for matching) may be interpreted as a measure of the degree to which (the content of) a proposition q covers (the content of) a proposition p. M is offered as an explicatum of the notion of evidential support.

$KF_1 (= KF)$ is based on a set S of propositions which is closed under negation, conjunction and disjunction. It contains five mappings, $C, M, P,$ V and T. C and M are defined by means of postulates which ensure that C and M in KF_1 are equivalent to C and M in KF_0 for every subset S_0 of S, such that the propositions in S_0 are individually and mutually consistent. C and M in KF_1 may accordingly be described as extensions of C and M in KF_0 from a set S_0 to a set S which in contrast to S_0 contains propositions which are mutually incompatible. C in KF_1 as well as in KF_0 may accordingly be interpreted as a content measure.

M in KF_1 (in contrast to M in KF_0, which is $\geqslant 0$), takes negative as well as non-negative values. When $M(p, q)$ in KF_1 takes non-negative values it is interpreted in the same way as $M(p, q)$ in KF_0. It is proved in KF_1 that $M(\bar{p}, q) > 0$ if $M(p, q) < 0$. This theorem ensures that $M(p, q)$ may be interpreted as (positive, zero or negative) agreement or matching between any two propositions for which M is defined.

There exist in addition to C and M three other mappings in KF_1 viz. P, V and T.

P is introduced as an auxiliary concept which helps to extend the C- and M-functions from a set S_0 to a set S of propositions. P has the properties of a probability and thus is a probability.

V is a probability function on S but it is defined so that it may only take two values 0 and 1. $V(p)$ is interpreted as the truth value of p. V is the basic concept in what will be called a first truth theory in $KF_1 = KF$.

T maps S on the set of real numbers. It is introduced by means of a definition of truth core which presupposes the V-concept and a postulate which links T to the concept of matching. T is interpreted as degree of truth. It is the basic concept in what I call the second truth theory in KF.

I proceed now to a presentation of KF_0.

IIIa. *The formalism KF_0*

S_0 is a set of propositions which are individually and mutually consistent. Three operators and three two-place predicates are introduced. S_0, the operators and the predicates form an algebra A_0.

$$A_0 = \langle S_0, \, \wedge, \, \times, \, -, \, \Rightarrow, \Leftrightarrow, \,)(\, \rangle$$

The operators will now be explained. If p and q are elements in S_0, then so are $p \wedge q$, $p - q$ and $p \times q$. $p \wedge q$ is a proposition, the content of which is the *union* of the contents of p and of q. $p \times q$ is a proposition, the content of which is the *intersection* of the contents of p and q. $p - q$ is a proposition, the content of which is the *difference* between the content of p and the content of q.

I proceed next to explain the three predicates in the algebra. $p \Rightarrow q$ (p entails q) asserts that the content of q is contained in the content of p. $p \Leftrightarrow q$ (p is equivalent with q) asserts that p and q have the same content. $p)(q$ (p excludes q) asserts that p and q do not have any common content. Their intersection is empty.

Here follows a number of valid assertions in the algebra A_0.

A_0 1.   The operators $\wedge$ and $\times$ are commutative. This means that $p \wedge q \Leftrightarrow q \wedge p$ and $p \times q \Leftrightarrow q \times p$.

A_0 2.   If $p \Rightarrow q$, then $p \times q \Leftrightarrow q$ and $p - q$ is empty. This means that $p - q$ has no content.

A_0 3.   If $p)(q$, then $p \times q$ is empty.

A_0 4. $p \wedge q \Leftrightarrow (p - q) \wedge q$.

A_0 5. $p \Leftrightarrow (p - q) \wedge (p \times q)$.

A_0 6. $p - q)(q$.

A_0 7. $p - q)(p \times q$.

A_0 8.   Every element in S_0 entails an empty proposition.

A_0 9.   If p is empty and if $p \Rightarrow q$, then q is also empty.

IIIa1. *A Restricted Content Theory*

Convention. I will label all postulates and theorems in KF consecutively as follows: $KF1$, $KF2$ etc. Postulates will be marked by asterisks.

C is a function which maps S_0 on the set of real numbers. C is to satisfy these postulates.

*KF1.   If p is empty, then $C(p) = 0$.
*KF2.   If $p \Rightarrow q$, then $C(p) \geqslant C(q)$.
*KF3.   If $p)(q$ then $C(p \wedge q) = C(p) + C(q)$.

From these postulates in conjunction with the assertions $A_0 1 - A_0 9$ it is very easy to establish the following theorems:

KF4.   If $p \Leftrightarrow q$, then $C(p) = C(q)$.
KF5. $C(p) \geqslant 0$ for all p.
KF6. $C(p) > 0$ if and only if p is non-empty.
KF7. $C(p \times q) = C(q \times p)$.
KF8. $C(p - q) = C(p \wedge q) - C(q)$.
KF9. $C(p \times q) = C(p) + C(q) - C(p \wedge q)$.

Proofs of KF8 and KF9. $p \wedge q \Leftrightarrow (p-q) \wedge q$ (according to $A_0 4$). Hence $C(p \wedge q) = C((p-q) \wedge q)$ (according to KF4) $= C(p-q) + C(q)$ (according to $A_0 6$ and *KF3). Hence $C(p-q) = C(p \wedge q) - C(q)$. This completes the proof of KF8.

$p \Leftrightarrow (p-q) \wedge (p \times q)$ (according to $A_0 5$). Hence $C(p) = C(p-q) + C(p \times q)$ (according to $A_0 7$ and *KF3) $= C(p \wedge q) - C(q) + C(p \times q)$ (according to KF8). Hence $C(p \times q) = C(p) + C(q) - C(p \wedge q)$. This completes the proof of KF9.

It is useful to introduce a new symbol (a stroke). This will be done by means of a definition.

DEFINITION. $C(p/q) = C(p \wedge q) - C(q)$.

The stroke symbol will be employed in subsequent theorems:

KF9. $C(p - q) = C(p/q)$
KF10. $C(p/q) \geqslant 0$
KF11.   If $q \Rightarrow p$, then $C(p/q) = 0$.
KF12.   If $p)(q$, then $C(p/q) = C(p)$.
KF13. $C(p \times q) = C(p) - C(p/q)$.
KF14. $C(p \times q) = C(q) - C(q/p)$.
KF15.   If $p)(q$, then $C(p \times q) = 0$.

Interpretation. C has such properties that it is plausible to regard C as a measure of the content of a proposition. The above presented part of KF may thus be described as a (to S_0) restricted content theory.

I proceed next to present a (to S_0) restricted matching theory.

IIIa2. *A Restricted Matching Theory*

My purpose of introducing the function M (for Matching) by means of a postulate linking it to the C-function as I will do in this section is to construct a measure of evidential support. The postulate and the theorems will ensure that M has such properties that it is reasonable to accept $M(p, q)$ as a measure of the support which the proposition q gives to the proposition p.

Here follows a postulate and a number of theorems.

Let M be a function which maps ordered pairs of elements in S_0 on the set of real numbers. M is to satisfy this postulate.

*KF*16.   Let p be a non-empty proposition in S_0 and let q be any proposition in S_0, then $M(p, q) = \dfrac{C(p \times q)}{C(p)}$

From this postulate in conjunction with previous postulates and theorems in KF_0 it is easy to prove the following theorems.

*KF*17. $M(p, q) = \dfrac{C(p) + C(q) - C(p \wedge q)}{C(p)}$.

*KF*18. $M(p, q) = \dfrac{C(p) - C(p/q)}{C(p)}$.

*KF*20. $M(p, q) \leqslant 1$.

*KF*21.   If $q \Rightarrow p$, then $M(p, q) = 1$.

*KF*22.   If $p \Rightarrow q$, then $M(p, q) = \dfrac{C(q)}{C(p)}$.

*KF*23.   If $p \,)(\, q$, then $M(p, q) = 0$.

The following theorem is a very important one in applications. It will be called the distribution theorem.

*KF*24.   (the distribution theorem).

$$M(p, q \wedge r) = M(p, q) + \frac{C(r)}{C(p)}(1 - M(r, q)) - \frac{C(r, p \wedge q)}{C(p)}$$

Proof of *KF*24.

$$C(p)(M(p, q \wedge r) - M(p, q)) = C(p/q) - C(p/q \wedge r),$$

(according to $KF17$)

$$= C(p \wedge q) - C(q) - C(p \wedge q \wedge r) + C(q \wedge r)$$

(according to $KF8$ and $KF9$)

$$= C(r/q) - C(r/p \wedge q) =$$
$$= C(r) - (C(r) - C(r/q)) - C(r/p \wedge q) =$$
$$= C(r)(1 - M(r, q)) - C(r/p \wedge q).$$

Hence

$$M(p, q \wedge r) = M(p, q) + \frac{C(r)}{C(p)}(1 - M(r, q)) - \frac{C(r/p \wedge q)}{C(p)}$$

Comment. If q and r are two bodies of evidence and p is an hypothesis, then the distribution theorem shows how the two bodies of evidence contribute to the support of p.

Interpretation. $M(p, q)$ which measures the degree to which (the content of) q overlaps with or covers (the content of) p can be described as a degree of covering. This notion is closely related to a notion which an historian employs when he tests a narrative. He compares the content of the narrative with that of (veridical) documents. If he finds that the information in the narrative agrees with that of the body of evidence he expresses his finding as follows: "The narrative is (well) covered by the documents".

The theorems above back up the contention that $M(p, q)$ may be taken as a measure of evidential support.

The presentation of the first part of KF, viz. the formalism KF_0 has now been completed. A main reason for proceding to KF_1 ($=KF$) is that KF_0 fails to cater for unfavorable evidence. It caters only for supporting or indifferent evidence.

IIIb. *The Formalism KF_1*

S is a set of propositions which is closed under negation, conjunction and disjunction. S is the 'set part' of a Boolean algebra.

IIIb1. *A Probability Theory*

P is a function which maps S on the set of real numbers. It is specified by the following postulates. p, q and r are elements in S.

KF25. $P(p) \leqslant 1$ for all p in S.
KF26. If p is empty ($=$ tautological), then $P(p) = 1$.
KF27. If $p \Rightarrow q$, then $P(p) \leqslant P(q)$.
KF28. If p)(q, then $P(p \wedge q) = P(p) P(q)$.
KF29. If p and q are mutually inconsistent, then
$$P(p \vee q) = P(p) + P(q).$$

A binary P-measure will also be introduced.

KF30. If $P(q) > 0$, then $P(p, q) = \dfrac{P(p \wedge q)}{P(q)}$.

Here follows a number of theorems:

KF31.   If $p \Leftrightarrow q$, then $P(p) = P(q)$.
KF32. $P(p \vee q) = P(p) + P(q) - P(p \wedge q)$.
KF33. $P(\bar{p}) = 1 - P(p)$.
KF34.   If p is inconsistent, then $P(p) = 0$.
KF35. $P(p) \geqslant 0$.

Proof of KF35. An inconsistent proposition q entails every proposition in S. It follows from *KF27* that $0 = P(q) \leqslant P(p)$. This concludes the proof.

It is presupposed in the following theorems that the binary P-function is defined.

KF36.   If $p \Rightarrow q$, then $P(q, p) = 1$.
KF37.   If p)(q, then $P(p, q) = P(p)$.
KF38. $P(p \vee q, r) = P(p, r) + P(q, r) - P(p \wedge q, r)$.

Proof of KF38.

$$P(r)(P(p \vee q, r)) = P(p \wedge r \vee q \wedge r) = P(p \wedge r \vee \bar{p} \wedge q \wedge r)$$
$$= P(p \wedge r) + P(\bar{p} \wedge q \wedge r) = P(p \wedge r) + P(\bar{p} \wedge q \wedge r) +$$
$$+ P(p \wedge q \wedge r) - P(p \wedge q \wedge r) = P(p \wedge r) + P(q \wedge r) -$$
$$- P(p \wedge q \wedge r)$$

(according to the rules of Boolean algebra in conjunction with *KF31*). As $P(r) > 0$, (otherwise $P(p \vee q, r)$ is not defined) it follows that $P(p \vee q, r) = P(p, r) + P(q, r) - P(p \wedge q, r)$. This concludes the proof of *KF38*.

KF39. $P(\bar{p}/q) = 1 - P(p/q)$.

These postulates show that P is a probability. No attempt at a general interpretation of P will be offered. P is assigned two tasks in KF_1: (1) It will be employed in extending the content and matching theories from the set S_0 to S. (2) It will be modified into a first truth theory by the simple device of adding one more postulate to the P-postulates.

IIIb2. *A First Truth Theory*

V is a function which maps S on the set of real numbers. V satisfies these postulates.

*KF40. $V(p) = 0$ or $V(p) = 1$.
*KF41. The function V satisfies the same postulates as the monadic function P.

Here follows a number of V-theorems:

KF42. If $p \Rightarrow q$, then $V(p) \leqslant V(q)$.
KF43. If $p \Leftrightarrow q$, then $V(p) = V(q)$.
KF44. $V(p \wedge q) = V(p) \, V(q)$.
KF45. $V(p \vee q) = V(p) + V(q) - V(p \wedge q)$.
KF46. $V(\bar{p}) = 1 - V(p)$

Interpretation. The V-postulates and -theorems warrant this interpretation. A proposition p is true if and only if $V(p) = 1$. A proposition p is false if and only if $V(p) = 0$.

IIIb3. *An Extended Content Theory*

I will now proceed to the task of extending the C- and M-functions from S_0 to S. The following theorem will be useful.

THEOREM. The function $- \log P$ satisfies the postulates *KF1, *KF2 and *KF3 for every subset S_0 of S, which consists of individually and mutually consistent propositions.

  Proof.
  (a) If p is empty, then $- \log P(p) = - \log 1 = 0$ (according to KF25). This shows that *KF1 is satisfied.
  (b) If $p \Rightarrow q$, then $P(p) \leqslant P(q)$ according to *KF26. It follows from the properties of the log-function, that $- \log P(p) \geqslant - \log P(q)$. This shows that *KF2 is satisfied.

(c) If p)(q, then $P(p \wedge q) = P(p) P(q)$ (according to *KF27). It follows from the properties of the log-function that $-\log P(p \wedge q) = (-\log P(p)) + (-\log P(q))$ This shows that *KF3 is satisfied. This concludes the proof of the theorem.

COROLLARY. All postulates and theorems from *KF1 → KF23 remain valid if the function C is replaced by the function $-\log P$ on every subset S_0 of S which consists of individually and mutually consistent propositions. The theorem and its corollary justify the introduction of the following postulate:

*KF47. Let p be any proposition in a set S of propositions which is closed under negation, conjunction and disjunction. Then $C(p) = -\log P(p)$.

Here follows a number of theorems in the extended content theory:

KF48.   If $P(q) > 0$, then $-\log P(p, q) = C(p \wedge q) - C(q)$

Proof. If $P(q) > 0$, then $P(p, q) = [P(p \wedge q)]/P(q)$ according to *KF30. It follows from this equality that $-\log P(p, q) = -\log P(p \wedge q) - (-\log P(q)) = C(p \wedge q) - C(q)$ (according to *KF47). Importing the definition of $C(p/q)$ from KF_0 into KF_1 we can rewrite the above theorem as follows:

KF49.   If $P(q) > 0$, then $C(p/q) = -\log P(p, q)$.
KF50.   If p is inconsistent, then $C(p) = \infty$.
KF51.   If $C(q)$ is finite, and if p is inconsistent with q, then $C(p/q) = \infty$

It was shown in KF_0 that $C(p \times q) = C(p) - C(p/q)$ and that $C(p \times q) \geqslant 0$. (Cf KF5 and KF13.) It can be shown in KF_1 that $C(p) - C(p/q)$ may be negative. It is evident that the proposition $p \times q$ is not defined in such a situation. The operator of content intersection can thus not be transferred from S_0 to S.

The next theorem is relevant here.

KF52.   If $C(p) - C(p/q) > 0$, then $C(\bar{p}) - C(\bar{p}/q) < 0$ and if $C(p) - C(p/q) < 0$, then $C(\bar{p}) - C(\bar{p}/q) > 0$

Proof. It is sufficient to prove the first part of KF52.

If $C(p) - C(p/q) > 0$, then $\log P(p) < \log P(p/q)$ (according to *KF47 and KF48).

Hence $P(p) < P(p/q)$. It follows from the subtraction theorems in the P-theory, that $P(\bar{p}) > P(\bar{p}/q)$. Hence $log\ P(\bar{p}) > log P(\bar{p}/q)$ so that $C(\bar{p}) < < C(\bar{p}/q)$ (according to *KF47 and KF48). This concludes the proof.

IIIb4. *An Extended Matching Theory*

The postulate *KF16 cannot be transferred from KF_0 into KF_1 because $p \times q$ is not defined for every pair of elements of S (of KF52). The theorem KF17, however, can be imported and 'raised to the dignity' of a postulate. I will do so.

$M(p, q)$ is a function on pairs of elements in a set S of propositions which is the 'set part' of a Boolean algebra. $M(p, q)$ satisfies this postulate:

*KF53. p is to satisfy this condition $0 < C(p) < \infty$

$$M(p, q) = \frac{C(p) + C(q) - C(p \wedge q)}{C(p)}$$

From this postulate in conjunction with postulates and theorems in the extended content theory it is easy to establish a number of theorems. Most of them have exactly the same form as theorems in the restricted matching theory. Owing to the importance of the matching theory I will nevertheless restate them here in order to have all assertions about matching collected within the same subsection. It will be presupposed throughout that $C(p)$ is positive and finite.

KF54. $M(p, q) = \dfrac{C(p) - C(p/q)}{C(p)} = \dfrac{C(q) - C(q/p)}{C(p)}$.

KF55. $M(p, q) \leqslant 1$.

KF56.   If $q \Rightarrow p$, then $M(p, q) = 1$.

KF57.   If $p \Rightarrow q$, then $M(p, q) = \dfrac{C(q)}{C(p)}$.

KF58.   If $p\ )(\ q$, then $M(p, q) = 0$.

KF59.   (new) $M(p, q) < 0$, if and only if (iff) $M(\bar{p}, q) > 0$ and $M(p, q) > 0$, iff $M(\bar{p}, q) < 0$.

*Proof of KF*59.

$C(p)\ M(p, q) = C(p) - C(p, q)$ (according to KF55) $= \log$ $P(p, q)/P(p)$ (according to *KF47 and KF49).

It follows that

$$C(p)\, M(p, q) \gtrless 0 \text{ iff } P(p, q) \gtrless P(p).$$

In the same way it is shown that

$$C(\bar{p})\, M(\bar{p}, q) \gtrless 0 \quad \text{iff} \quad P(\bar{p}, q) \gtrless P(\bar{p}) \quad \text{and} \quad P(p, q) \gtrless P(p)$$
$$\text{iff } P(\bar{p}, q) \lesssim P(\bar{p}) \text{ (according to the subtraction theorems in the } P\text{-theory).}$$

We conclude that:

$$C(p)\, M(p, q) \gtrless 0 \text{ iff } C(\bar{p})\, M(\bar{p}, q) \lesssim 0$$

It remains to prove that:

$$0 < C(\bar{p}) < \infty \text{ if } 0 < C(p) < \infty$$
$$0 < C(p) < \infty \rightarrow 0 < P(\bar{p}) < 1 \rightarrow 0 < 1 - P(p) < 1 \rightarrow 0 <$$
$$< P(\bar{p}) < 1 \rightarrow 0 < C(\bar{p}) < \infty$$

As both $C(p)$ and $C(\bar{p})$ have positive and finite values we conclude that $KF59$ is valid.

$KF60$.   (new) If q contradicts p, then $M(p, q) = -\infty$

Proof:

$$M(p, q) = \frac{C(p) - C(p/q)}{C(p)} = 1 - \frac{C(p/q)}{C(p)} = 1 + \frac{\log P(p, q)}{C(p)}.$$

q contradicts $p \rightarrow P(p, q) = 0 \rightarrow \log P(p, q) = -\infty \rightarrow M(p, q) = -\infty$. This concludes the proof of $KF60$.

$KF61$ (the distribution theorem).

$$M(p, q \wedge r) = M(p, q) + \frac{C(r)}{C(p)}(1 - M(r, q)) - \frac{C(r/p \wedge q)}{C(p)}.$$

$KF62$.   If $M(p, q \geqslant 0$, then $M(p, q) = C(p \times q)/C(p)$.

Proof of KF62. $M(p, q) \geqslant 0 \rightarrow p$ and q are mutually consistent $\rightarrow p$ and q belong to a subset S_0 of S for which $p \times q$ is defined. $C(p \times q) = C(p) - -C(p/q)$ (according to $KF13$) $= C(p)\, M(p, q)$ (according to $KF54$) $\rightarrow$ $\rightarrow M(p, q) = (C(p \times q))/C(p)$ because $C(p)$ has a positive finite value. This concludes the proof.

I have now completed the presentation of the extended matching theory = the matching theory within *KF*.

Interpretation of the M-function in KF.

(a) If $M(p, q) > 0$, then $M(p, q) = [C(p \times q)]/C(q)$ (according to *KF*62) = the degree to which (the content of the proposition) q covers (the content of the proposition) p. If p is an hypothesis and q is evidence, we may accordingly interpret $M(p, q)$ as a measure of evidential support.

(b) If $M(p, q) = 0$, then q is indifferent to p (in accordance with *KF*58).

(c) If $M(p, q) < 0$, then $M(\bar{p}, q) > 0$ (according to *KF*59). If p stands for an hypothesis and q for evidence we may in accordance with (a) conclude that q supports the negation of p. It follows that q is unfavorable evidence for p.

IIIb5. *A Theory of Degree of Truth*

It occurs often in the history of science that an hypothesis h, which has been confirmed at some occasion, will nevertheless be replaced by another hypothesis h', which is incompatible with h. h' will be confirmed at a later occasion, in spite of its incompatibility with h. Example: Planetary laws before Kepler were replaced by Kepler's laws, which in their turn have been replaced by other planetary laws within Newton's theory of gravitation. Sequences of confirmed hypotheses such as these suggest that confirmed hypotheses may not be perfectly true. It is evident therefore that we need a notion of degree of truth in a theory of knowledge-formation. I will present a theory of degree of truth in this subsection. A concept of degree of truth T will be introduced in two steps. First a notion of truth-core is defined, then T is introduced by means of a postulate.

Definition of truth core. Let p be an element in S. The truth-core of p is a proposition p^* with these properties:

(1) $p \Rightarrow p^*$.

(2) $V(p^*) = 1$.

(3) It holds for every proposition q, such that (a) $p \Rightarrow q$ and (b) $V(q) = 1$, that $p^* \Rightarrow q$. In other words, the content of p^* is the true part of the content of p.

*KF*63. Let p be a proposition in S, such that $C(p) > 0$. Let p^* be the truth-core of p. Then $T(p) = M(p, p^*)$.

Remark. The condition $C(p) > 0$, which must be satisfied in order for $M(p, p^*)$ to be defined, implies that only a proposition with a content is assigned a T-value, i.e. a degree of truth. A number of theorems follow from this postulate in conjunction with previous postulates and theorems in KF.

KF64. $T(p) = \dfrac{C(p^*)}{C(p)}$.

KF65. $0 \leqslant T(p) \leqslant 1$.

KF66.   If $C(p) > 0$, then a) $T(p) = 1$ if and only if $V(p) = 1$, and b) $V(p) = 0$ if and only if $T(p) < 1$.

Proof. p and p^* are consistent. Hence $p - p^*$ and $p \times p^*$ are defined. $p \times p^* \Leftrightarrow p^*$ (according to $A_0 2$) because $p \Rightarrow p^*$. It follows that $p \Leftrightarrow (p - p^*) \wedge p^*$ (according to $A_0 5$). Application of $*KF3$ and $KF44$ to this formula yields the formulas:

$$\left. \begin{array}{l} C(p) = C(p - p^*) + C(p^*) \text{ and} \\ V(p) = V(p - p^*)\, V(p^*) = V(p - p^*) \end{array} \right\} \quad (*)$$

because $V(p^*) = 1$ according to the definition of truth-core.

(I) Suppose that $T(p) = 1$, then $C(p) = C(p^*)$ (according to $KF66$). Hence $C(p - p^*) = 0$ (according to $(*)$). $p - p^*$ is thus empty. It follows from $*KF41$ in conjunction with $*KF26$, that $V(p - p^*) = 1$. Hence $V(p) = 1$ (according to $(*)$).

I have now shown that $T(p) = 1 \Rightarrow V(p) = 1$ (I)

(II) Suppose that $V(p) = 1$. Then $p^* \Rightarrow p$ (according to the definition of truth-core). It holds also according to that definition that $p \Rightarrow p^*$. Hence $p \Leftrightarrow p^*$. It follows from $KF4$ that $C(p) = C(p^*)$. Hence $T(p) = 1$.

I have now shown that $V(p) = 1 \Rightarrow T(p) = 1$.   (II)

It follows from (I) and (II) that $V(p) = 1$ if and only if $T(p) = 1$. This concludes the proof of $KF66(a)$.

$KF66(b)$ follows from $KF66(a)$ as follows:
$T(p) = 1$ iff $V(p) = 1 \rightarrow T(p) \neq 1$ iff $V(p) \neq 1$ (by contraposition). $T(p) \neq 1 \Leftrightarrow T(p) < 1$ (according to $KF65$) and $V(p) \neq 1 \Leftrightarrow V(p) = 0$ (according to $*KF40$).

We may thus conclude that $T(p) < 1$ if and only if $V(p) = 0$. This concludes the proof of $KF66b$) and thus of $KF66$.

KF67.   Let q^* be the truth-core of a proposition q which is compatible with the proposition p.
Then $T(p) \geqslant M(p, q^*) \geqslant 0$

Proof. p and q^* are compatible. Hence $p \times q^*$ is defined. $q^* \Rightarrow p \times q^*$. Hence $V(q^*) = 1 \leqslant V(p \times q^*) \leqslant 1$. Thus $V(p \times q^*) = 1$. $p \Rightarrow p \times q^*$.

The last two formulas together with the definition of truth-core entail that $p^* \Rightarrow p \times q^*$ where p^* is the truth-core of p.

From this relation it follows that:

$$C(p^*) \geqslant C(p \times q^*) = C(p) - C(p/q^*) = C(p) \, M(p, q^*).$$

Hence $T(p^*) \geqslant M(p, q^*)$.

As p and q^* are compatible, it follows that $M(p, q^*) \geqslant 0$. We conclude that $T(p^*) \geqslant M(p, q^*) \geqslant 0$. This ends the proof of *KF*67.

> *KF*68.   If $M(p, q^*) < 0$ where q^* is the truth-core of a proposition q, then $T(p) < 1$.

Proof: The assertion of this theorem is equivalent to the assertion: If $T(p) = 1$, then $M(p, q^*) \geqslant 0$. If $T(p) = 1$, then $V(p) = 1$ according to *KF*66. It follows that p and q^* must be compatible as both of them are true, so that $p \times q^*$ is defined. Hence

$$M(p, q^*) = \frac{C(p \times q^*)}{C(p)} \geqslant 0,$$

because $C(p \times q^*) \geqslant 0$ according to *KF*5.

I have now shown that $T(p) = 1$ entails $M(p, q^*) \geqslant 0$, which is equivalent to the assertion in *KF*68. This concludes the proof.

The last three theorems lead to the following important conclusion: Suppose that an hypothesis h has been first confronted with a body e of evidence, which is completely true and which supports h to the degree $M(h, e)$ and that h afterwards meets another body of completely true evidence e' which is unfavorable to h. The last theorems warrant this assessment of the situation.

(1) The degree of truth of h is at least as large as $M(h, e)$, but (2) the degree of truth of h is less than 100% because of the negative true evidence e'. Hence (3) h is false in the sense that $V(h) = 0$.

An hypothesis may thus have a very high degree of truth *even* if it meets negative evidence which is perfectly true. It seems obvious that this conclusion has a strong bearing on the rationale of adopting the decision to reject an hypothesis in the face of undermining evidence. More about this later on (Cf IVd).

I have now completed the construction of *KF*. It will be used in dealing with those problems about piecemeal knowledge-formation which I have raised in a previous section.

IV. APPLICATION OF THE FORMALISM *KF* TO PROBLEMS CONCERNING PIECEMEAL KNOWLEDGE-FORMATION

IVa. *Criteria on Knowledgehood*

Under what conditions does an hypothesis deserve to be confirmed? I will consider two answers to this question.

(A) An hypothesis *h* deserves to be confirmed (=is a piece of knowledge) if and only if (1) *h* is true in the sense that $V(h)=1$; and if (2) it has been proved that $V(h)=1$.

The second condition is satisfied if and only if a body of evidence *e* has been gathered such that

(2a) $M(h, e)=1$; and

(2b) $V(e)=1$; and

(2c) it is proved that $V(e)=1$.

The last three conditions must be satisfied in order for (2) to be satisfied in accordance with *KF*67, which asserts that $T(h) \geqslant M(h, e^*)$.

This criterion on knowledge-hood is much too severe. It rarely happens that a body of evidence completely covers an hypothesis. The condition (2b) and (2c) generate a nasty regress: *e* must satisfy the same conditions as *h*. Hence there must exist another body of evidence *e'* which satisfies the same conditions as *h*. Hence there must exist another body of evidence *e''* which satisfies the same conditions as *h* and so on. I conclude that the criterion *A* on knowledgehood must be rejected; and I propose another criterion on knowledgehood.

(B) An hypothesis *h* deserves to be confirmed *to the extent that*

(1) $T(h)$ is high; and

(2) there are good grounds to hold that $T(h)$ is high.

(2) is satisfied *to the extent that* a body *e* of evidence has been collected such that

(2a) $M(h, e)$ is large, and that

(2b) $T(e)$ is large.

If (2a) and (2b) are satisfied then it is plausible to conclude that (2) also is satisfied in accordance with the theorem $KF67$, which asserts that $T(h) \geqslant M(h, e^*)$

Comments on (B). (B) but not (A) is compatible with the historical fact that a confirmed hypothesis h is frequently replaced by another confirmed hypothesis h' such that h' is incompatible with h. Such a transition amounts according to (B) to an improvement in knowledge status if $T(h') > T(h)$ and if there are good grounds to hold that this is the case. (B) implies that knowledgehood is not a provable but an 'improvable' property. Once confirmed hypotheses may be disconfirmed later on in favor of other hypotheses with a higher degree of truth. This happens regularly in synthesizing research.

IVb. *Relation Between Evidence and Hypotheses*

This relation is construed as follows:

(1) e is favorable evidence for h, if (the content of) e covers (the content of) h. The evidential strength is measured by the function $M(h, e)$.

(2) e is indifferent to h, if e)(h, in which case $M(h, e) = 0$.

(3) e is unfavorable evidence for h, if e is favorable evidence for h. $M(h, e) < 0$ in such a case. The degree to which e is unfavorable to h may be measured by the function $-M(h, e)$.

IVc. *Two Kinds of Testing*

A researcher has proposed several hypothses h, h', h'',... as tentative solutions to a cognitive problem. Surveying them he may arrive at the opinion that some of them are doubtful (= have low T-values) and that others are credible (= have high T-values). A working hypothesis about an hypothesis h to the effect that $T(h)$ is low provides a motive for designing a testing plan aiming at its refutation. I will call such a plan an elimination plan.

If on the other hand a researcher adopts the working hypothesis that $T(h)$ is large, then he is motivated to design a testing plan aiming at its ultimate confirmation. I will call such a plan a supporting plan.

IVc1. *Attempted Elimination*

A working hypothesis to the effect that a particular hypothesis h has a low degree of truth is presumably based on the opinion that h disagrees with already confirmed hypotheses. It is therefore natural for a researcher who holds that $T(h)$ is low to confront h with already accepted hypotheses.

Suppose that he is able to bring together a body k of confirmed hypotheses such that $M(h, k) < 0$. Does this condition entitle him to refute h? The condition that $M(h, k) < 0$ does not exclude the possibility that $T(h)$ is large. There may exist a body e of evidence such that $M(h, e)$ is positive and large and that $T(e)$ is near 100%, in which case $T(h)$ also would be large.

I will take up the question about the rationale of hypothesis refutation in a later subsection (Cf.IVd).

IVc2. *Attempted Confirmation*

A researcher believes that an hypothesis h has a high T-value. He bases this belief on the feeling that h is in good agreement with already confirmed hypotheses. His working hypothesis provides a motive for him to design a testing plan, aiming at the ultimate confirmation of h. A plan guides him in the process of bringing together evidence. One kind of evidence consists of hypotheses which have already been confirmed. Another kind of evidence consists of data in the form of results of measurement or in the form of documents as in historiography or in the form of reports from interviews as in social anthropology. How should the researcher go about in bringing together the evidence? The distribution theorem in KF is relevant here.

Let h be the hypothesis to be tested, let k be a body of confirmed hypotheses and let d stand for *virtual* data, i.e. information of data type which the tester is looking for. The distribution theorem (D) applied to the triple (h, k, d) asserts that

$$M(h, k \wedge d) = M(h, k) + \frac{C(d)}{C(h)}(1 - M(d, k)) - \frac{C(d/h \wedge k)}{C(h)} \quad \text{(D)}$$

$$\text{(I)} \qquad\qquad\qquad \text{(II)} \qquad\qquad\quad \text{(III)}$$

(Cf. *KF*61).

h has a high T-value if (a) $M(h, k \wedge d)$ is large and if (b) $T(k)$ and $T(d)$ are large (according to *KF*67). The tester wants $M(h, d \wedge k)$ to have a large value. The formula (D) shows that this is the case if (I) and (II) which both may be positive are large and that (III) which can never be positive is near 0. In order to reach his goal the tester should try to bring together a body k of confirmed hypotheses which covers k to a large extent, for (I) is large in that case. He should, moreover, look for data of such a kind that the data adds as little content as possible to h in conjunction with k, because (III) is then near 0. (III) is 0 if d is deduced from h in conjunction with k. The data d should, moreover, satisfy the following condition: $C(d)$ is multiplied by a factor in (II) which cannot be negative. If data have the form of results of measurement it is desirable on this score that they are precise, because precise results of measurement entail less precise ones and have thus a larger C-value according to *KF2. d should, moreover, have the property that $1 - M(d, k)$ is large. $M(d, k)$ may be negative in which case this condition is well satisfied. However, it is not desirable for other reasons that $M(d, k) < 0$, because this condition is satisfied only if d is incompatible with k, in which case d would presumably have an unsatisfactory T-value. The last requirement on d should therefore be that d is nearly independent of k in which case $M(d, k) \sim 0$.

I have now dealt with requirements on the testing plan, which is under human control. When the testing plan is implemented actual data d' are collected. What happens when h is confronted with these data depends on the T-value of h which is outside human control. If replacement of d by d' in the formula (D) leads to the result that the sum of the terms II and III is negative, then data are unfavorable to h even if $M(h, k \wedge d) > 0$. The working hypothesis that $T(h)$ is large is shaken in such a case, and it is doubtful whether it would pay off to proceed to attempt to confirm h (Cf.IVd). If on the other hand d' gives a positive contribution to the covering of h in the sense that $M(h, k \wedge d') > M(h, k)$, then the researcher has to make up his mind whether he should judge that h is acceptable or whether he should bring together further evidence.

Suppose that he opts for the latter alternative. How should the new evidence e' be related to the previous evidence e? It ought to have the property that $M(h, e \wedge e') > M(h, e)$. This means according to (D) that

$$\frac{C(e')}{C(h)} (1 - M(e', e)) - \frac{C(e'/e \wedge h)}{C(h)}$$

ought to be positive and preferably large. This relation shows that entirely new evidence is desirable. If $M(h, e \wedge e') > M(h, e)$, then the researcher faces again the two alternatives stated above, but now it is more reasonable than before to opt for the first of them. What should he do, if he finds on the contrary that $M(h, e \wedge e') < M(h, e)$? This problem will be dealt with in the next subsection.

IVd. *Problems Concerning Negative Evidence*

$M(h, e)$ may be large. Then as $T(h) \geqslant M(h, e^*)$, it follows that h would have a high T-value under the condition that $T(e)$ is large. Should the researcher then simply ignore negative evidence e'? Practicians of research would protest: Negative evidence should be taken seriously! Is such an attitude reasonable or not?

The criteria (A) and (B) on knowledgehood (of Section IVa) are relevant here. An hypothesis h should be rejected if it is confronted with negative true evidence according to A, no matter how well it may be supported by other evidence (Cf *KF*66 and *KF*68).

The condition B recommends that the situation should be handled in another manner. Suppose that h is supported by evidence e. Then h and e are compatible so that $h - e$ and $h \times e$ are both defined. The negative evidence e' is incompatible with $h \times e$, only if e' is incompatible with e. This possibility is ruled out by the truth requirement on the two bodies of evidence e and e'. I conclude that e' is incompatible with h, only if it is incompatible with $h - e$. This means that the part of (the content of) h which is not covered by e may have a low degree of truth.

To reject h altogether is to throw out the baby ($=$ the proposition $h \times e$) with the bath water ($=$ the proposition $h - e$). A more commendable procedure is to employ both e and e' as clues for a modification of h into an hypothesis h' which should satisfy these two conditions:

(1) h' ought to be supported by e to at least the same extent as the hypothesis h.

(2) The evidence e' which is unfavorable to h ought to be either neutral or else favorable to h'.

Under these conditions it is reasonable to believe that h' has a higher degree of truth than h.

It is desirable to follow up the working hypothesis $T(h') > T(h)$ by means of a testing plan for h' of the supporting kind. Such testing might be called discriminatory testing. I will call the operation described above of modifying h into h' instead of rejecting h when h meets negative evidence a refinement if it is successful.

Using this term I will summarize the discussion of negative evidence as follows. A major role which negative evidence should play in knowledge-formation is to provide a motive for as well as a clue for possible refinement-operations which may have the effect of raising the T-level at which hypotheses are confirmed, prior to their acceptance. It may not pay off to try to refine hypotheses which have low T-values to begin with. Such hypotheses deserve to be eliminated (Cf. Section IVc1).

IVe. *Long Run Trends to be Expected from Piecemeal Knowledge-Formation Unaccompanied by Attempts at Synthesis*

In the confrontations between a new hypothesis h and a confirmed hypothesis k, k runs no risk of refutation. I will show this by considering the two (exhaustive) possibilities which have been treated in the Subsections (IVc1) and (IVc2).

Case 1. Suppose that $M(h, k) < 0$. This condition is met only in tests of the eliminative type (Cf IVc1) motivated by the working hypothesis that $T(h)$ is low. k will of course not be rejected in such a case.

Case 2. Suppose that $M(h, k) > 0$. This condition is met in tests of the supporting kind (Cf IVc2). The only risk of refutation that k could run in such a situation is that data d are brought together which satisfy this condition: $M(h, k \wedge d) < M(h, k)$.

But this condition will not shed doubt on k. Unfavorable data are judged not to affect the part $h \times k$ of (the content of) h but instead the part $h - k$, which goes beyond (the content of) k. The body k of previously confirmed hypotheses is thus safe from refutation. The immunity from refutation of once confirmed hypotheses in piecemeal knowledge-formation brings about that this long run trend is to be expected, provided that no attempts at systematization of accepted pieces of knowledge are made.

Knowledge is accumulated without replacement in stores of public knowledge. Refinements may take place, but only prior to confirmation (Cf IVc2). Accumulation means that more and more 'covering material' will be available as evidence in hypothesis testing. Absence of replacements means that the quality (= the T-value) of the 'covering material' may not be raised as time goes on. Quantitative growth is thus unaccompanied by qualitative improvements. A machinery for the improvement of the degree of truth of public knowledge is absent from piecemeal knowledge-formation, but not from research in which syntheses of knowledge

(= theories as in physics or narratives as in historiography) are produced. I conclude that researchers who are concerned with positive knowledge ought not to be satisfied with theory weak research. They ought not to frown at speculations which are essential in synthesizing research. Radical positivism is not a viable ideal of science.

University of Göteborg

BIBLIOGRAPHY

Albert, H.: *Traktat über kritische Vernunft*, Tübingen, 1968.
Bunge, M.: *Method, Model and Matter*, Dordrecht, 1972.
Bunge, M.: *Philosophy of Physics*, Dordrecht, 1973.
Bunge, M.: *Scientific Research*, New York, 1967.
Churchman, C. W.: *The Design of Inquiring Systems*, Cambridge, Mass., 1968.
Danielson, A. and Törnebohm, H.: 'On Complex Systems with Human Components', A. Elzinga (ed.), *FOA P Report C 8212-10*, Stockholm, 1968.
Elzinga, A.: *On a Research Program in Early Modern Physics*, Göteborg, 1973.
Hanson, N. R.: *Patterns of Discovery*, Cambridge, 1958.
Klaus, G.: *Kybernetik und Erkenntnistheorie*, Berlin, 1966.
Kuhn, T. S.: *The Structure of Scientific Revolutions*, Chicago, 1962.
Lakatos, I. and Musgrave, A. (eds.): *Criticism and the Growth of Knowledge*, Cambridge, 1970.
Laszlo, E.: *Introduction to Systems Philosophy*, New York, 1972.
Lindström, J.: 'Vetenskapsteori för något empiriskt forskningsområde', del 1: *Perspektiv och Vetenskapsideal*, No. 29 in RDTS (Reports from the Department of Theory of Science), Göteborg, 1972.
Nagel, E.: *The Structure of Science*, London, 1961.
Popper, K. R.: *Conjectures and Refutations*, London, 1963.
Popper, K. R.: *The Logic of Scientific Discovery*, London, 1959.
Popper, K. R.: *Objective Knowledge*, Oxford, 1972.
Radnitzky, G.: 'Reflections on the Theory of Research', in *RDTS*, No. 34, Göteborg, 1972.
Spinner, H.: *Theorien und Metatheorien* (mimeo), 1973.
Toulmin, S.: *Human Understanding*. I, Oxford, 1972.
Törnebohm, H.: 'Perspectives on Inquiring Systems', in RDTS, No. 54, Göteborg, 1973.
Törnebohm, H.: 'A Systems Approach to Inquiring Systems', in RDTS, No. 54, Göteborg, 1973.
Törnebohm, H.: 'Aggregerade undersökande system i vetenskapens värld', in *RDTS*, No. 57, Göteborg, 1974.
Törnebohm, H.: 'Scientific Knowledge-Formation', in *RDTS*, No. 58, Göteborg, 1974.
Törnebohm, H.: 'Paradigm i vetenskapernas värld och i vetenskapsteorin', in *RDTS*, No. 59, Göteborg, 1974.
Törnebohm, H.: 'An Essay on Knowledge-Formation', in *RDTS*, No. 61, Göteborg, 1974.
Wallén, G. (ed.): *Om Komplexa System*, FOA P Report C 8280-10, Stockholm, 1971.

RAIMO TUOMELA

CONFIRMATION, EXPLANATION,
AND THE PARADOXES OF TRANSITIVITY*

I

The purpose of this paper is to discuss the transitivity paradoxes of (qualitative) confirmation (or support) by connecting the concept of confirmation with that of explanation. We cannot here discuss in any detail the general philosophical reasons for making this connection. But let it be said that even a brief consultation of scientific examples should convince one of 'facts' such as that when a theory explains some statements (explananda) it gives confirmatory credence to these statements, and these explananda can also be taken to support or confirm the theory. What we shall do below is to spell out this in some detail and show that even by means of such small, intuitively obvious, observations one can show that many of the transitivity paradoxes of confirmation can be avoided or blocked. In other words, the main idea in our paper is, so to speak, to 'localize' some general principles of confirmation in the sense of restricting their range of applicability to nomological explanation situations.

Let us start our discussion by stating some of the conditions of confirmation which have proved to be problematic. We write '$h I g$' for 'h confirms g'. Here h and g are statements in some scientific language.

(C1)   For all h there is a g such that not $h I g$.   (Nonuniversalizability)

(C2)   If $h \vdash g$ then $h I g$.   (Entailment)

(C3)   If $h \vdash g$ then $g I h$.   (Converse entailment)

(C4)   If $h I g$ and $k \vdash g$ then $h I k$.   (Converse consequence)

(C5)   If $h I g$ and $g \vdash k$ then $h I k$.   (Special consequence)

Of these conditions (C4) and (C5) state weak transitivity properties for the relation I. To gain some insight into the paradoxes of transitivity we shall present three 'classical' examples of such paradoxes (cf. Hempel (1945) and Skyrms (1966)). We prove, first, that the set $\{C1, C3, C5\}$ is

R. J. Bogdan (ed.), Local Induction, 319–328. All rights reserved.
Copyright © 1976 by D. Reidel Publishing Company, Dordrecht-Holland

inconsistent. Secondly and thirdly, we prove that the sets $\{C1, C4, C5\}$ and $\{C1, C2, C4\}$ are inconsistent.

(I)   (1) $h \& g \vdash h$ for all h, g, by propositional logic
     (2) $h I h \& g$ by (C3) from (1)
     (3) $h \& g \vdash g$ by propositional logic
     (4) $h I g$ from (2) and (3) by (C5), contradicting (C1)

(II) (1) $e I h$ assume for some e, h
     (2) $h \& g \vdash h$ for any g, by propositional logic
     (3) $e I h \& g$ from (1) and (2) by (C4)
     (4) $h \& g \vdash g$ by propositional logic
     (5) $e I g$ from (3) and (4) by (C5), contradicting (C1)

(III) (1) $h \vdash h \vee g$ for any h, g, by propositional logic
     (2) $h I h \vee g$ from (1) by (C2)
     (3) $g \vdash h \vee g$ by propositional logic
     (4) $h I g$ from (2) and (3) by (C4), contradicting (C1)

Our first two proofs have been reconstructed from Hempel (1945) and the third is a proof by Skyrms (1966). To us these proofs concretely illustrate that there is something wrong with conditions (C2)–(C5). Consider the first proof. If g is completely irrelevant (whatever that is taken to mean) to h, then step (2) seems to be unwarranted. The same can be said of step (4). Therefore, we may conclude, rules (C3) and (C5) do not qualify without restrictions, if at all. In proof (II), if h and g are irrelevant to each other, steps (3) and (5) are not intuitively acceptable. The same remark applies to steps (2) and (4) in (III). These observations indicate that in addition to (C3) and (C5), (C2) and (C4) also have to be rejected in their above form.

However, we wish to suggest that there is something acceptable in these rules: if a relation of *explanation* holds between two statements one of which implies the other, then these statements convey confirmation to each other.[1] In other words, it is philosophically incorrect to connect confirmation with deducibility *per se*. The connection sought after holds between confirmation and explanatory deducibility only. The acceptance of this suggestion immediately disqualifies proofs (I)–(III) above, because deductions of the kind $h \& g \vdash g$ and $h \vdash h \vee g$ do not qualify as explanations. The former is completely circular and the latter is partially circular and does not show the relevance of h to g.

Let us write '$h\,E\,g$' for 'h explains g'. (We shall here be concerned with nomological deductive explanation only.) Then we suggest that the following necessary conditions of confirmation might be regarded as severally and perhaps also jointly acceptable:

(C2*)   If $h\,E\,g$ then $h\,I\,g$.
(C3*)   If $h\,E\,g$ then $g\,I\,h$.
(C4*)   If $h\,I\,g$ and $k\,E\,g$ then $h\,I\,k$.
(C5*)   If $h\,I\,g$ and $g\,E\,k$ then $h\,I\,k$.

There is a number of other well-known necessary conditions for I. Here we shall not deal with them except for suggesting that the substitution of '⊢' by 'E' never makes a condition less acceptable.

Conditions (C2*)–(C5*) at least formally disqualify the above proofs of the paradoxes of transitivity. However, they may not be very interesting as such unless supplemented with specific explicates for relations I and E.

II

We shall concentrate on a situation in which we have a statement (or a set of statements) which is said to deductively explain another statement. Our relation 'E' will rely on the idea that the explanans must contain a law or theory which together with some initial conditions will have to logically imply the explanandum statement. However, as recent discussion has shown, that is far from sufficient. We need a few other requirements, which will enable us to pick out explanatory deductive arguments from the set of all deductive arguments. (It suffices to be concerned only with the 'logico-inferential' conditions of explanation for the purposes of this paper.)

One central requirement for deductive explanations is that they should not be more circular than what is 'necessary'. By this we mean the following. In order to have a deductive relationship between a potential explanans and an explanandum there must be some common content in them. However, to avoid excessive self-explanation and circularity, such as found in the deductions in proofs (I)–(III) above, some precautions have to be taken. As recent discussion on explanation has shown, the

following kind of general condition has to be accepted: In an explanation
the components of the explanans and the explanandum should be in a
sense noncomparable. We say that two components or statements p and q
are *noncomparable* exactly when not $p \vdash q$ and not $q \vdash p$. (See e.g. Ackermann (1965) and Tuomela (1972) for a discussion leading to the acceptance of this general condition.) Actually our analysis of noncomparability needs some refinement, mainly because of the vagueness of the
notion of a component. To accomplish this we use the notions of a
sequence of truth functional components of an explanans and of a set
of ultimate sentential conjuncts of an explanans (cf. Ackermann and
Stenner (1966)).

A sequence of statemental well formed formulas $\langle w_1, w_2, ..., w_n \rangle$ of a
scientific language $\mathscr{L}$ is a *sequence of truth functional components* of an
explanans t if and only if t may be built up from the sequence by the
formation rules of $\mathscr{L}$, such that each member of the sequence is used
exactly once in the application of the rules in question. The w_i's are thus
to be construed as tokens. The formation rules of $\mathscr{L}$ naturally have to be
specified in order to see the exact meaning of the notion of a sequence of
truth functional components of a theory finitely axiomatized by a sentence
t. A *set of ultimate sentential conjuncts tc* of a sentence t is any set whose
members are the well formed formulas of the longest sequence $\langle w_1, w_2,$
$..., w_n \rangle$ of truth functional components of t such that t and w_1 & w_2
& ... & w_n are logically equivalent. If t is a set of sentences then the set
tc of ultimate conjuncts of t is the union of the sets of ultimate sentential
conjuncts of each member of t. We may notice here that although by
definition the tc-sets of two logically equivalent theories are logically
equivalent they need not be the same.

Now we are ready to state a better version of the noncomparability
requirement for a tc of a theory t constituting an explanans (cf. Tuomela
(1975a)): For any tc_i in the largest set of truth functional components
of t, tc_i is noncomparable with the explanandum.

In addition to this condition we require that the explanans and the
explanandum of an explanation are consistent, that the explanans
logically implies the explanandum, and that the explanans contains some
universal laws.

Finally, there is a nontrivial logical condition for E which guarantees
that an explanans provides a proper amount of relevant information.

This is condition (5) below. (The reader is referred to Tuomela (1972) and (1975a) for a discussion of its acceptability.) Now we can state the logico-inferential conditions for our model of explanation (termed the *weak DEL model* in Tuomela (1972)). Let t be a statement, tc a set of ultimate sentential components of t (or actually a conjunction of components in the context $tcEl$), and l a statement to be explained. Then we say that the relation $tcEl$ satisfies the logico-inferential conditions of adequacy for the deductive explanation of (singular or general) scientific statements if and only if

(1) $\{l, tc\}$ is consistent;

(2) $tc \vdash l$;

(3) tc contains at least some universal laws;

(4)   for any tc_i in the largest set of truth functional components of l, tc_i is noncomparable with l;

(5)   it is not possible, without contradicting any of the previous conditions for explanation, to find sentences $s_i, ..., s_r (r \geqslant 1)$ at least some of which are essentially universal such that for some $tc_j, ..., tc_n$ $(n \geqslant 1)$:

        tc_j & ... & $tc_n \vdash_p s_i$ & ... & s_r;

        not s_i & ... & $s_r \vdash tc_j$ & ... & tc_n;

        $tc_s \vdash l$,

where tc_s is the result of the replacement of $tc_j, ..., tc_n$ by $s_i, ..., s_r$ in tc, and '$\vdash_p$' means 'deducible by means of predicate logic but not by means of universal or existential instantiation only'.

The reader is referred to Tuomela (1972) and (1975a) for a detailed discussion of the formal properties of the notion of explanation that this model generates. Here it must suffice to make the following general remarks only.

In the above model of explanation an explanandum may have several explanantia differing in their quantificational strength (depth). On each quantificational level, however, only the weakest explanans-candidate qualifies. Our model thus generates an explanation-tree for each explanandum such that the explanantia in different branches may be incompatible whereas the explanatia within the same branches are compatible and increasingly stronger.

III

As we already saw above, Hempel's and Skyrms' paradoxical proofs do not apply in the case of conditions (C2*)–(C5*), provided that E satisfies the noncomparability condition (condition (4) of our model). It is still quite possible, of course, that paradoxes of transitivity may arise in some other ways. It would not be surprising if some specific probabilistic interpretation of I would lead into a paradox. Let us briefly examine this latter possibility.

We consider the *high probability* (HP) and *positive relevance* (PR) criteria:

(HP) $h \, I \, g$ if and only if $p(g/h) > 1 - \varepsilon$ $(\varepsilon < 0.50)$
(PR) $h \, I \, g$ if and only if $p(g/h) > p(g)$.

It is well known that these two interpretations satisfy different conditions of confirmation. Here we shall only be concerned with the critical and important conditions (C2)–(C5) and especially with their modified versions (C2*)–(C5*). First we may notice that (HP) satisfies (C2) and (C5) but not (C3) and (C4). The positive relevance criterion (provided that $p(g) < 1$) satisfies (C2) and (C3) but fails to satisfy (C4) and (C5). This well-known result can be taken to indicate that conditions (C2)–(C5) are not in general jointly acceptable, even if they may be so considered severally. The viewpoint of this paper is, however, that all of these conditions should be rejected. Instead we should consider the conditions (C2*)–(C5*). But here again, even if they may be severally acceptable, they are not, generally and without qualifications, jointly acceptable, as we shall soon see.

It has been argued by Smokler (1968) that we have two fundamentally conflicting intuitive notions of confirmation. One is represented by 'abductive inference' and the other is called 'enumerative inference'. Smokler claims that of the conditions being investigated here abductive inference satisfies (C3) and (C4) but fails to satisfy (C2) and (C5). Enumerative inference satisfies (C2) and (C5) but does not satisfy (C3) and (C4). On the other hand, Smokler seems to take the high probability criterion to be representative of enumerative inference and the positive relevance criterion to illustrate abductive inference (cf. Smokler (1968), pp. 310–312). But as we just saw, positive relevance does not satisfy the same criteria as Smokler's abductive inference. This may suggest that we have a multiplicity of conflicting intuitive notions of confirmation rather than just two of them. Or, alternatively, it may be taken to suggest that the dichotomy has been incorrectly established.

On the basis of the fact that within our model of explanation $h E g$ implies $h \vdash g$ it is easily seen that, considering the matter *in general* and without qualifications on the probability measures used, the high probability criterion satisfies (C2*) and (C5*) but not (C3*) and (C4*), whereas positive relevance satisfies (C2*) and (C3*) but not (C4*) and (C5*) (cf. however, Tuomela (1973), Ch. VIII). Our remarks on abductive and enumerative inference carry over analogously.

Now it seems natural to argue that, in slight contrast with Smokler (1968), at least conditions (C1), (C2*), (C3*) and the well-known consistency, and perhaps symmetry, conditions implicitly characterize abductive inference. (Notice that our conditions (C2*) and (C3*) imply that I is symmetrical, provided its arguments are in explanatory relationship.) Positive relevance satisfies these conditions and is hence at least partially representative of abductive inference, whereas, for instance, high probability represents enumerative inference (characterized primarily by conditions (C1), (C2*) and (C5*)).

Our results do not prove that conditions (C2*)–(C5*) cannot be jointly satisfied (together with (C1)). But what we have shown is that Hempelian and Skyrmsian type of proofs will not be found to prove that. Still, the fact that neither positive relevance nor high probability criteria generally satisfy all of (C2*)–(C5*) strongly indicates that these conditions contribute to the formalization of at least two different notions of confirmation.

We have above concentrated on the scientifically central situations in which the concern is the confirmatory relations between theories (laws), on the one hand, and singular statements or theories, on the other hand. Our main observations carry to the cases of singular explanation and confirmation as well. If we let h and g be singular statements we may define a relation of singular explanation E^* for them as follows: $h E^* g$ if and only if there is a nomological theory t (which possibly contains some singular statements, too) such that t jointly with h deductively explains g in the sense of our explanatory relation E. Of course now if $h E^* g$ then h alone need not entail g, and this makes some of our above comments for E false in the case of E^*. But what we have said against Hempel's and Skyrms' proofs of course comes out true here as well. We cannot here go into any a detailed investigation of the properties of E^* (see Tuomela (1975b) for that). Suffice it to remark only that in this case we have a really clear case of relativization to a body of knowledge, i.e. to the theory backing the singular explanation.

IV

We shall end this paper by discussing briefly an almost classical example, which purports to show the inadequacy of contemporary confirmation theory by means of the nontransitivity of the notions of confirmation. This example in a sense goes beyond our above cases. But we claim that a correct analysis of this situation is complementary, and far from contradictory, to what we have said earlier in this paper.

Let us thus recall the argument concerning the 'consilience of inductions' discussed by Putnam (1963) and Hesse (1970). Before the atomic theory was tested by means of atomic bomb explosions it had received plenty of physical and chemical confirmation from some of its consequences (call them e_1). Let us call the atomic theory (plus whatever initial conditions are needed) t, and let e_2 denote the statement for atomic bomb explosions. We can agree that the prediction of e_2 from t and e_1 could be made with great inductive assurance. However, Putnam and Hesse argue, current confirmation theory is unable to handle this situation without paradoxes. By this they mean that current confirmation theory is not able to explicate why e_1 confirms e_2 in this situation. To see why, we present the wanted argument as follows (we assume that t potentially explains both e_1 and e_2):

(1)	$t\,E\,e_1$	assume
(2)	$t\,E\,e_2$	assume
(3)	$e_1\,I\,t$	from (1) by (C3*)
(4)	$e_1\,I\,e_2$	from (2) and (3) by (C5*).

But as we saw above, this argument does not seem acceptable, as none of the usual explicates of I satisfy both (C3*) and (C5*) for all, or even for all 'interesting', probability measures. Let us here admit this.

However, we may argue that the whole situation has been analyzed incorrectly. What we want is a confirmational justification for e_2, given e_1 *and* the theory t. We may analyze this to mean that we want the direct relationship

$$(e_1 \,\&\, t)\, I \, e_2$$

rather than transitivity of confirmation as above. That is, we want for e_2 *direct* empirical *cum* theoretical support on the basis of $e_1 \,\&\, t$ rather than *indirect* empirical support by e_1 *via* theory t. Thus, for instance, taking I to be positive probabilistic relevance we have

$$(e_1 \,\&\, t)\, I \, e_2,$$

while it need not be the case that

$$e_1 \, I \, e_2.$$

As is shown by Niiniluoto and Tuomela (1973), our present suggestion can indeed be worked out in detail at least for some interesting cases. These authors show, among other things, that the above idea for avoiding inductive transitivity paradoxes can also be applied to *inductive* explanation (by interpreting 'E' as positive probabilistic relevance) and, more generally, to what these authors call *hypothetico-inductive inference*.

University of Helsinki

NOTES

* This is a revised version of a paper read at the XVth World Congress of Philosophy, Varna, Bulgaria, 1973, and forthcoming in its *Proceedings*.
[1] This suggestion has in effect been made by Brody (1968), but he does not go on to clarify explanation in any detail. For critical remarks against his suggestion as far as non-deductive explanation is concerned see Niiniluoto and Tuomela (1973).

BIBLIOGRAPHY

Ackermann, R.: 1965, 'Deductive Scientific Explanation', *Philosophy of Science* **32**, 155–167.
Ackermann, R. and Stenner, A.: 1965, 'A Corrected Model of Explanation', *Philosophy of Science* **33**, 168–171.
Brody, B. A.: 1968, 'Confirmation and Explanation', *Journal of Philosophy* **65**, 282–299.
Hempel, C. G.: 1945, 'Studies in the Logic of Confirmation', *Mind* **54**, 1–26, 97–121.

Hesse, M. B.: 1970, 'Theories and the Transitivity of Confirmation', *Philosophy of Science* 37, 50–63.

Niiniluoto, I. and Tuomela, R.: 1973, *Theoretical Concepts and Hypothetico-Inductive Inference*, D. Reidel, Dordrecht-Holland.

Putnam, H.: 1963, 'Degrees of Confirmation and Inductive Logic', in *The Philosophy of Rudolf Carnap* (ed. by P. A. Schilpp), pp. 761–784, Open Court, La Salle.

Skyrms, B.: 1966, 'Nomological Necessity and the Paradoxes of Confirmation', *Philosophy of Science* 33, 230–249.

Smokler, H.: 1968, 'Conflicting Concepts of Confirmation', *The Journal of Philosophy* LXV, 300–312.

Tuomela, R.: 1972, 'Deductive Explanation of Scientific Laws', *Journal of Philosophical Logic* 1, 369–392.

Tuomela, R.: 1973, *Theoretical Concepts*, Springer-Verlag, Wien.

Tuomela, R.: 1975a, 'Morgan on Deductive Explanation: A Rejoinder'; forthcoming in *Journal of Philosophical Logic*.

Tuomela, R.: 1975b, 'Causes and Deductive Explanation', forthcoming in *PSA* 1974, D. Reidel, Dordrecht.

RADU J. BOGDAN

A SELECTED BIBLIOGRAPHY OF LOCAL INDUCTION

This bibliography attempts to include the most important works on, or related to, local induction. However, its scope is far from being exhaustive. For more comprehensive bibliographies the reader is referred to R. L. Slaght's in Swain [8] (which also mentions significant reviews) and of course Kyburg's [15] and [16].

The first section contains anthologies including several studies on local induction and related problems, technical books often referred to in the literature of local induction, and historical surveys. The second section covers the mainstream of contemporary local induction, i.e. those works dealing with cognitive decision theory, epistemic utilities, and criteria and rules of acceptance. The third comprises works stressing the formal and linguistic parameters of inductive inferences and probability assignments. The fourth section is devoted to other relevant approaches to local induction, and the last section mentions some recent attempts to uncover the local dimensions of inductive systematizations and explanations.

I. REFERENCE WORKS

Anthologies

[1] R. J. Bogdan and I. Niiniluoto (eds.), *Logic, Language, and Probability*, D. Reidel, Dordrecht-Holland, 1973.

[2] J. Hintikka and P. Suppes (eds.), *Aspects of Inductive Logic*, North-Holland, Amsterdam, 1966.

[3] J. Hintikka and P. Suppes (eds.), *Information and Inference*, D. Reidel, Dordrecht-Holland, 1970.

[4] H. E. Kyburg, Jr. and H. E. Smokler (eds.), *Studies in Subjective Probability*, Wiley and Sons, New York, 1964.

[5] I. Lakatos (ed.), *The Problem of Inductive Logic*, North-Holland, Amsterdam, 1968.

[6] S. A. Luckenbach (ed.), *Probabilities, Problems, and Paradoxes*, Dickenson, Encino, Calif., 1972.

[7] P. A. Schilpp (ed.), *The Philosophy of Rudolf Carnap*, Open Court, La Salle, 1963.

[8] M. Swain (ed.), *Induction, Acceptance, and Rational Belief*, D. Reidel, Dordrecht-Holland, 1970.

Technical Books

[9] H. Chernoff and L. Moses, *Elementary Decision Theory*, Wiley and Sons, New York, 1959.

[10] I. J. Good, *The Estimation of Probabilities*, MIT Press, Cambridge, Mass., 1965.

[11] D. Luce and H. Raiffa, *Games and Decisions*, Wiley and Sons, New York, 1957.

[12] J. von Neumann and O. Morgenstern, *Theory of Games and Economic Behavior*, Princeton University Press, Princeton, 1944.

R. J. Bogdan (ed.), Local Induction, 329–336. All rights reserved.
Copyright © 1976 by D. Reidel Publishing Company, Dordrecht-Holland.

[13] H. Raiffa, *Decision Analysis*, Addison Wesley, Reading, Mass., 1968.
[14] L. J. Savage, *The Foundations of Statistics*, Wiley and Sons, New York, 1954.

Surveys

[15] H. E. Kyburg, Jr., 'Recent Works in Inductive Logic', *American Philosophical Quarterly* **1** (1964), 1–39.
[16] H. E. Kyburg, Jr., *Probability and Inductive Logic*, MacMillan, London, 1970.
[17] I. Lakatos, 'Changes in the Problem of Inductive Logic', in Lakatos [5].
[18] A. C. Michalos, *The Popper-Carnap Controversy*, Nijhoff, The Hague, 1971.
[19] E. Nagel, *Principles of the Theory of Probability*, University of Chicago Press, Chicago, 1939.
[20] J. Pietarinen, *Lawlikeness, Analogy, and Inductive Logic* (Acta Philosophica Fennica, XXVI), North-Holland, Amsterdam, 1972.
[21] W. C. Salmon, *The Foundations of Scientific Inference*, University of Pittsburgh Press, Pittsburgh, 1967.
[22] F. Schick, 'Three Logics of Belief', in Swain [8], pp. 6–26.

II. DECISION THEORY, EPISTEMIC UTILITY, AND ACCEPTANCE

Monographs

[23] R. Ackermann, *Nondeductive Inference*, Dover, New York, 1966.
[24] A. W. Burks, *Cause, Chance, and Reason*, Forthcoming.
[25] C. W. Churchman, *Prediction and Optimal Decision*, Prentice-Hall, Englewood Cliffs, N.J., 1961.
[26] R. Hilpinen, *Rules of Acceptance and Inductive Logic* (Acta Philosophica Fennica, XXII), North-Holland, Amsterdam, 1968.
[27] R. C. Jeffrey, *The Logic of Decision*, McGraw-Hill, New York, 1965.
[28] K. Lehrer, *Knowledge*, Oxford University Press, Oxford, 1974.
[29] I. Levi, *Gambling with Truth*, A. Knopf, New York, 1967.
[30] I. Niiniluoto and R. Tuomela, *Theoretical Concepts and Hypothetico-Inductive Inference*, D. Reidel, Dordrecht-Holland, 1973.

Articles

[31] R. B. Braithwaite, 'The Role of Values in Scientific Inference', in H. E. Kyburg, Jr. and E. Nagel (eds.), *Induction: Some Current Issues*, Wesleyan U.P., Middletown, 1963, pp. 180–193.
[32] R. Carnap, 'The Aim of Inductive Logic', in E. Nagel, P. Suppes, and A. Tarski (eds.), *Logic, Methodology, and Philosophy of Science I*, Stanford U.P., Stanford, 1962, pp. 303–318. Also in [6].
[33] J. H. Fetzer, 'Elements of Induction', *this volume*, p. 145.
[34] H. A. Finch, 'Confirming Power of Observations Metricized for Decisions Among Hypotheses', *Philosophy of Science* **27** (1960), 293–307, and 391–404.
[35] I. J. Good, 'Weight of Evidence, Corroboration, Explanatory Power, and the Utility of Experiments', *Journal of the Royal Statistical Society* **22** (1960), 319–331.
[36] W. K. Goosens, 'A Critique of Epistemic Utilities', *this volume*, p. 93.
[37] C. G. Hempel, 'Inductive Inconsistencies', *Synthese* **12** (1960), 439–469.
[38] C. G. Hempel, 'Deductive-Nomological vs Statistical Explanation', in H. Feigl

and G. Maxwell (eds.), *Minnesota Studies in Philosophy of Science III*, University of Minnesota Press, Minneapolis, 1962, pp. 98–169.

[39] C. G. Hempel, 'Recent Problems of Induction', in R. G. Colodny (ed.), *Mind and Cosmos*, University of Pittsburgh Press, Pittsburgh, 1966, pp. 112–134. Also in Luckenbach [6].

[40] R. Hilpinen (see [43]).

[41] R. Hilpinen, 'On the Information Provided by Observations', in Hintikka and Suppes [3], pp. 97–122.

[42] R. Hilpinen, 'Decision-Theoretic Approaches to Rules of Acceptance', in R. E. Olson and A. M. Paul (eds.), *Contemporary Philosophy in Scandinavia*, The John Hopkins Press, Baltimore, 1972, pp. 147–168.

[43] J. Hintikka and R. Hilpinen, 'Knowledge, Acceptance, and Inductive Logic', in Hintikka and Suppes [2], pp. 1–20.

[44] J. Hintikka and J. Pietarinen, 'Semantic Information and Inductive Logic', in Hintikka and Suppes [2], pp. 96–113.

[45] J. Hintikka, 'The Varieties of Information and Scientific Explanation', in van Rootselaar and Staal (eds.), *Logic, Methodology, and Philosophy of Science III*, North-Holland, Amsterdam, 1968, pp. 311–331.

[46] R. C. Jeffrey, 'Valuation and Acceptance of Scientific Hypotheses', *Philosophy of Science* 23 (1956), 237–246.

[47] R. C. Jeffrey, 'New Foundations for Bayesian Decision Theory', in Y. Bar-Hillel (ed.), *Logic, Methodology, and Philosophy of Science II*, North-Holland, Amsterdam, 1964, pp. 289–300.

[48] R. C. Jeffrey, 'Probable Knowledge', in Lakatos [5], pp. 166–180.

[49] R. C. Jeffrey, 'The Whole Truth', *Synthese* 18 (1968), 24–27.

[50] R. C. Jeffrey, 'Dracula Meets Wolfman: Acceptance vs Partial Belief', in Swain [8], pp. 157–185.

[51] R. C. Jeffrey, 'Preferences Among Preferences', *Journal of Philosophy* 71 (1974), 377–391.

[52] H. E. Kyburg, Jr., 'Probability, Rationality, and a Rule of Detachment', in Y. Bar-Hillel (ed.), *Logic, Methodology, and Philosophy of Science II*, North-Holland, Amsterdam, 1964, pp. 301–310.

[53] H. E. Kyburg, Jr., 'Probability and Decision', *Philosophy of Science* 33 (1966), 250–261.

[54] H. E. Kyburg, Jr., 'The Rule of Detachment in Inductive Logic', in Lakatos [5], pp. 98–119.

[55] H. E. Kyburg, Jr., 'Conjunctivitis', in Swain [8], pp. 55–82.

[56] H. E. Kyburg, Jr., 'Local and Global Induction', *this volume*, p. 191.

[57] K. Lehrer, 'Induction: A Consistent Gamble', *Noûs* 3 (1969), 285–297.

[58] K. Lehrer, 'Induction, Reason, and Consistency', *The British Journal for the Philosophy of Science*, 21 (1970), 103–114.

[59] K. Lehrer, 'Justification, Explanation, and Induction', in Swain [8], pp. 100–133.

[60] K. Lehrer, 'Induction and Conceptual Change', *Synthese* 23 (1971), 206–25.

[61] K. Lehrer, 'Evidence and Conceptual Change', *Philosophia* 2 (1972), 273–281. Also in Bogdan and Niiniluoto [1], pp. 100–107.

[62] K. Lehrer, 'Evidence, Meaning, and Conceptual Change: A Subjective Approach', in G. Pearce and P. Maynard (eds.), *Conceptual Change*, D. Reidel, Dordrecht-Holland, 1973, pp. 94–122.

[63] K. Lehrer, Truth, Evidence, and Inference', *American Philosophical Quarterly* 11 (1974), 79–92.

[64] K. Lehrer, 'Induction, Consensus, and Catastrophe', *this volume*, p. 115.

[65] K. Lehrer, 'Social Consensus and Rational Agnoiology', forthcoming.

[66] K. Lehrer, 'When Rational Disagreement is Impossible', forthcoming.

[67] I. Levi, 'Must the Scientist Make Value Judgments', *Journal of Philosophy* 57 (1960), 345–357.

[68] I. Levi, 'Decision Theory and Confirmation', *Journal of Philosophy* 58 (1961), 614–625.

[69] I. Levi, 'On the Seriousness of Mistakes', *Philosophy of Science* 29 (1962), 47–65.

[70] I. Levi, 'Corroboration and Rules of Acceptance', *The British Journal for the Philosophy of Science* 13 (1963), 307–313.

[71] I. Levi, 'Belief and Action', *The Monist* 48 (1964), 306–316.

[72] I. Levi, 'Deductive Cogency in Inductive Inference', *Journal of Philosophy* 62 (1965), 68–77.

[73] I. Levi, 'On Potential Surprise', *Ratio* 8 (1966), 107–129.

[74] I. Levi, 'Information and Inference', *Synthese* 17 (1967), 369–391.

[75] I. Levi, 'Utility and Acceptance of Hypotheses', in S. Morgenbesser (ed.), *Philosophy of Science Today*, New York, 1967.

[76] I. Levi, 'Probability Kinematics', *The British Journal for the Philosophy of Science* 18 (1967–68), 197–209.

[77] I. Levi, 'Probability and Evidence', in Swain [8], pp. 134–156.

[78] I. Levi, 'Certainty, Probability, and the Correction of Evidence', *Nous* 5 (1971), 299–312.

[79] I. Levi, 'On Indeterminate Probabilities', *Journal of Philosophy* 71 (1974), 391–418.

[80] I. Levi, 'Acceptance Revisited', *this volume*, p. 1.

[81] J. Marschak, 'Economics of Information Systems', *Journal of American Statistical Association* 66 (1971), 192–219.

[82] J. Marschak, 'Optimal Systems for Information and Decision', in A. Balakrishnan (ed.), *Techniques of Optimization*, Academic Press, New York, 1972, pp. 355–369.

[83] J. Marschak, 'Prior and Posterior Probabilities and Semantic Information', in G. Menges (ed.), *Information, Inference, and Decision*, D. Reidel, Dordrecht-Holland, 1974, pp. 167–180.

[84] J. Marschak, 'Information, Decision, and the Scientist', in C. Cherry (ed.), *Pragmatic Aspects of Human Communication*, D. Reidel, Dordrecht-Holland, 1974, pp. 145–178.

[85] G. Menges, 'On Subjective Probability and Related Problems', *Theory and Decision* 1 (1970), 44–60.

[86] G. Menges, 'Inference and Decision', *Selecta Statistica Canadiana* 1 (1973), 1–14.

[87] G. Menges, 'Elements of An Objective Theory of Inductive Behavior', in G. Menges (ed.), *Information, Inference, and Decision*, D. Reidel, Dordrecht-Holland, 1974, pp. 3–49.

[88] G. Menges and E. Kofler, 'Cognitive Decisions under Partial Information', *this volume*, p. 183.

[89] A. C. Michalos, 'Cost-Benefit vs Expected Utility Acceptance Rules', in Michalos [18].

[90] I. Niiniluoto, 'Inquiries, Problems, and Questions', *this volume*, p. 263.
[91] F. P. Ramsey, 'Truth and Probability', in Kyburg and Smokler [4], pp. 61–92.
[92] N. Rescher, 'Notes on Preference, Utility, and Cost', *Synthese* **16** (1966), 332–343.
[93] R. Rosenkrantz, 'Experimentation as Communication with Nature', in Hintikka and Suppes [3], pp. 58–93.
[94] R. Rosenkrantz, 'Cognitive Decision Theory', *this volume*, p. 73.
[95] R. S. Rudner, 'The Scientist Qua Scientist Makes Value Judgments', *Philosophy of Science* **20** (1953), 1–6.
[96] L. J. Savage, 'The Foundations of Statistics Reconsidered', in Kyburg and Smokler [4], pp. 175–188.
[97] J. D. Sneed, 'Strategy and the Logic of Decision', *Synthese* **16** (1966), 270–283.
[98] J. D. Sneed, 'Entropy, Information, and Decision', *Synthese* **17** (1967), 392–407.
[99] P. Suppes, 'The Philosophical Significance of Decision Theory', *Journal of Philosophy* **58** (1961), 605–614.
[100] P. Suppes, 'Concept Formation and Bayesian Decision', in Hintikka and Suppes [2], pp. 21–48.
[101] P. Suppes, 'Information Processing and Choice Behavior', in I. Lakatos and A. Musgrave (eds.), *Problems in the Philosophy of Science*, North-Holland, Amsterdam, 1968, pp. 278–299.
[102] P. Suppes, 'The Role of Subjective Probability and Utility in Decision Making', in R. Luce *et al.* (eds.), *Readings in Mathematical Psychology*, New York, 1965.
[103] K. Szaniawski, 'Some Remarks Concerning the Criterion of Rational Decision Making', *Studia Logica* **9** (1960), 221–239.
[104] K. Szaniawski, 'The Value of Perfect Information', *Synthese* **17** (1967), 408–424.
[105] K. Szaniawski, 'Questions and Their Pragmatic Value', in Bogdan and Niiniluoto [1], pp. 121–123.
[106] K. Szaniawski, 'Two Concepts of Information', *Theory and Decision* **5** (1974), 9–22.
[107] K. Szaniawski, 'On Sequential Inference', *this volume*, p. 171.
[108] P. M. Williams, 'The Structure of Acceptance and Its Evidential Basis', *The British Journal for the Philosophy of Science* **19** (1969), 325–344.

III. INDUCTION, PROBABILITY, AND LANGUAGE

Monographs

[109] R. Carnap, *The Logical Foundations of Probability*, University of Chicago Press, Chicago, 1950.
[110] R. Carnap, *The Continuum of Inductive Methods*, University of Chicago Press, Chicago, 1952.
[111] R. Carnap and W. Stegmüller, *Inductive Logik und Wahrscheinlichkeit*, Springer, Wien, 1959.
[112] N. Goodman, *Fact, Fiction, and Forecast*, Harvard U.P., Cambridge, 1955.
[113] R. Hilpinen (see [26]).
[114] H. E. Kyburg, Jr., *Probability and the Logic of Rational Belief*, Wesleyan U.P., Middletown, 1961.
[115] I. Niiniluoto and R. Tuomela (see [30]).

Articles

[116] R. Carnap and Y. Bar-Hillel, 'An Outline of a Theory of Semantic Information' (1952), in Y. Bar-Hillel, *Language and Information*, Addison Wesley and Jerusalem Academic Press, Reading, Mass. and Jerusalem, 1964, pp. 221–274.

[117] R. Carnap, 'The Basic System of Inductive Logic, Part I', in R. Carnap and R. C. Jeffrey (eds.), *Studies in Inductive Logic and Probability*, University of California Press, Berkeley and Los Angeles, 1971, pp. 33–165.

[118] I. Hacking, 'Linguistically Invariant Inductive Logic', *Synthese* 20 (1969), 25–47.

[119] R. Hilpinen, 'On Inductive Generalization in Monadic First-Order Logic with Identity', in Hintikka and Suppes [2], pp. 133–154.

[120] R. Hilpinen, 'Relational Hypotheses and Inductive Inference', *Synthese* 23 (1971), 266–286.

[121] R. Hilpinen, 'Carnap's New System of Inductive Logic', *Synthese* 25 (1973), 307–333.

[122] J. Hintikka, 'Toward a Theory of Inductive Generalization', in Y. Bar-Hillel (ed.), *Logic, Methodology, and Philosophy of Science II*, North-Holland, Amsterdam, 1965, pp. 274–288.

[123] J. Hintikka, 'On a Combined System of Inductive Logic', *Acta Philosophica Fennica* 18 (1965), 21–30.

[124] J. Hintikka, 'A Two-Dimensional Continuum of Inductive Methods', in Hintikka and Suppes [2], pp. 113–132.

[125] J. Hintikka, 'Induction by Enumeration and Induction by Elimination', in Lakatos [5], pp. 191–216.

[126] J. Hintikka, 'On Semantic Information', in Hintikka and Suppes [3], pp. 3–27.

[127] J. Hintikka, 'Surface Information and Depth Information', in Hintikka and Suppes [3], pp. 263–297.

[128] J. Hintikka, 'Unknown Probabilities, Bayesianism, and de Finetti's Representation Theorem', in R. Buck and R. S. Cohen (eds.), *Boston Studies in the Philosophy of Science VIII*, D. Reidel, Dordrecht-Holland, 1971, pp. 325–341.

[129] K. Lehrer, 'Descriptive Completeness and Inductive Methods', *Journal of Symbolic Logic* 28 (1963), 157–160.

[130] I. Levi, 'Confirmation, Linguistic Invariance, and Conceptual Innovation', *Synthese* 20 (1969), 48–55.

[131] W. C. Salmon, 'Carnap's Inductive Logic', *Journal of Philosophy* 64 (1967), 725–739.

[132] R. Tuomela, 'Inductive Generalization in an Ordered Universe', in Hintikka and Suppes [2], pp. 155–174.

IV. OTHER RELEVANT APPROACHES

[133] R. J. Bogdan, 'Hume and the Problem of Local Induction', *this volume*, p. 217.

[134] L. J. Cohen, 'A Logic of Evidential Support', *The British Journal for the Philosophy of Science* 17 (1966), 21–43 and 105–126.

[135] L. J. Cohen, *The Implications of Induction*, Methuen, London, 1970.

[136] L. J. Cohen, 'The Inductive Logic of Progressive Problem-Shifts', *Revue Internationale de Philosophie* (1971), 62–77.

[137] L. J. Cohen, 'A Note on Inductive Logic', *Journal of Philosophy* 70 (1973), 27–40.

[138] L. J. Cohen, 'The Paradox of Anomaly', in Bogdan and Niiniluoto [1], pp. 78–82.
[139] L. J. Cohen, 'A Conspectus of the Neo-Classical Theory of Induction', *this volume*, p. 235.
[140] L. J. Cohen, *The Probable and the Provable*, forthcoming.
[141] B. de Finetti, 'Foresight: Its Logical Laws, Its Subjective Sources', in Kyburg and Smokler [4], pp. 93–158.
[142] B. de Finetti, 'Initial Probabilities: A Prerequisite for Any Valid Induction', *Synthese* 20 (1969), 2–16.
[143] J. G. Greeno, 'Evaluation of Statistical Hypotheses Using Information Transmitted', *Philosophy of Science* 37 (1970), 279–293.
[144] I. Hacking, *Logic of Statistical Inference*, Cambridge University Press, Cambridge, 1965.
[145] G. H. Harman, 'Induction', in Swain [8], pp. 83–99.
[146] M. Hesse, *The Structure of Scientific Inference*, MacMillan, London, 1974.
[147] H. E. Kyburg, Jr., *The Logical Foundations of Statistical Inference*, D. Reidel, Dordrecht-Holland, 1974.
[148] I. Lakatos (see [17]).
[149] A. Musgrave, 'Logical vs Historical Theories of Confirmation', *The British Journal for the Philosophy of Science* 25 (1974), 1–23.
[150] H. Putnam, 'The "Corroboration" of Theories', in P. A. Schilpp (ed.), *The Philosophy of Karl Popper*, Book I, Open Court, La Salle, 1974, pp. 221–239.
[151] R. Rosenkrantz, 'Inductivism and Probabilism', *Synthese* 23 (1971), 167–205.
[152] R. Rosenkrantz, 'Simplicity', forthcoming.
[153] R. Rosenkrantz, 'Popper and Bayes', forthcoming.
[154] A. Shimony, 'Scientific Inference', in R. Colodny (ed.), *The Nature and Function of Scientific Theories*, University of Pittsburgh Press, Pittsburgh, 1970.
[155] R. C. Stalnaker, 'Probability and Conditionals', *Philosophy of Science* 37 (1969), 64–80.
[156] H. Törnebohm, 'Two Measures of Evidential Strength', in Hintikka and Suppes [2], pp. 81–95.
[157] H. Törnebohm, 'On the Confirmation of Hypotheses About Regions of Existence', *Synthese* 18 (1968), 28–45.
[158] H. Törnebohm, *Scientific Knowledge-Formation*, Department of Theory of Science, University of Göteborg, Report No. 58, April 1974.
[159] H. Törnebohm, 'On Piecemeal Knowledge-Formation', *this volume*, p. 297.

V. INDUCTIVE SYSTEMATIZATION AND EXPLANATION

[160] J. A. Coffa, 'Hempel's Ambiguity', *Synthese* 28 (1974), 141–164.
[161] J. H. Fetzer, 'Statistical Explanations', in K. F. Schaffner and R. S. Cohen (eds.), *Boston Studies in the Philosophy of Science XX*, D. Reidel, Dordrecht-Holland, 1974, pp. 337–347.
[162] J. H. Fetzer, 'A Single Case Propensity Theory of Explanation', *Synthese* 28 (1974), 171–198.
[163] C. G. Hempel, 'The Theoretician's Dilemma' (1958), in C. G. Hempel, *Aspects of Scientific Explanation*, The Free Press, New York, 1965, pp. 173–228.
[164] C. G. Hempel, 'Aspects of Scientific Explanation', in the same volume as above, pp. 331–496.

[165] C. G. Hempel, 'Maximal Specificity and Lawlikeness in Probabilistic Explana-tion', *Philosophy of Science* 35 (1968), 116–134.

[166] R. C. Jeffrey, 'Statistical Explanation vs Statistical Inference' (1969), in Salmon, Jeffrey and Greeno, *Statistical Explanation and Statistical Relevance*, University of Pittsburgh Press, Pittsburgh, 1971, pp. 19–28.

[167] K. Lehrer, 'Theoretical Terms and Inductive Inference', *American Philosophical Quarterly*, Monograph Series No. 3 (1969), 30–41.

[168] I. Niiniluoto, 'Inductive Systematization', *Synthese* 25 (1972), 25–81.

[169] I. Niiniluoto, 'Empirically Trivial Theories and Inductive Systematization', in Bogdan and Niiniluoto [1], pp. 108–114.

[170] I. Niiniluoto and R. Tuomela (see [30]).

[171] J. Pietarinen, 'Quantitative Tools for Evaluating Scientific Systematizations', in Hintikka and Suppes [3], pp. 123–147.

[172] W. C. Salmon, 'Statistical Explanation', in the same volume as [166], pp. 29–88.

[173] R. Tuomela, *Theoretical Concepts*, Springer, Wien, 1973.

[174] R. Tuomela, 'Confirmation, Explanation, and the Paradoxes of Transitivity', *this volume*, p. 319.

ADDENDA

[175] J. Leach, R. Butts, and G. Pearce (eds.), *Science, Decision, and Value*, D. Reidel, Dordrecht-Holland, 1973 (papers by Suppes, Braithwaite, Fishburn, Good, Levi, etc.).

[176] *Theory and Decision* 1 (1970) – 6 (1975).

[177] 'Methodologies: Bayesian and Popperian', *Synthese* 30 (1975).

Stanford University

INDEX OF NAMES

INDEX OF SUBJECTS

SYNTHESE LIBRARY

Monographs on Epistemology, Logic, Methodology,
Philosophy of Science, Sociology of Science and of Knowledge, and on the
Mathematical Methods of Social and Behavioral Sciences

Managing Editor:

JAAKKO HINTIKKA (Academy of Finland and Stanford University)

Editors:

ROBERT S. COHEN (Boston University)
DONALD DAVIDSON (The Rockefeller University and Princeton University)
GABRIËL NUCHELMANS (University of Leyden)
WESLEY C. SALMON (University of Arizona)

1. J. M. BOCHEŃSKI, *A Precis of Mathematical Logic.* 1959, X + 100 pp.
2. P. L. GUIRAUD, *Problèmes et méthodes de la statistique linguistique.* 1960, VI + 146 pp.
3. HANS FREUDENTHAL (ed.), *The Concept and the Role of the Model in Mathematics and Natural and Social Sciences, Proceedings of a Colloquium held at Utrecht, The Netherlands, January 1960.* 1961, VI + 194 pp.
4. EVERT W. BETH, *Formal Methods. An Introduction to Symbolic Logic and the Study of Effective Operations in Arithmetic and Logic.* 1962, XIV + 170 pp.
5. B. H. KAZEMIER and D. VUYSJE (eds.), *Logic and Language. Studies Dedicated to Professor Rudolf Carnap on the Occasion of his Seventieth Birthday.* 1962, VI + 256 pp.
6. MARX W. WARTOFSKY (ed.), *Proceedings of the Boston Colloquium for the Philosophy of Science, 1961–1962,* Boston Studies in the Philosophy of Science (ed. by Robert S. Cohen and Marx W. Wartofsky), Volume I. 1973, VIII + 212 pp.
7. A. A. ZINOV'EV, *Philosophical Problems of Many-Valued Logic.* 1963, XIV + 155 pp.
8. GEORGES GURVITCH, *The Spectrum of Social Time.* 1964, XXVI + 152 pp.
9. PAUL LORENZEN, *Formal Logic.* 1965, VIII + 123 pp.
10. ROBERT S. COHEN and MARX W. WARTOFSKY (eds.), *In Honor of Philipp Frank,* Boston Studies in the Philosophy of Science (ed. by Robert S. Cohen and Marx W. Wartofsky), Volume II. 1965, XXXIV + 475 pp.
11. EVERT W. BETH, *Mathematical Thought. An Introduction to the Philosophy of Mathematics.* 1965, XII + 208 pp.
12. EVERT W. BETH and JEAN PIAGET, *Mathematical Epistemology and Psychology.* 1966, XII + 326 pp.
13. GUIDO KÜNG, *Ontology and the Logistic Analysis of Language. An Enquiry into the Contemporary Views on Universals.* 1967, XI + 210 pp.
14. ROBERT S. COHEN and MARX W. WARTOFSKY (eds.), *Proceedings of the Boston Colloquium for the Philosophy of Science 1964–1966, in Memory of Norwood Russell Hanson,* Boston Studies in the Philosophy of Science (ed. by Robert S. Cohen and Marx W. Wartofsky), Volume III. 1967, XLIX + 489 pp.

15. C. D. BROAD, *Induction, Probability, and Causation. Selected Papers.* 1968, XI + 296 pp.
16. GÜNTHER PATZIG, *Aristotle's Theory of the Syllogism. A Logical-Philosophical Study of Book A of the Prior Analytics.* 1968, XVII + 215 pp.
17. NICHOLAS RESCHER, *Topics in Philosophical Logic.* 1968, XIV + 347 pp.
18. ROBERT S. COHEN and MARX W. WARTOFSKY (eds.), *Proceedings of the Boston Colloquium for the Philosophy of Science 1966–1968*, Boston Studies in the Philosophy of Science (ed. by Robert S. Cohen and Marx W. Wartofsky), Volume IV. 1969, VIII + 537 pp.
19. ROBERT S. COHEN and MARX W. WARTOFSKY (eds.), *Proceedings of the Boston Colloquium for the Philosophy of Science 1966–1968*, Boston Studies in the Philosophy of Science (ed. by Robert S. Cohen and Marx W. Wartofsky), Volume V. 1969, VIII + 482 pp.
20. J. W. DAVIS, D. J. HOCKNEY, and W. K. WILSON (eds.), *Philosophical Logic.* 1969, VIII + 277 pp.
21. D. DAVIDSON and J. HINTIKKA (eds.), *Words and Objections: Essays on the Work of W. V. Quine.* 1969, VIII + 366 pp.
22. PATRICK SUPPES, *Studies in the Methodology and Foundations of Science. Selected Papers from 1911 to 1969.* 1969, XII + 473 pp.
23. JAAKKO HINTIKKA, *Models for Modalities. Selected Essays.* 1969, IX + 220 pp.
24. NICHOLAS RESCHER *et al.* (eds.). *Essay in Honor of Carl G. Hempel. A Tribute on the Occasion of his Sixty-Fifth Birthday.* 1969, VII + 272 pp.
25. P. V. TAVANEC (ed.), *Problems of the Logic of Scientific Knowledge.* 1969, XII + 429 pp.
26. MARSHALL SWAIN (ed.), *Induction, Acceptance, and Rational Belief.* 1970. VII + 232 pp.
27. ROBERT S. COHEN and RAYMOND J. SEEGER (eds.), *Ernst Mach; Physicist and Philosopher*, Boston Studies in the Philosophy of Science (ed. by Robert S. Cohen and Marx W. Wartofsky), Volume VI. 1970, VIII + 295 pp.
28. JAAKKO HINTIKKA and PATRICK SUPPES, *Information and Inference.* 1970, X + 366 pp.
29. KAREL LAMBERT, *Philosophical Problems in Logic. Some Recent Developments.* 1970, VII + 176 pp.
30. ROLF A. EBERLE, *Nominalistic Systems.* 1970, IX + 217 pp.
31. PAUL WEINGARTNER and GERHARD ZECHA (eds.), *Induction, Physics, and Ethics, Proceedings and Discussions of the 1968 Salzburg Colloquium in the Philosophy of Science.* 1970, X + 382 pp.
32. EVERT W. BETH, *Aspects of Modern Logic.* 1970, XI + 176 pp.
33. RISTO HILPINEN (ed.), *Deontic Logic: Introductory and Systematic Readings.* 1971, VII + 182 pp.
34. JEAN-LOUIS KRIVINE, *Introduction to Axiomatic Set Theory.* 1971, VII + 98 pp.
35. JOSEPH D. SNEED, *The Logical Structure of Mathematical Physics.* 1971, XV + 311 pp.
36. CARL R. KORDIG, *The Justification of Scientific Change.* 1971, XIV + 119 pp.
37. MILIČ ČAPEK, *Bergson and Modern Physics*, Boston Studies in the Philosophy of Science (ed. by Robert S. Cohen and Marx W. Wartofsky), Volume VII, 1971, XV + 414 pp.
38. NORWOOD RUSSELL HANSON, *What I do not Believe, and other Essays*, ed. by Stephen Toulmin and Harry Woolf, 1971, XII + 390 pp.

39. ROGER C. BUCK and ROBERT S. COHEN (eds.), *PSA 1970. In Memory of Rudolf Carnap*, Boston Studies in the Philosophy of Science (ed. by Robert S. Cohen and Marx W. Wartofsky), Volume VIII. 1971, LXVI + 615 pp. Also available as a paperback.

40. DONALD DAVIDSON and GILBERT HARMAN (eds.), *Semantics of Natural Language*. 1972, X + 769 pp. Also available as a paperback.

41. YEHOSUA BAR-HILLEL (ed।)., *Pragmatics of Natural Languages*. 1971, VII + 231 pp.

42. SÖREN STENLUND, *Combinators, λ-Terms and Proof Theory*. 1972, 184 pp.

43. MARTIN STRAUSS, *Modern Physics and Its Philosophy*. Selected Papers in the Logic, History, and Philosophy of Science. 1972, X + 297 pp.

44. MARIO BUNGE, *Method, Model and Matter*. 1973, VII + 196 pp.

45. MARIO BUNGE, *Philosophy of Physics*. 1973, IX + 248 pp.

46. A. A. ZINOV'EV, *Foundations of the Logical Theory of Scientific Knowledge (Complex Logic)*, Boston Studies in the Philosophy of Science (ed. by Robert S. Cohen and Marx W. Wartofsky), Volume IX. Revised and enlarged English edition with an appendix, by G. A. Smirnov, E. A. Sidorenko, A. M. Fedina, and L. A. Bobrova. 1973, XXII + 301 pp. Also available as a paperback.

47. LADISLAV TONDL, *Scientific Procedures*, Boston Studies in the Philosophy of Science (ed. by Robert S. Cohen and Marx W. Wartofsky), Volume X. 1973, XII + 268 pp. Also available as a paperback.

48. NORWOOD RUSSELL HANSON, *Constellations and Conjectures*, ed. by Willard C. Humphreys, Jr. 1973, X + 282 pp.

49. K. J. J. HINTIKKA, J. M. E. MORAVCSIK, and P. SUPPES (eds.), *Approaches to Natural Language*. Proceedings of the 1970 Stanford Workshop on Grammar and Semantics. 1973, VIII + 526 pp. Also available as a paperback.

50. MARIO BUNGE (ed.), *Exact Philosophy – Problems, Tools, and Goals*. 1973, X + 214 pp.

51. RADU J. BOGDAN and ILKKA NIINILUOTO (eds.), *Logic, Language, and Probability*. A selection of papers contributed to Sections IV, VI, and XI of the Fourth International Congress for Logic, Methodology, and Philosophy of Science, Bucharest, September 1971. 1973, X + 323 pp.

52. GLENN PEARCE and PATRICK MAYNARD (eds.), *Conceptual Chance*. 1973, XII + 282 pp.

53. ILKKA NIINILUOTO and RAIMO TUOMELA, *Theoretical Concepts and Hypothetico-Inductive Inference*. 1973, VII + 264 pp.

54. ROLAND FRAÏSSÉ, *Course of Mathematical Logic – Volume I: Relation and Logical Formula*. 1973, XVI + 186 pp. Also available as a paperback.

55. ADOLF GRÜNBAUM, *Philosophical Problems of Space and Time*. Second, enlarged edition, Boston Studies in the Philosophy of Science (ed. by Robert S. Cohen and Marx W. Wartofsky), Volume XII. 1973, XXIII + 884 pp. Also available as a paperback.

56. PATRICK SUPPES (ed.), *Space, Time, and Geometry*. 1973, XI + 424 pp.

57. HANS KELSEN, *Essays in Legal and Moral Philosophy*, selected and introduced by Ota Weinberger. 1973, XXVIII + 300 pp.

58. R. J. SEEGER and ROBERT S. COHEN (eds.), *Philosophical Foundations of Science. Proceedings of an AAAS Program, 1969*. Boston Studies in the Philosophy of Science (ed. by Robert S. Cohen and Marx W. Wartofsky), Volume XI. 1974, X + 545 pp. Also available as a paperback.

59. ROBERT S. COHEN and MARX W. WARTOFSKY (eds.), *Logical and Epistemological*

Studies in Contemporary Physics, Boston Studies in the Philosophy of Science (ed. by Robert S. Cohen and Marx W. Wartofsky), Volume XIII. 1973, VIII + 462 pp. Also available as paperback.

60. ROBERT S. COHEN and MARX W. WARTOFSKY (eds.), *Methodological and Historical Essays in the Natural and Social Sciences. Proceedings of the Boston Colloquium for the Philosophy of Science, 1969–1972*, Boston Studies in the Philosophy of Science (ed. by Robert S. Cohen and Marx W. Wartofsky), Volume XIV. 1974, VIII + 405 pp. Also available as paperback.

61. ROBERT S. COHEN, J. J. STACHEL, and MARX W. WARTOFSKY (eds.), *For Dirk Struik. Scientific, Historical and Political Essays in Honor of Dirk J. Struik*, Boston Studies in the Philosophy of Science (ed. by Robert S. Cohen and Marx W. Wartofsky), Volume XV. 1974, XXVII + 652 pp. Also available as paperback.

62. KAZIMIERZ AJDUKIEWICZ, *Pragmatic Logic*, transl. from the Polish by Olgierd Wojtasiewicz. 1974, XV + 460 pp.

63. SÖREN STENLUND (ed.), *Logical Theory and Semantic Analysis. Essays Dedicated to Stig Kanger on His Fiftieth Birthday*. 1974, V + 217 pp.

64. KENNETH F. SCHAFFNER and ROBERT S. COHEN (eds.), *Proceedings of the 1972 Biennial Meeting, Philosophy of Science Association*, Boston Studies in the Philosophy of Science (ed. by Robert S. Cohen and Marx W. Wartofsky), Volume XX. 1974, IX + 444 pp. Also available as paperback.

65. HENRY E. KYBURG, JR., *The Logical Foundations of Statistical Inference*. 1974, IX + 421 pp.

66. MARJORIE GRENE, *The Understanding of Nature: Essays in the Philosophy of Biology*, Boston Studies in the Philosophy of Science (ed. by Robert S. Cohen and Marx W. Wartofsky), Volume XXIII. 1974, XII + 360 pp. Also available as paperback.

67. JAN M. BROEKMAN, *Structuralism: Moscow, Prague, Paris*. 1974, IX + 117 pp.

68. NORMAN GESCHWIND, *Selected Papers on Language and the Brain*, Boston Studies in the Philosophy of Science (ed. by Robert S. Cohen and Marx W. Wartofsky), Volume XVI. 1974, XII + 549 pp. Also available as paperback.

69. ROLAND FRAÏSSÉ. *Course of Mathematical Logic – Volume II: Model Theory*. 1974, XIX + 192 pp.

70. ANDRZEJ GRZEGORCZYK, *An Outline of Mathematical Logic. Fundamental Results and Notions Explained with all Details*. 1974, X + 596 pp.

71. FRANZ VON KUTSCHERA, *Philosophy of Language*. 1975, VII + 305 pp.

75. JAAKKO HINTIKKA and UNTO REMES, *The Method of Analysis. Its Geometrical Origin and Its General Significance*. 1974. XVIII + 144 pp.

76. JOHN EMERY MURDOCH and EDITH DUDLEY SYLLA, *The Cultural Context of Medieval Learning. Proceedings of the First International Colloquium on Philosophy, Science, and Theology in the Middle Ages – September 1973*. Boston Studies in the Philosophy of Science (ed. by Robert S. Cohen and Marx. W. Wartofsky), Volume XXVI. 1975, X + 566 pp. Also available as paperback.

77. STEFAN AMSTERDAMSKI, *Between Experience and Metaphysics. Philosophical Problems of the Evolution of Science*, Boston Studies in the Philosophy of Science (ed. by Robert S. Cohen and Marx W. Wartofsky), Volume XXXV. 1975, XVIII + 193 pp. Also available as paperback.

SYNTHESE HISTORICAL LIBRARY

Texts and Studies
in the History of Logic and Philosophy

Editors:

N. KRETZMANN (Cornell University)
G. NUCHELMANS (University of Leyden)
L. M. DE RIJK (University of Leyden)

1. M. T. BEONIO-BROCCHIERI FUMAGALLI, *The Logic of Abelard.* Translated from the Italian. 1969, IX + 101 pp.

2. GOTTFRIED WILHELM LEIBNITZ, *Philosophical Papers and Letters.* A selection translated and edited with an introduction, by Leroy E. Loemker. 1969, XII + 736 pp.

3. ERNST MALLY, *Logische Schriften*, ed. by Karl Wolf and Paul Weingartner. 1971, X + 340 pp.

4. LEWIS WHITE BECK (ed.), *Proceedings of the Third International Kant Congress.* 1972, XI + 718 pp.

5. BERNARD BOLZANO, *Theory of Science*, ed. by Jan Berg. 1973, XV + 398 pp.

6. J. M. E. MORAVCSIK (ed.), *Patterns in Plato's Thought. Papers arising out of the 1971 West Coast Greek Philosophy Conference.* 1973, VIII + 212 pp.

7. NABIL SHEHABY, *The Propositional Logic of Avicenna: A Translation from al-Shifā': al-Qiyās*, with Introduction, Commentary and Glossary. 1973, XIII + 296 pp.

8. DESMOND PAUL HENRY, *Commentary on 'De Grammatico'. The Historical-Logical Dimensions of a Dialogue of St. Anselm's.* 1974, IX + 345 pp.

9. JOHN CORCORAN, *Ancient Logic and Its Modern Interpretations.* 1974. X + 208 pp.

10. E. M. BARTH, *The Logic of the Articles in Traditional Philosophy.* 1974, XXVII + 533 pp.

11. JAAKKO HINTIKKA, *Knowledge and the Known. Historical Perspectives in Epistemology.* 1974, XII + 243 pp.

12. E. J. ASHWORTH, *Language and Logic in the Post-Medieval Period.* 1974, XIII + 304 pp.

13. ARISTOTLE, *The Nicomachean Ethics.* Translated with Commentaries and Glossary by Hippocrates G. Apostle. 1975, XXI + 372 pp.

14. R. M. DANCY, *Sense and Contradiction: A Study in Aristotle.* 1975, XII + 184 pp.

15. WILBUR RICHARD KNORR, *The Evolution of the Euclidean Elements. A Study of the Theory of Incommensurable Magnitudes and Its Significance for Early Greek Geometry.* 1975, XI + 374 pp.

16. AUGUSTINE, *De Dialectica*, Translated with Introduction and Notes by B. Darrell Jackson. 1975, XII + 151 pp.